Structural Biology Using Electrons and X-Rays

Structural Biology Using Electrons and X-Rays

An Introduction for Biologists

Michael F. Moody

AMSTERDAM • BOSTON • HEIDELBERG • LONDON
NEW YORK • OXFORD • PARIS • SAN DIEGO
SAN FRANCISCO • SINGAPORE • SYDNEY • TOKYO
Academic Press is an imprint of Elsevier

Academic Press is an imprint of Elsevier
32 Jamestown Road, London NW1 7BY, UK
30 Corporate Drive, Suite 400, Burlington, MA 01803, USA
525 B Street, Suite 1800, San Diego, CA 92101-4495, USA

First edition 2011

Notice

British Library Cataloguing-in-Publication Data
A catalogue record for this book is available from the British Library

Library of Congress Cataloging-in-Publication Data
A catalog record for this book is available from the Library of Congress

ISBN: 978-0-12-370581-5

Typeset by MPS Limited, a Macmillan Company, Chennai, India
www.macmillansolutions.com

Printed and bound in United States of America

11 12 13 14 15   10 9 8 7 6 5 4 3 2 1

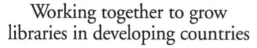

Contents

Preface

The aim of this book is to provide a comprehensible introduction to diffraction- and image-based methods for finding the structures of complex biological macromolecules. The reader is presumed to be a biologist or biochemist, interested in how these methods have revealed the extraordinary sub-cellular world of molecular machines, yet lacking the physical and (especially) the mathematical background to follow the primary literature. Fortunately, this common obstacle of higher mathematics can be circumvented, in the case of Fourier transform theory. For there exists an *intuitive* version of the theory, in which symbolism and calculation are replaced by pictures and pictorial rules, and yet sound conclusions can be drawn. This theory, covered in Part I, provides the key to understanding the applications of Fourier transforms described in later parts. It also forms the basis for keeping all the mathematics in this book (outside Part V) at the level of simple algebra, despite presenting some highly mathematical topics.

The applications of Fourier transforms include Part II (light- and electron-optics and diffraction theory), and (in Part IV) the more 'crystallographic' methods of image processing, used in X-ray crystallography and the analysis of electron micrographs of (especially) crystalline sheets and helices. However, structural analysis also uses other methods. *Symmetry theory* is needed to understand the organization of natural aggregates of identical subunits (relevant to viruses and muscle) and also those artificial aggregates, like protein crystals, constructed to facilitate structural analysis. *Matrix methods* are also needed for data-analysis (especially by 'single particle' methods). All these extra techniques are explained in Part III, which also discusses the general problem of turning 2D images into 3D structures.

All detailed structural analyses fall into different classes through the specimen's inherent symmetry (crystals, helices, polyhedral viruses and asymmetrical particles). These alternatives are covered in Part IV, which deals with those classes in declining order of symmetry-order. Naturally, the methods required to analyze highly symmetrical crystals (Chapters 13 and 14) differ from those appropriate for isolated, randomly-oriented, asymmetrical particles (Chapter 17). Thus two different kinds of method are in use. 'Crystallographic' methods, based on Fourier transforms, are suitable for structures with translational symmetry (crystals and helices) whereas 'single particles' need correlation and statistical techniques.

To accommodate newcomers, Parts I–IV (inclusive) are a (higher) mathematics–free zone. However, some readers may eventually develop an interest in the native mathematical 'language' of structure research; so an elementary introduction is provided for them in Part V. But this isn't required for understanding other chapters, though it would be required to understand most other books or papers on the subject.

There is a wide range of structural techniques, and a comprehensive coverage would produce an encyclopedic tome whose contents would also lack overall coherence. Therefore the topics chosen here have been restricted to molecular structure determination and to image-based methods. The main topic is high-resolution electron microscopy, but the framework also includes X-ray crystallography, which also produces images. However, NMR is excluded, despite its importance, because it is a very different kind of method, and a proper account of it must include quantum-mechanical spin-operators. Another important exclusion is electron cryo-tomography, a major new bridge between light microscopy and molecular structures. However, it lies rather more on the non-molecular side of this gap. Nevertheless, its promise and its use of three-dimensional reconstruction techniques makes this omission a matter of regret, justified mostly by practical considerations. Yet another regretted omission is the reconstruction of sections of thick crystals (applied most notably to muscle), a topic included in the long review article (Moody, 1990, Image analysis of electron micrographs, pp. 145–287. In: *Biophysical Electron Microscopy* (P.W. Hawkes & U. Valdrè, eds.). London: Academic Press.) on which this book was originally based.

I am grateful to Drs. R. A. Crowther, R. Henderson, E. Orlova, J.J. Ruprecht and Professor H. Saibil for their comments on drafts of this book.

CHAPTER 1

Overview

Chapter Outline

1.1 Role of Structural (Molecular) Biology

'Structural biology' (Liljas et al., 2009) (perhaps more precisely titled 'structural *molecular* biology') covers those macromolecular structures with dimensions between the small molecules of chemistry and the much larger features of cell structure visible in the light microscope. Chemistry has already shown how to explain the behavior of small molecules in terms of the mechanics of atoms, and molecular biology extends this on a larger scale and with a focus on living processes. Just as chemical explanations of molecules are based on molecular structures, so too must explanations of sub-cellular processes be based on the structures of their much larger components, the various protein and nucleic acid macromolecules. Indeed, as the components' size becomes more macroscopic, so the mechanics shifts from the quantum to the more familiar 'classical' level, and we approach the situation in mechanical engineering, where a detailed knowledge of any machine is the basis of understanding how it works.

In biology, that understanding at present lags far behind its level in mechanical engineering, so the unified subject of 'machines' is split into a (purely) descriptive branch, analogous to anatomy but expressed as molecular structures; a functional branch, analogous to physiology but mostly based on organic reactions and the physical chemistry of macromolecules in solution; and a third branch, analogous to embryology but following the transfer of genetic information from its DNA stores to the construction of functional macromolecules.

At present this third branch, under the title of 'bio-informatics', dominates. Description of structures is complemented by the acquisition of familiarity with the use of data-bases, including those describing structures. However, the methods by which the structures are determined also deserve study in structural biology, just as they do in organic chemistry. There the dominant techniques are spectroscopies (mass, IR, UV and NMR), which find the covalent framework more efficiently than did the earlier chemical methods. In structural biology, that framework is more easily determined (from DNA sequences) but also less important. Instead, structural studies focus on an aspect of only minor importance in organic chemistry: **conformation**, the three-dimensional structures consistent with free rotation about single bonds; their description, enumeration and energetics. In a small molecule, these structures are few and easily interconverted. In a protein, they are effectively infinite in number and there are too many to interconvert under physiological conditions. Consequently, the dominant techniques must determine shapes, and they are based on image-formation (microscopy, where electrons are the appropriate probe at these dimensions), and diffraction (X-ray crystallography).

As explained in the Preface, this book is intended to introduce these techniques to biologists who lack a strong background in physical chemistry and mathematics. It starts with a short history of the techniques, including the developments in NMR that allow the calculation of 'images' of proteins.

1.2 A Short History of Structural (Molecular) Biology

1.2.1 The Nature of the Problem

The study of macromolecule structure is a special subject because macromolecules have a special kind of structure. The ordinary small molecules of biochemistry (e.g. glucose) have a carbon–carbon skeleton that must be built up, step by step and almost atom by atom, using many different enzymes; they are thus limited to around 10–20 non-hydrogen atoms. By contrast, the macromolecules (proteins and nucleic acids) are made from many copies of smaller units (differing only in side-chains or **residues**), using special methods of fabrication. The units join together in the same standard way, using the same enzymes. This system has the important consequence that the product consists of a linear chain, in which the sequence is the only variable. That sequence, specifically controlled by cellular information (originally in the DNA) is maintained by covalent bonds, so the chain is specific and strong. Thus it can be analyzed by the methods of organic chemistry, aided by natural enzymes (proteases, nucleases, etc).

This is as far as organic chemistry takes either the cell itself, or our efforts to analyze its macromolecule structures. But this is not far; a long-chain molecule of *random* shape is of little use, whatever its sequence. It needs a *specific* shape to bring specific residues to the right places, and the specific sequence itself results in a spontaneous folding of the chain. This process of self-assembly yields a detailed structure which emerges through a complex physical process of translating information. It also requires a great increase of information for the final *detailed* structure, since a protein structure file of atomic coordinates is far longer than the few lines needed to carry its sequence. However, there is a more interesting question: how much more information is needed to convey the *essential* features of the structure, from which most of the detail could be obtained by computer-fitting the polypeptide chain? This is interesting because the information content of these essential features measures the minimum we need to extract from experiment.

We get more insight into this by considering the stereochemistry of polypeptide chains. Two successive C_α-atoms in a polypeptide are linked through two atoms (C, N), and therefore through three bonds. In principle, these are single bonds about which rotation is free, allowing three adjustable angles per residue. However, Pauling et al (1951) discovered that the —CO—NH— peptide bond that links amino-acids, though naively seen as a single bond about which rotation is free, is actually partly a double bond which keeps the six atoms involved (carbonyl-C, O, N, H and the two C_α-atoms) in the same plane (more or less). Thus there are actually only *two* adjustable angles per residue, which allows all possible configurations of two successive residues to be represented in a plane diagram (the Ramachandran plot). If the actual energy of all these configurations is calculated, it is found that most of the plot is 'forbidden' to a polypeptide chain with only ordinary thermal energies. Indeed, there are essentially only two 'allowed' regions of the plot: a small but deep one labelled 'α', and a broader, shallower one labelled 'β'. (A chain with all residues fitting the 'α' region forms an α-helix, and one with all residues fitting the 'β' region forms a β-strand.) Thus the most essential structural information, concerning the conformation of the polypeptide chain, allows only about 4 alternatives per residue. This is less than the number of alternative residues (20), suggesting some redundancy in the sequence-coding of structure (see Schröder et al., 2010).

We can draw the following two conclusions from the foregoing discussion. (i) The essential structural information needed to find a polypeptide chain's conformation is of the same order of magnitude as the information in its sequence. This is a substantial quantity, requiring many complicated experiments, especially when directly sequencing the polypeptide chain, but also when the relevant DNA is sequenced. Any useful structural technique has to yield a comparable quantity of data. (ii) Given the polypeptide chain's conformation, there is a second phase of structure-determination ('refinement'), where a chain of the known sequence is fitted to the conformation so as to minimize the energy. In other words, structure-determination is a two-phase process, the first experimental and the second computer-intensive. Of these, the first is the more uncertain, challenging and crucial, since it is essential to reach a basically correct main-chain conformation before we can draw any useful conclusions.

We now consider the kinds of experimental measurements that could give us the necessary data. (i) As we saw, they must yield a rich harvest of information; many otherwise useful techniques, like labeling reactive groups or measuring sedimentation rate, have no chance of getting us to the crucial threshold. (ii) The self-assembly of a polypeptide chain is controlled by many weak, reversible bonds and steric constraints. The final structure, maintained by many weak forces, therefore depends on the cooperative support of all parts and thus cannot survive the rough treatment imposed by chemical analysis. Structure-determination must depend on physical methods which leave macromolecules' delicate structures undisturbed.

1.2.2 'Imaging' Techniques

Next we consider the physical methods found to fit these conditions. The earliest studies of biological structure were based on images, originally ocular but, starting in the 17th century, with the help of the microscope. After two centuries of improving its resolution, convenience and contrast, light microscopy reached a block at a resolution of around 1 micron, nearly 10,000 times too big to reveal molecular structures. In the late 19th century this block was shown (Abbe) to be the wavelength of light and, at the century's end, the radiations were discovered that would overcome this block: electrons (Thomson) and X-rays (Röntgen). Electrons can be deflected and therefore focused, but it was only 30 years later that they were used in a microscope (Ruska) and the early electron microscope started in resolution near where the light microscope left off, though with many disadvantages in relation to specimen handling and visualization. X-rays, however, could not be focused, so images were excluded and it was only through curiosity about their physical nature that von Laue discovered, nearly 20 years after their discovery, that X-rays give diffraction patterns from crystals. These patterns corresponded to Abbe's intermediate stage in a light microscope, and an X-ray lens could have converted them into images. Unfocused, they remained puzzles from which image-like information could be derived by guesswork of increasing sophistication. Starting at the structural basement (NaCl), more and more complex molecules were solved by this approach, culminating in Pauling et al's α-helix (1951) and Watson & Crick's (1953) DNA structures.

Globular protein crystals under natural conditions had already been found to give very detailed diffraction patterns (Bernal & Crowfoot, 1934), thus satisfying both conditions of the previous section. However, the level of detail precluded 'guessing' the solution, stimulating Perutz's long search for a direct experimental method to find the missing phase data. He discovered it in 1953, and intensive work over the next few years led to the creation of a complicated methodology (effectively a semi-computational 'X-ray microscope') that extracted the desired images from myoglobin crystals (Kendrew et al., 1960). For the first time the threshold was crossed, and enough detail was visible to start refinement (when the sequencing had caught up). Of course, the successful methodology has since been greatly improved, both computationally and experimentally (see Chapter 13).

In 1934 a race was effectively started between the first protein diffraction patterns and the first experimental electron microscope (which had recently surpassed the light microscope's resolution). Any observer who then compared these two competitors would have been most impressed by the apparently insoluble puzzle of the diffraction pattern. By contrast, it might have seemed that the nascent electron microscope had only to follow the light microscope's example to achieve direct images of biological structures at all levels. However, by 1960 X-rays had yielded clear images of α-helices in myoglobin, and it would take more than twice as long for the electron microscope to reach a (barely) comparable resolution with bacteriorhodopsin (Henderson et al., 1990).

How had the tortoise so far overtaken the hare? About half of the delay was consumed by learning to maintain a thin specimen in a high vacuum, so that its high-resolution structure was preserved, as judged by electron diffraction patterns. But an image – that great merit of a microscope with lenses – lost about half the resolution. So why didn't the microscopists then forget their images and simply analyse the diffraction patterns by crystallographic techniques? Unfortunately, those would not work with electrons; while X-rays are scattered much more strongly by heavy atoms, that advantage is much reduced with electrons (see Glaeser et al., 2007, their §9.2). Thus physics forced the microscopists to remain microscopists and to improve their images, a task that occupied the remaining half of the delay.

Electron microscopy endured its long struggle, as had the early phase of protein crystallography. However, whereas that is just history to a modern crystallographer, one crucial part of microscopy's problems remains as an inescapable burden: **radiation damage**, the destructive effect of high-energy electrons that interact strongly with biological specimens. This can be minimized by exposing them at very low temperatures, but the only remedy is to use very low radiation exposures, taking many similar data-sets which are averaged to regain the required exposure. Thus all high-resolution electron microscope techniques need to record many thousands of images and then – the most difficult part – to align them all extremely accurately so that they can be averaged without blurring.

A few years ago it began to seem that electron microscopy's 'race' with X-rays might be over, and it had settled into a niche, acting as a bridge between light microscopy's cell-structures and crystallography's molecule-structures. Electron microscopy does indeed perform that role, and even its pursuit of higher resolution is consistent with it, since a 'bridge' should produce molecular structures that can be recognized well enough to insert the better X-ray versions in the correct places. But races between techniques can restart (in the 1950s, who would have predicted the present flowering of fluorescence microscopy?). The apparent 'barrier' around 4 Å resolution has recently been broken by studies on aquaporin (Gonen et al., 2004; Hite et al., 2010) and on a reovirus (Zhang et al., 2010) which have reached resolutions comparable with those of X-ray diffraction. The last paper is particularly encouraging, showing that molecular structures can be obtained from icosahedral symmetry. Thus specimens of all major symmetry-types (Part IV) have now yielded molecular structures. This is not yet true of the low-symmetry 'single particles', which suffer from two main problems: exposure-dependent specimen-movement (Rosenthal & Henderson, 2003) and the need to record, align and average thousands of images by hand. But at least that problem should soon be removed by using sophisticated image-recognition software to process images from electronic detectors, leaving only enough challenges to keep the subject interesting. Perhaps the next few decades will see the roles of hare and tortoise interchanged again.

Before leaving the 'imaging' techniques, it should be noted that radiation damage is not some curse peculiar to electron microscopy. It also afflicts X-ray diffraction, especially with very small crystals of very big structures like viruses (the relevant parameter being how few unit cells contribute to the recorded diffraction pattern). Thus the most sensitive specimens require each photon to contribute to the record, from which the crystal orientation is also found. This is essentially what is used in electron microscopy (§7.2.2). Even light microscopy is not entirely immune: fluorescent specimens, that need damaging UV, have a 'photon budget'. The uncertainty principle might suggest a naïve argument, that all high-resolution structures must find accurate positions and therefore leave momenta uncertain, raising speeds or quantum energies and thereby causing radiation damage. But that has already been upset by a technique for finding *molecular* structures using radiation of *centimeter* wavelengths.

1.2.3 Nuclear Magnetic Resonance

Images are not the only way to get a molecular structure. When light microscopy was stuck at the micron level, organic chemistry evolved a way to get *molecular* structures through strong-bond connectivities; and, although atomic bonds cannot define macromolecule structures, they might be defined by atomic *proximities*. A third structure method of this type was developed in the 1980s. Nuclear magnetic resonance (NMR), unlike the diffraction-based methods, is a form of spectroscopy, and is thus based on exploring the energy levels observed with radiation. Most nuclei possess small magnetic moments and, when such a nucleus is placed in a strong magnetic field, it occupies one of several energy levels. Even with the strongest laboratory magnets, these energy differences are small, corresponding only to thermal energies. So the energy levels are all partly filled at ambient temperature, and no damaging high energies are needed[1], as with other structural techniques (see above). Moreover, transitions between these energy levels emit or absorb radio waves, for which an efficient technology was developed in the early 20th century. Consequently, the spectroscopy was discovered in 1946 (Bloch and also Purcell), and was soon applied to organic chemistry after the structure-dependent energy-level shifts (**chemical shifts**) were discovered (Arnold et al., 1951).

But three major advances were needed to reach the resolution for analysing even the smallest biological macromolecules. First, the spectrum (or energy-level gaps) needed to be stretched out with higher magnetic fields of great stability, for which the development of superconducting magnets was essential. (They are also of potential value in electron microscopy.) Second, the universal need to average data to increase the signal:noise ratio led, in NMR, to the development of pulse techniques, from which the spectra could be calculated by Fourier transforms (Part I) (Ernst & Anderson, 1966). The resulting rich spectral data needed clarification and, like the clarification of gel electrophoresis patterns, this has achieved by extension into two (and more) dimensions (Aue et al., 1976).

---

[1]   This advantage has also allowed the development of NMR for medical imaging (MRI and f-MRI).

Finally, the necessary details could be measured to allow structure determination (Wüthrich, 1986) using **sequential assignment**. The mature methodology uses proton spectra obtained with a wide range of possible pulse-sequences (all known by acronyms like 'NOESY'), from which 2D (etc.) Fourier transforms convert the spectra into the form of peaks. Depending on the chosen pulse-sequence, these can yield 'through bond' peaks characteristic of each amino-acid, allowing the residue type to be identified. (This assignment process is sometimes aided by using deuterated proteins that lack proton spectra.) Other pulse-sequences give 'through space' peaks indicating the proximity of different protons, giving **distance constraints**. Since the closest amino-acids are adjacent in the sequence, a combination of both data sources identifies specific amino-acids. At this stage, we know (in principle) the origin of each proton-peak in the spectra, so we also know that certain amino-acids, though distant in the sequence, are close in the folded structure. These are the essential data for determining structures. Unfortunately, the relative sparseness of this information means that the calculated structures are not precisely determined. However, unlike the parallel situation with low-resolution X-ray data, the positions of specific residues will be known, allowing the main-chains configuration to be found. This is because NMR differs radically from the other methods in not generating an image; instead, it gives structural information in terms of local interactions, rather like organic chemistry. (However, like ordinary chemical methods, it provides no information about overall chirality (handedness).)

The complexities of pulse-sequence 'spin choreography' exclude NMR from this book, but a few aspects of it can be included. Resonance, spectra and the underlying Fourier transform theory find places in the first two Parts, and the interesting methodology for structure-determination from distance data comes into Chapter 11. These methods may well have some relevance in analysing electron microscope data.

1.2.4 Fundamental Limitations to Finding Macromolecule Structures

Despite the successes of NMR, there are still the same two fundamental problems in *microscopy*: resolution and radiation damage, and these are complementary. Good resolution requires a small wavelength, and hence a substantial photon energy that can generate active free radicals (either from the specimen or its surroundings) that then react with specimen molecules. Most visible light, which has insufficient energy to generate such radicals, has too long a wavelength for good resolution; though the problem is slightly different when light is emitted by a fluorescent specimen. However, improving the resolution by using UV (for fluorescence) starts to cause direct specimen damage; and X-rays, with far shorter wavelengths, can create more radicals. Electrons have even shorter wavelengths and they are potentially very destructive through their high speed and charge. However, Henderson (1995) found that the available data show them to be 1000 times less damaging than ordinary crystallographic X-rays (wavelength 1.5 Å), if evaluated in terms of total damage per useful elastic event.

The naïve argument from the uncertainty principle can be circumvented by fastening the specimen to a solid substrate. Then the big momentum change, required for accurate positions, can be absorbed without causing much movement. A similar advantage accompanies the use of heavier probes, especially the molecules that 'discriminate' each others' shapes, as in enzyme–substrate interactions. And it applies *a fortiori* to the probe of an atomic force microscope.

Note Added in Proof

Since the text of this book was completed, two important applications of electron cryo-microscopy have been published (see p. 357 and p. 384).

Fourier Transforms

CHAPTER 2

Correlations and Convolutions

Chapter Outline

2.1 Introducing Correlations

All image analysis uses either **Fourier transforms** (FTs) or **cross-correlations** (CCs); and FTs are based on CCs, which are usually calculated with FTs. We shall focus on FTs in the next chapter. Here we introduce the more fundamental concept of a **correlation function**, referring to situations familiar in molecular biology.

2.1.1 Cross-Correlations

Correlation is used in sequence comparisons. Suppose we are comparing two nucleotide sequences (X and Y), but distinguish only purines from pyrimidines, as black or white squares. The familiar *dot plot* (or *dot matrix*) comparison of the sequences is shown in the big square at the bottom left of Fig. 2.1 (where X and Y are respectively horizontal and vertical lines of black/white squares at the bottom and left of the big square). In order to find all comparisons with a given nucleotide in X, we follow the vertical column above it. A light or dark grey square indicates a match, a white square a mismatch. (Five or more consecutive matches along a 45° diagonal are marked with darker shading.) We count all the non-white squares along each 45° diagonal and plot the numbers on the diagonal histogram at the top-right.

Each 45° diagonal of the big square corresponds to a different *relative shift* of the X and Y sequences. Most striking is the 20-base-long dark-shaded diagonal connecting the X and

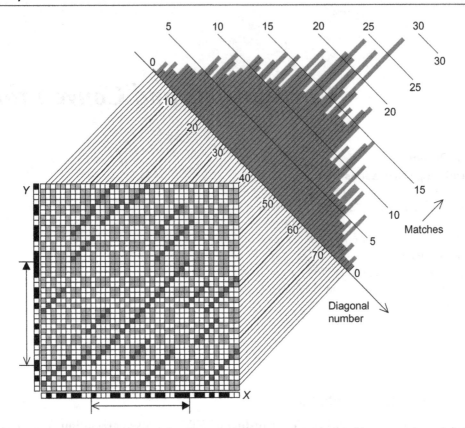

Figure 2.1: Converting the dot plot (bottom left) into a correlation histogram (top right).

Y segments arrowed at the bottom and side. In Fig. 2.2 (a) it is isolated from the dot plot, showing its connection with the common 20-base sequence in *X* and *Y* (arrowed at the left and bottom). This starts at base 11 in *X* and base 6 in *Y*, so the relative shift is $11 - 6 = +5$. Different 45° diagonals, with different lengths, correspond to different shifts; the longest has a zero shift. Diagonals below it have *Y* to the right of *X*, giving a positive shift; those above it give a negative shift. In Fig. 2.2 (b), sequence *Y* is shifted 5 bases to the right of sequence *X*, and matching nucleotides are indicated by 28 successive vertical double-headed arrows.

Using W and B to symbolize white and black squares (respectively), and '~' to represent blanks, the comparison can be represented in just three lines:

```
~~~~~WWBWBBWWWBBWBWWWBWBBWWBBBBBWWBBBWWBBWWWB  Y
WBWBBWBBBWWBWWWBBWBWWWBWBBWWBBBWBBWBWBBWW~~~~~  X
00000101101111111111111111111010110101000000
```

In the top two lines, spaces are shown by '~' in the bottom line, matches are shown by 1's and mis-matches by 0's. The matches follow the rules W/W = B/B = 1: W/B = B/W = 0: and W/~ = B/~ = 0. The tally is 28 1's (28 matches).

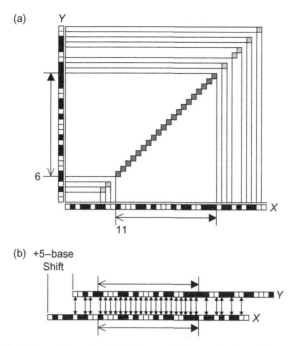

Figure 2.2: (a) A 20-base-long shaded diagonal from Fig. 2.1. **(b)** Sequence Y shifted 5 bases to the right of sequence X.

Here the matches or non-matches are digital but the sequences are alphabetic. It is more logical to make all digital, so we simply encode every W as +1 (written as 1) and every B as −1 (written as **1**). Then we just multiply the two strings of digits, using 'no-carriage' multiplication (i.e., without carrying to the next column). Multiplication of digits follows the usual rule of $1 \times 1 = 1$, $\mathbf{1} \times \mathbf{1} = (-1) \times (-1) = 1$ and $1 \times \mathbf{1} = \mathbf{1}$ (so identical digits give 1 and different digits give −1). Then the above matching process can be written as:

```
~~~~~11111111111111111111111111111111111111111   Y
111111111111111111111111111111111111111111~~~~~   X
0000011111111111111111111111111111111111100000
```

The tally is again 28 1's, but now there are seven −1's giving $28 − 7 = 21$ net matches.

2.1.2 Cross-Correlation Functions

This net match is called the **cross-correlation** (CC) of the (numerically encoded) X and Y sequences. It obviously depends on their relative shift. The +5 shift shown above gives the most matches, because it brings into register the common 20-base sequence BWWWBBWBWWWBWBBWWBBB in both X and Y. We found that by comparing all the relative shifts in Fig. 2.1. The comparisons are summarized at the top-right as a correlation histogram, plotting the number of sequence matches (shaded squares). A plot of the net match

(CC) as a function of shift is called the cross-correlation function (CCF). This function's highest peak marks the relative shift producing the best match.

This has introduced the concept of CCFs, but it is much easier to study them with a very short sequence; so we look at the CCF of WBB with itself. Thus we are briefly returning to the alphabetic sequence notation with 'W', 'B' and ';', but using the digital notation for matches. Thus W/W = B/B = 1: W/B = B/W = 1= −1: and W/; = B/; = 0. But now the CCF is generated by a relative movement of the sequences, keeping the lower one fixed and moving the upper one stepwise to the right. We start and end at the smallest possible overlaps, in between generating all grades of overlap:

(B~~/WBB) → (100) → −1; (BB~/WBB) → (110) → 0; (WBB/WBB) → (111) → 3; (~WB/WBB) → (011) → 0; (~~W/WBB) → (001) → −1.

Using the symbol ✪ to mean correlation, the cross-correlation function (CCF) is (WBB)✪(WBB) = (1|**1**|1)✪(1|**1**|1) = (−1|0|3|0| −1), which has its peak (3) where the sequence exactly matches itself. However, since the two lines are the same (WBB), we need not worry about their order with respect to ✪ (see below).

So far we have considered sequences with only two alternatives, W and B, encoded as 1 and 1 = −1. But suppose we wished to compare the hydrophobicity patterns of two amino-acid chains. Using some formula, we could convert each amino-acid type into a hydrophobicity number, so each chain would give a line of numbers. Then we could compare these lines by calculating their CCF. As an extremely simple example, we might calculate the CCF (31)✪(12). Now the two lines differ, so we need to distinguish between (31)✪(12) and (12)✪(31). As the lines approach each other, the first contact is between the right end of the first line and the left end of the second. So (12)✪(31) starts with $2 \times 3 = 6$, and ends with $1 \times 1 = 1$. In the middle we have (12) matching (31), giving $1 \times 3 + 2 \times 1 = 5$. Thus (12)✪(31) = (651), whereas (31)✪(12) = (156).

We could write each chain's numbers, (12) and (31), at equal distances along a paper strip (but with the second group reversed, ready for folding), as in Fig. 2.3 (a), and fold it at the middle. Then, in (b)–(d), we slide the upper numbers past the lower ones, (12) moving right relative to (31). We multiply the number-pairs and add the products, as shown. So the three positions (b), (c) and (d) give us respectively 6, 5 and 1; and (12)✪(31) = (651).

As an alternative to this paper-strip comparator, we could use plates with the numbers physically encoded as transmittances in transparent squares, but opaque elsewhere, as in Fig. 2.4 (a). Two successive transmittances multiply (approximately), so that the transmitted light's brightness is proportional to (transmittance 1) × (transmittance 2). Also, the sliding process can now be done automatically. We illuminate the upper plate with diffuse light, so that it transmits light in all possible directions. The light transmitted through the lower plate gives (approximately) the CCF (6,5,1). (This analogue device, the optical correlator, was invented by Robertson (1943)

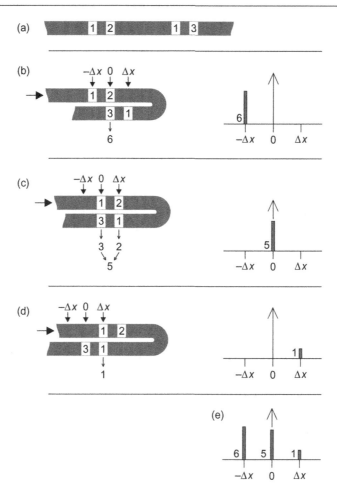

Figure 2.3: A paper strip **(a)** containing two number series. After folding, the upper strip's number [1] starts in **(b)** at $-\Delta x$, then moves in **(c)** to 0, and finally in **(d)** to Δx. So these are the coordinates at which the CCs are plotted on the right summarized in **(e)**.

to reduce the calculation burden of pre-computer crystallographers[1]). We can write the top two lines of Fig. 2.4 (a) in one line with the symbol '//' to represent the arrangement of $(1|2)$ above $(3|1)$, thus: $(1|2)\otimes(3|1) = (1|2)//(3|1) = (2 \times 3|1 \times 3 + 2 \times 1|1 \times 1) = (6|5|1)$. In general, $(a,b)\otimes(A,B) = (a|b)//(A|B) = (A|baA + bB|aB)$.

As often happens with mathematical operations, the CCF can be considered from two different viewpoints. Although they give identical answers, they differ in their intuitive significance, and one viewpoint may be more appropriate than another for certain applications. The viewpoint implicit in Fig. 2.3 might be called the *matching viewpoint*,

---

[1] See Buerger (1959) and Hosemann & Bagchi (1962) for more details of optical correlators.

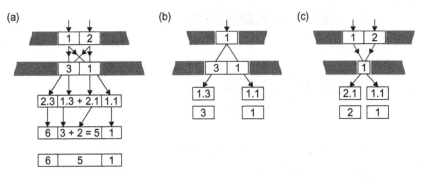

Figure 2.4: (a) The process illustrated in Fig. 2.3, is here accomplished automatically through transmittance and a gap between the strips, allowing light to pass in all possible directions. That gives the peak diagram from Fig. 2.3 (e) in one step. **(b)** A point source of illumination gives a copy in the same sequence. **(c)** A pinhole camera gives a copy in the reversed direction.

since it sees the CCF as a map of all the matches between the two sequences with different shifts.

The other viewpoint on the CCF can be approached through shrinking one of the functions into a point. Figure 2.4 (b) and (c) show the effects of constricting one plate to a pinhole. When the upper plate is a pinhole (b), the output below is like a shadow of the lower plate, and has the numbers in the same order. However, when the pinhole is in the lower plate (c), the output resembles the image cast by a pinhole camera, so the order of the (upper-plate) numbers is reversed. This suggests a different viewpoint on the CCF, the *copy viewpoint*. Each point on the upper plate generates a copy of the whole lower plate, in its existing orientation. Also, each point on the lower plate generates a copy of the whole upper plate, in the reversed orientation. The 'copy viewpoint' involves a reversal of one of the functions when applied to the CCF.

2.1.3 Correlation by Multiplication

These two point-simplifications can also be obtained from our one-line version of Fig. 2.4:

$$(1|2)\otimes(3|1) = (2 \times 3|1 \times 3 + 2 \times 1|1 \times 1) = (6|5|1)$$

First, when $2 \to 0$, $(1|0)\otimes(3|1) = (0|3|1) = (3|1)$. That is, $(1)\otimes(3|1) = (3|1)$, reminding us of:

$$\begin{array}{r} 1 \\ \times\,31 \\ \hline 31 \end{array}$$

Next, when $3 \to 0$, $(1|2) \odot (0|1) = (0|2|1) = (2|1)$. That is, $(1|2) \odot (1) = (2|1)$, reminding us of:

$$
\begin{array}{r}
21 \\
\times\ 1 \\
\hline
21 \\
\hline
\end{array}
$$

This suggests that the CCF might possibly be obtained by (almost) ordinary multiplication, if we *reverse* the order of the *first* multiplicand (12) to become (21):

$$
\begin{array}{r}
21 \\
\times\ 31 \\
\hline
21 \\
63 \\
\hline
651 \\
\hline
\end{array}
$$

which is indeed the correct answer. Thus $(1|2) \odot (3|1) \to 21 \times 31 = 651$. The multiplication is '*almost* ordinary' because we must *avoid* carrying tens to the column on the left. In the CCF calculation, the powers of ten only serve to keep the digits apart for sorting them[2], and that would be spoilt by carrying. To separate the columns, we can draw lines on paper, and in print we can use vertical bars or commas in vertical lines.

2.1.4 Convolution and Correlation

We noted that, to obtain the CCF, the first number in the multiplication must be reversed. But what would we get if it were *not* reversed? The first case gives us:

$$
\begin{array}{r}
12 \\
\times\ 31 \\
\hline
12 \\
36 \\
\hline
372 \\
\hline
\end{array}
$$

which is quite different. This is not the CCF but an even more important operation called **convolution**, which will occur again and again throughout this book. We shall write it as

---

[2] Our decimal representation of numbers is an example of a *power series*: the digits are to be multiplied by successive diminishing powers of ten. The convolution rule applies whenever two power series are multiplied: Bracewell (2000).

$(1|2)\star(3|1) = (1 \times 3|1 \times 1 + 2 \times 3|2 \times 1) = (3|7|2)$. We can summarize the two processes by the equations

$$(a,b)\mathbf{O}(A,B) = (a|b)//(A|B) = (Ab|aA + bB|aB) \tag{2.1}$$

$$(a,b)\mathbf{O}(A,B) \rightarrow ba \times AB \tag{2.2}$$

$$(a,b)\star(A,B) = (b|a)//(A|B) = (aA|aB + Ab|bB) \tag{2.3}$$

$$(a,b)\star(A,B) \rightarrow ab \times AB \tag{2.4}$$

Convolution $\star$ is best understood through the 'copy viewpoint', as a process of replacing every element of one function by an entire copy of the other function. Thus, in the last multiplication, each line is a copy of the first number, multiplied by successive digits of the second number[3]. Alternatively, it is a copy of the second number, multiplied by successive digits of the first number, since equation (2.3) is symmetrical in a and A (and also in b and B), so $(a,b)\star(A,B) = (A,B)\star(a,b)$. Thus convolution is commutative like ordinary multiplication (e.g. $2 \times 3 = 3 \times 2$). (However, correlation involves reversing the *first* number, which destroys this symmetry, so it is **non-commutative**.)

The connection between correlation and convolution is important. $(a,b)\star(A,B) = (aA|aB + Ab|bB)$ implies $(b,a)\star(A,B) = (bA|bB + Aa|aB)$ (if we simply interchange a and b). But equation (2.1) is $(a,b)\mathbf{O}(A,B) = (Ab|aA + bB|aB) = (b,a)\star(A,B)$, as we have just found. Thus, if $f(-)$ means $f(-x)$, i.e. the function f reversed, then:

$$f\mathbf{O}g = f(-)\star g \tag{2.5}$$

Because of the importance of reversing functions and their occasional symmetry under reversal, we devote the next few sections to this topic.

2.2 Function Parity

2.2.1 Even and Odd Functions

We start by considering a simple symmetry that is applicable to all 1D functions. Parity is the difference between symmetric and anti-symmetric functions. When a function is reversed, so that it runs from right to left, **symmetric** or **even** functions are unchanged. But **anti-symmetric** or **odd** functions have their signs reversed (Fig. 2.5).

---

[3] Thus long multiplication is a process of convolution, so convolution is taught in primary school! This connection is also used when we want computers to do very long multiplications quickly: Press et al. (1992), p. 909. Although unfamiliar, these calculations have been known for a considerable time; see Bracewell (2000). Similar calculations have also been long used in numerical analysis; see Abramowitz & Stegun (1964), formulae 25.3.21 and 25.3.23.

The simplest examples of even and odd functions are pairs of peaks; see Fig. 2.5 (a).
If we add the left and right pairs, the peaks at $-a$ cancel, leaving only those at $+a$. Reversing
this, a peak can be expressed as the sum of an even peak-pair and an odd peak-pair.
Now any function can be divided into isolated peaks, and each of those peaks can be split
into even and odd peak-pairs. So we can take any function, divide it into its constituent peaks,
split each peak into a pair of even peaks and a pair of odd peaks, and finally combine all the
even peak-pairs to get an even function, while combining all the odd peak-pairs to get an
odd function. These two functions are the even and odd *components* or *parts* of the original
function, which will be regenerated if we add them together. Perhaps this is clearer with
simple algebra:

$$f = \tfrac{1}{2}[f + f(-)] + \tfrac{1}{2}[f - f(-)] = f_e + f_o$$

where f_e and f_o are (respectively) the even and odd parts of f.

The cross-correlation (CC) of an *even* (symmetric) curve with an odd (anti-symmetric) curve
must be zero; for, as they match on both sides, but with opposite signs, the sum of the two
sides' contributions (and hence the CC) must be zero. We shall call this result the *CC parity
rule*. However, one should recall the distinction between CC and CCF (the CC being only one
point in the CCF). The CC parity rule only concerns the CC, and tells us little about the CCF.

If we have a general function, and calculate its CC with an even function, the CC will give
zero with the general function's odd component; it will preserve only the even component

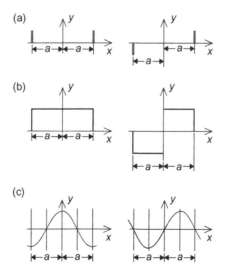

Figure 2.5: Even (left) and odd (right) functions. One-dimensional functions of y against x.
(a) pairs of peaks. **(b)** Left, rectangle function. **(c)** Left, part of cosine; right, part of sine.

of the general function, and 'filter out' the odd component. (We could alternatively arrange, *mutatis mutandis*, to 'filter out' the even component.)

2.2.2 Some Even Functions and Their Cross-Correlation Functions

The simplest positive functions are even, if the origin is put at their center of gravity. Thus very low-resolution approximations to a structure are likely to be even functions. They have the simplifying feature that the CCF with another even function is the same as the convolution with it, since $f(-) = f$ in equation (2.5).

The most important even functions are related to the rectangle. The rectangle function has two extreme forms, depending on its width. As it shrinks, a rectangle approaches the 'peak' function. Usually, a 'peak' means any function whose width is smaller than the *resolution limit* of the physical system used. (**Resolution** measures the size of the smallest detail that can be interpreted reliably.) However, in mathematics, where there is no *resolution limit*, the 'peak' must be a special function of infinitesimal width: the **delta-function** (δ-function). But it is not merely infinitesimal; unlike the Euclidean point, the δ-function has position *and* magnitude. For, if the rectangle is not to lose significance as its width shrinks, its height must increase to maintain a constant area (§3.4.1).

The CCF of two peaks is another peak. Its x-coordinate is the sum of the x-coordinates of the two peaks; its height is the product of the two peaks' heights; and its width is the sum of their widths. Convoluting a function with a unit peak leaves its shape unchanged; but it *shifts* it so that its original origin gets moved to the peak's position, and also *re-scales* it (multiplying it by the peak's height).

No name seems to have been given to the other extreme form of the rectangle: when its width expands towards infinity. This function, which we shall call the *plateau*, is the opposite of the peak. Whereas correlation or convolution with the peak leaves other functions unchanged, the same process using the plateau only leaves the plateau unchanged. The plateau is universally 'dominant' or indestructible, while the peak is universally 'recessive' or transformable.

The regular *comb function* (or simply *comb*; also called the 'shah' function by Bracewell (1965–2000) is a kind of combination of the peak and the plateau. It is an infinite series of equidistant, identical peaks, one of the peaks being at the origin. The distance between peaks is the *period* of the comb. As Fig. 2.6 (a) shows, this function is symmetric (even). Of course, such symmetry can be destroyed by shifting, but a half-period shift also gives a symmetrical comb, as shown in Fig. 2.6 (c).

To make an *odd* function from the regular comb, we must obviously have as many negative as positive peaks, and the function must be zero at the origin. If we alternate the peaks, we get the '*alternating comb*': Fig. 2.6 (e) and (g).

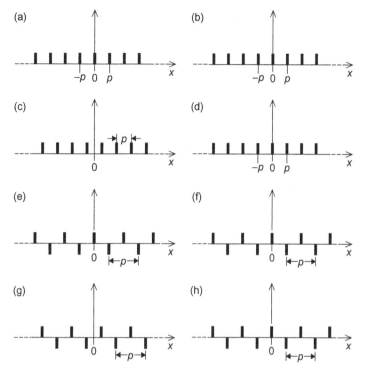

Figure 2.6: ACFs (right) of some periodic combs (left). **(a)** Simple even comb ('simple comb'). **(b)** ACF is exactly the same. **(c)** Shifted simple comb (also even). **(d)** The same as (b); shifts are lost in the ACF. **(e)** 'Alternating comb' (even). **(f)** Its ACF is the same as (e). **(g)** Simple odd comb; this is (e) shifted. **(h)** Similar but even, like all ACFs); same as (f).

2.3 Auto-Correlation Function

Having seen how the CCF and convolution might be calculated, we now consider what they mean. What is the effect on an image of convolution with some function (call it the 'convolutor')? Obviously, the effect must depend on the nature of the convolutor, which is an additional variable. So we start by considering the simplest case, where there is no new function because the image is correlated or convoluted with itself. (There will be no difference between correlation and convolution in practice, as we are considering only even functions.) Correlation of a function with itself gives us its **auto-correlation function** (ACF).

2.3.1 Interpretations of the Auto-Correlation Function

The CCF measures the matching of two different curves, as a function of their overlap; so the ACF must measure this between two identical curves. At zero displacement, a function obviously matches itself perfectly, giving every ACF a high peak at the origin. From this maximum, the ACF must decline (perhaps erratically), reaching zero when the displacement equals the function's width. The ACF is always an even function, because the later stages of overlap exactly mirror the early stages, except that the top and bottom curves are exchanged. (See the ACF of (WBB) in §2.1.2.)

In electron microscopy, this should apply to two exact copies of a particle in the same orientation. And, although differing noise contributions destroy the identity of such images, the ACF is still relevant to methods for aligning them prior to averaging them (Chapter 12, section §12.2).

In X-ray crystallography the ACF is called the **Patterson function**, which can be calculated directly from the experimental intensity data, without the arduous and uncertain search for the phases of the diffraction pattern. Consequently, much effort has gone into finding its uses. Here the most useful viewpoint sees the ACF as a map of all vectors joining dense regions within the image to each other. The vectors are weighted according to the product of the densities they join.

Figure 2.7 shows the simplest case, a double peak and its ACF whose prominent central peak represents zero vectors. This connects with the other two peaks through copies of the vector (length Δx) joining the original peaks.

The ACF of a rectangle function is shown in Fig. 2.8 to be a triangular function of twice the breadth, since the area of overlap increases and then decreases linearly with displacement. (This is the reason for the triangular-shaped CCF in the top-right of Fig. 2.1.) The same

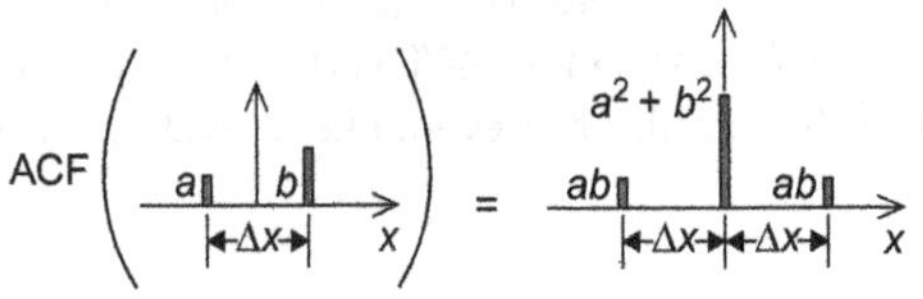

$$ \text{ACF}\left(\begin{array}{c} a \hspace{0.3cm} b \\ \overset{|\!\leftarrow\!\Delta x\!\rightarrow\!|}{} \end{array} \right) = \begin{array}{c} a^2 + b^2 \\ ab \hspace{1cm} ab \\ \overset{|\!\leftarrow\!\Delta x\!\rightarrow\!|\!\leftarrow\!\Delta x\!\rightarrow\!|}{} \end{array} $$

Figure 2.7: ACF of a double peak. Note its symmetry and big central peak.

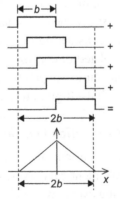

Figure 2.8: The ACF of a rectangle, width b, is a triangle, width $2b$.

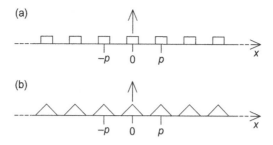

Figure 2.9: (a) Regular repeat of rectangles, width b, period p. **(b)** ACF of (a). The period remains p, but the rectangles have become triangles of width $2b$.

result is obtained numerically by multiplying by itself (i.e., squaring) one of the numbers $1111\ldots111$. For example, $(11111111)^2 = 123456787654321$, where the shape and length of the answer approximate to the geometrical result (though tens-carriage must be avoided in the calculation).

Another set of functions with simple ACFs are the comb functions, illustrated on the left of Fig. 2.6. As these functions are of infinite extent, so also are their ACFs, shown on the right.

The ACF is closely connected with convolution. From equation (2.5):

$$\mathrm{CCF}(f) = f \,❂\, f = f(-)\bigstar f$$

In Fig. 2.9, ACF (rectangle $\bigstar$ comb) = ACF [(a)] = (b) = triangle $\bigstar$ comb = ACF (rectangle) $\bigstar$ ACF (comb). That is an example of a useful rule: the ACF of the convolution of two functions is the convolution of their ACFs.

2.3.2 Locating the Auto-Correlation Function's Central Peak

When in electron microscopy the ACF is used to locate different particles with the same image, the location depends on a strong sharp central peak in the ACF. That is not always present. For example, Fig. 2.6 shows that the ACFs of 'combs' consist of a long series of equal peaks. To ensure that most of the ACF gets concentrated in the central peak, the function should lack prominent internal vectors that will make subsidiary ACF peaks. Thus the function should lack obvious structure; in a word, it should be complicated. And, to a superficial observer, a complicated structure is indistinguishable from noise (see Chapter 11.1.2). So the ACF is most useful with such dull or even ugly structures, and is not so well adapted for beautifully symmetrical structures, which are better analyzed by Fourier methods (see Part IV).

Given a known complicated structure, the ACF can be remarkably powerful. It is said that an early attempt to measure the distance of the planet Venus sent it a radar signal and tried to detect the echo. Because of the great distance, the echo was extremely weak and thus (to all appearances) lost in the noise. But then they sent a very long complicated signal, whose ACF consequently had a strong central peak. The radar echo was cross-correlated with a record of the original signal, and a clear peak could be seen in the CCF. This corresponded to the center of the signal's ACF, and showed the exact time required to receive the echo, and hence the exact distance of Venus at the time.

This story is instructive in several respects. The peak was found in a cross-correlation function, not an ACF. The calculation gave not $f \odot f'$, but $f \odot (f' + n)$ where n is a large noise component. (f and f' are copies of the original signal at different times.) Now it is convenient that the operation $\odot$ acts rather like multiplication (see § 2.1.3), so that $f \odot (f' + n) = f \odot f' + f \odot n = \text{ACF}(f) + \text{CCF}(f,n)$. Since f and n are quite uncorrelated, complicated functions, $\text{CCF}(f,n)$ is simply noise and hence very unlikely to have a very strong peak that could obscure that of the ACF.

In one respect, the situation was specially favorable to the astronomers: they knew the original signal and could use it in the CCF. But suppose that they had lost it, and had to compare several different radar echoes of the same unknown signal f after different times. Then the calculations would give not $f \odot (f' + n)$, but $(f + m) \odot (f' + n)$, where m is a different large noise component. This time we get:

$$(f + m) \odot (f' + n) \; = \; (f + m) \odot (f + n) \; = \; \text{ACF}(f) + \text{CCF}(f,[m + n]) + \text{CCF}(m,n)$$

So we still get a contribution from the ACF, but now in the presence of $\text{CCF}(f,[m + n])$ and $\text{CCF}(m,n)$. Because the noises m and n are quite uncorrelated, they more or less add together in $[m + n]$ to make a bigger noise. Moreover, their CCF is also noise, so $\text{CCF}(m,n)$ is yet another noise source. So the ACF has now to compete with much more noise. Competing noise consists of random peaks of various sizes, among which there is a typical size, while bigger peaks are progressively rarer. When the noise is strong enough, its typical peaks may approach the height of the central ACF peak and thus obscure it.

This is the situation when we try comparing faint, noisy images of the same particle type. For we are comparing 'weak' data with 'weak' data, rather than 'weak' data with 'strong' data as when the astronomers cross-correlated the ('weak') echo with the ('strong') copy of the original signal. But in microscopy we cannot know the particle image beforehand. Eventually, particle averaging may give us a reliable average, i.e. a 'strong' signal, and we can then add more data using particle locations obtained from 'strong' with 'weak' comparisons. But all *initial* ACF data must come from 'weak' with 'weak' comparisons. This is avoidable only if we can use *theoretical* data that is 'strong', as we discuss in the next chapter.

Fourier Fundamentals

Fourier analysis and Fourier transforms are needed for most of the applications described in this book, and this chapter introduces the subject.

There is a very long history of using sines and cosines (sinusoids) for analyzing periodic phenomena, dating back to the Eudoxan – Ptolemaic system for approximating planetary orbits as a series of epicycles (i.e., essentially 'circular' or trigonometric functions). The mathematical analysis of such approximations had to wait for Fourier's book on heat flow in 1815, but its major applications in that century were to periodic phenomena such as tides, whose level was approximated by a sum of sinusoids (a **Fourier series**), then calculated by complicated analogue computers. The 20th century saw two major developments. First, a sound mathematical basis was developed for extending the method to non-repeating curves (called the **Fourier transform** (FT) as it was originally suggested by Fourier in 1822). Second, the development of computers led to Fourier analysis being very widely used (the **digital FT**, here introduced in Chapter 4).

Fourier analysis is only one example of the *general* mathematical techniques for analyzing curves in terms of simpler component functions[1]. We look at their principles before focusing on the sines and cosines (odd and even sinusoids) used in basic Fourier analysis. However, these sinusoids are mostly used in special 'complex' combinations (here called **phasors**), a complication tackled in section §3.3.2.

This account aims to introduce the non-mathematical reader to the parts relevant to X-ray diffraction, image analysis and optics; it will use almost entirely intuitive methods. This chapter and the two succeeding ones are confined to 1D FTs for simplicity.

A largely intuitive account of Fourier transforms, aimed at crystallographers, is the book by Lipson & Taylor (1958), and Lipson's other books supplement this with optical diffraction patterns. Of the many mathematical accounts of Fourier transforms, Bracewell's (1965–2000) book is specially recommended for comprehensibility (see also his history of the FT: Bracewell (1989)). Bricogne's (2001) modern mathematical account is oriented towards crystallographers.

Fourier transforms are useful for both image analysis and also for describing diffraction and wave-optical imaging. All waves involve repeated vibrations, the simplest being sinusoidal; so every signal carried by waves is composed of sinusoids (see Part II). In this Part we concentrate on image analysis.

3.1 Component Functions

The high noise level of electron micrographs means that image information must be extracted through averaging. However, that requires a precise knowledge of the relative positions of the images or signals that must be averaged. As shown in the last chapter, precise relative positions can be found from the cross-correlation functions (CCFs) of two signals, provided one of them is near-perfect. To get adequate precision, we need one of the signals to be 'strong'. But what could we do when all available signals are imperfect or 'weak'?

---

[1] A simple example is the familiar *power series* such as $e^x = 1 + x + x^2/2! + \ldots$

One way to get a 'strong' image for comparison is to use entirely theoretical or mathematical functions as models for comparison, just as we often describe an unfamiliar shape by comparing it with known ones. Since the unfamiliar shape might be anything, our comparison must be able to achieve generality. So we need, not one comparison function, but a whole series, designed so that an appropriate mixture could fit *any* possible image, and to do so in the presence of noise.

This process is similar to a chemical analysis of mineral water. The main components (ions like sodium, chloride, etc.) are listed, but trace substances are ignored. Then, using the analytical data, a satisfactory copy of the original solution could be produced, differing only in trace substances. By analyzing only for the main components, we used *data compression* to obtain the essential analytical data. Similarly, we need to use data compression to extract important data from an image, ignoring the kind of fine detail that would be swamped by noise.

If we are to apply these principles to images or signals, we must first find a mathematical analog to chemical analysis. The analysis must satisfy two conditions. The component list should be *comprehensive*, omitting nothing significant; and the analytical tests for *pure* components should not be affected by the presence of contaminants. Similarly, the component functions, while providing a comprehensive description of any signal, should be 'pure' and not interfere with the tests for other components.

The validity of both the chemical and mathematical analyses depends on the quality of the analytical tests and the existence of pure components that (like chemical elements) are fundamentally different. We therefore need mathematical functions that are also 'fundamentally different' (called **orthogonal**), and a mathematical 'analytical test' that distinguishes them.

That test is the cross-correlation (CC) (§2.1.2), so different component functions are orthogonal if their CC is zero. As for suitable 'orthogonal' functions, there are many possible sets, but the most generally useful are the sinusoids, shown in Fig. 3.1. These curves are the simplest and most symmetrical of all component sets. Besides endlessly repeating, the repeat unit consists of four identical sectors or motifs (shaded) which are related by rotation and reflection within the period. One end of each motif is a node where the function is zero (excluding the bounds of the 'fundamental repeat' p, indicated). Thus a complete period contains two 'nodes' (indicated by small squares). Figure 3.1 sets up a fundamental repeat p within which sinusoids are shown repeating once, twice, etc., the number of repeats forming the basis for ordering the sinusoids. Their ordinal number $n = 0, 1, 2,...,$ where n is the number of nodes within p. ($n = 0$ is a constant.) Sinusoids with even n are even functions (cosines); those with odd n are odd functions (sines).

This simple numerical ordering is possible because our components all repeat after the distance p. The number of nodes within p orders the hierarchy of sinusoids so that they form a hierarchy

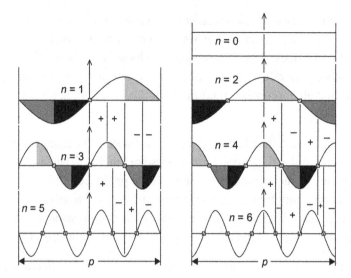

Figure 3.1: All sinusoids are constructed from the same basic motif (shown in white, light grey, dark grey and black), and ordered by the number n of internal nodes within the fundamental repeat p (see text). Left: sines; right: cosines. ($+$ and $-$ indicate signs of the products of adjacent sinusoids.)

of oscillations with *increasing detail*. The lowest members, slowest oscillations, can represent an experimental curve's *low-resolution* features (corresponding to blurred, large-scale features of images). We need the increasingly fine detail to make the sinusoids comprehensive (*complete*). The restricted detail in any experimental curve limits the number of component functions needed; after a point, further components include noise rather than information.

Sinusoids are useful components because any two of them with different n are orthogonal, i.e. give zero CC over p. We can see this by a few examples in Fig. 3.1. Both $n = 1$ and $n = 2$ have equal sizes of positive and negative sectors, and therefore give a zero CC with $n = 0$ (i.e., are orthogonal). The left curves (sines) are odd functions and therefore (§2.2.1) give a zero CC with the right curves (cosines). The even curves $n = 2$ and $n = 4$ are orthogonal: comparing the right sides, we see that the positive sectors exactly balance the negative sectors; and the left sides are identical as cosines are even. The same applies to the $n = 1$ and $n = 3$ sines (left); although they are anti-symmetrical, the zero right side still implies a zero left side. Similar symmetry considerations show the orthogonality of $n = 3$ with $n = 5$, and of $n = 4$ with $n = 6$. All other pairs must be orthogonal, by a mathematical argument that makes use of the high symmetry of sinusoids.

3.2 Fourier Analysis of Periodic Even Functions

The early Fourier analysis expressed periodic functions in terms of sinusoids. As a simple introduction, we start with periodic *even* functions. Of course, almost all images are *general* functions (neither even nor odd), but any general function can be expressed as the sum of an even function and an odd function (§2.2.1). Thus an analysis of even functions accomplishes half the requirement; the other half will be discussed in section §3.3.1.

3.2.1 Fourier Components and Diagrams

So we start with some chosen periodic even function, and find its components by calculating its CC with all the different sinusoid components. Since the CC of an even function with an odd function is zero (§2.2.1), our even function cannot have components that are odd functions; all its components must be even sinusoids, i.e. cosines. Since our function is also periodic (period p), these components' periods must fit with p: they must be p, or $p/2$, $p/3$, ...; or infinity (i.e. a constant).

We therefore need to find the cosine components with a period taken from the list: infinity, p, $p/2$, $p/3$, ... We select the components from the list by considering each candidate cosine and calculating its CC with the experimental curve. The CC gives the size of the candidate's contribution to the Fourier-series. Obviously, most give a zero CC and are therefore omitted. Also, we cannot test all the cosines from this infinite list; in practice, we keep testing until we find a few reasonably large CCs, and we stop when the size of additional CCs has dropped consistently to the noise level.

We must not only find all the components, but also represent them. Figure 3.2 shows how we will represent a single component compactly. Compactness is possible since an entire cosine curve is defined by just three parameters. First is the period or its reciprocal, the **spatial frequency**, which is more convenient as it includes a *constant* 'curve' as one of infinite period (i.e., zero spatial frequency). Moreover, submultiples of the period appear as equidistant points. Second, the curve's amplitude is represented as the length of a line. Third, the sign of the line corresponds to the sign of the curve at its origin, equivalent to the

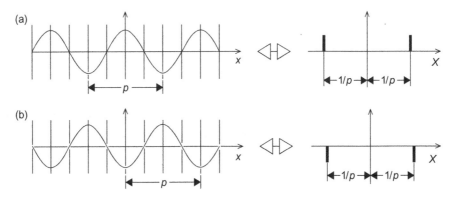

Figure 3.2: The compact representation of a cosine component. (The transform arrows are shown double-headed to maintain uniformity throughout all diagrams, although they imply a reversibility that has not yet been discussed.)

displacement of the cosine's positive peak, normally at its origin. This displacement is called **phase**, here shown by the sign of the line.

The two mirror-related sides of our representation may seem redundant, but a general curve lacks symmetry, and therefore needs both sides (the positive and negative axes). Because of the symmetry of even functions, we connect the curve (a function of x) with its representation (using a different axis, X) with a symmetrical symbol, linking the curve with its Fourier diagram. Of the two arrow-heads, one leads (in analysis) from the curve to its representation, the other from the representation (a kind of 'recipe' for the curve) back to the curve.

As noted, the function (left) is plotted on the x-axis, while the representation (right) is plotted on the X-axis. These axes differ, since x contains the period p, whereas X contains its reciprocal $1/p$. Thus x is called 'real space' and X is called 'reciprocal space'.

The representation acts as a 'recipe' by combining the peak-pairs for every cosine component of the curve. This is illustrated in Fig. 3.3. On the left, three cosine components (a)–(c) are added to give an even curve (d) at the bottom. Each component has its representation on the right.

Component (a), a constant, is a cosine with infinite period, so its representation has the two peaks coinciding at the origin $X = 0$ and has a double amplitude (Fig. 5.3); (b) is just Fig. 3.2(a); (c) is similar, but with three times the spatial frequency and a negative amplitude at the origin.

Each representation is added to give the composite representation on the right of (d), a compact summary of the components (achieving data-compression), which is a Fourier diagram, related to the 'Fourier Transform' (§3.4). Notice that, if p and all the periods based on it were increased, then $1/p$ and all the peak positions based on *it* would be moved closer to the origin. That is, stretching the x-axis shrinks the X-axis. This is called the *scale rule*.

3.2.2 Fourier Analysis of a Comb

The last section suggests the experiment of discovering what function emerges from an equal series of identical in-phase cosines, differing only their periods (Fig. 3.4). (This is a series of cosines taken from the right series in Fig. 3.1). On the right, we shall get the simplest possible representation: equal peaks equally spaced along the X-axis at multiples of $1/p$. This is a 'unit comb' (§2.2.2), but in reciprocal space (X). At the bottom left we have added all these cosines together, and we get another unit comb, though of course in real space (x). Thus the Fourier series of a unit comb is another (reciprocal) unit comb.

3.3 Sines and Phasors

3.3.1 Fourier Series for Odd Functions

We decided to consider first the Fourier analysis of even functions and, having done that, it is logical to continue with odd functions. Just as even functions needed even sinusoids (cosines),

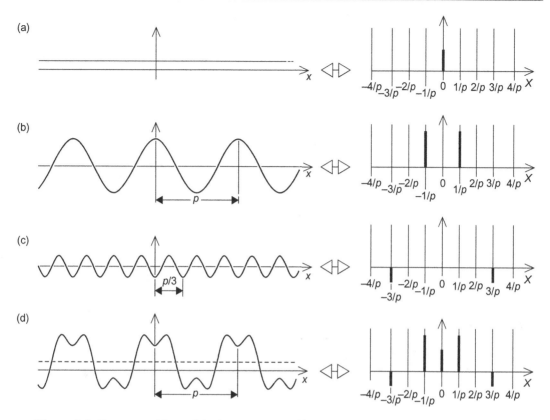

Figure 3.3: Decomposition of the bottom curve (d) into its Fourier components. The right transform arrow-heads indicate going from a curve to its Fourier diagram. **(a)** Left: cosine curve with period ∞; right: Fourier diagram peak at $X = 1/\infty = 0$. **(b)** Left: cosine curve with period p; right: Fourier diagram peak at $X = 1/p$. **(c)** Left: cosine curve with period $p/3$; right: Fourier diagram peak at $X = 3/p$. **(d)** Left: cosine curve with its Fourier diagram (right).

so odd functions need odd sinusoids (sines). Expressing an odd function as the sum of sines is quite simple. Indeed, we can derive an example from Fig. 3.3 simply by shifting the axis. We have to omit the first component (a), the constant function, as it is even. But (b) and (c), though even functions, become odd when we make a small axis shift, as shown in Fig. 3.5 (a) and (b). When these are added, they of course yield the same function they yielded in Fig. 3.3 (d), except for the omitted constant.

This much is simple, but Fig. 3.5 has an important omission: the representations on the right side. This is because the rules we have so far used fail to give a valid representation for a sine curve. The right of Fig. 3.3 contained representations for a positive cosine (b) and a negative cosine (c). These kinds of cosine differ by a half-period x-shift, and their representations are positive and negative lines. Now a sine needs a *quarter*-period x-shift, so it would seem to be intermediate between positive and negative (though certainly not zero!).

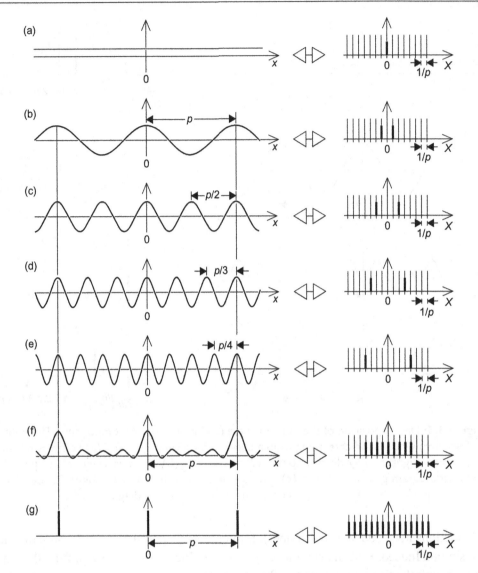

Figure 3.4: (a)–(e) On the left, cosine curves with the following periods: **(a)**: ∞; **(b)**: p; **(c)**: $p/2$; **(d)**: $p/3$; and **(e)**: $p/4$. (a)–(e) On the right, the FT of the cosine curve on the left. **(f)** Left: sum of the cosine curves in (a)–(e); right: sum of the FTs in (a)–(e). **(g)** The appearance of (f) if *all* the components (with periods $5/p$, $6/p$, etc.) were included. Note the symmetrical transform arrows, used with cosines (see Fig. 3.6).

3.3.2 Representation of Sines: Phasors

The solution to this problem is shown in Fig. 3.6. On the left we have a cosine curve in (a) that moves progressively, by eighth periods, to the right in (b)–(e), which is the negative of (a). The representations on the right start in (a) with a different representation for the

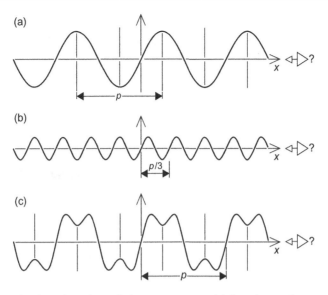

Figure 3.5: Fig. 3.3 displaced so that all the curves are odd functions. The transform arrows are asymmetric because sines are odd functions (see Fig. 3.1).

cosine: the lines are horizontal instead of vertical (as in previous figures). Actually, they are vectors, since we need their directions to distinguish positive lines (for a positive cosine in (a)) from negative lines (for a negative cosine in (e)).

Now the sine (c) fits the scheme by having a vector in an intermediate *direction* between right-pointing and left-pointing. Since the sine curve is an odd function, it is reasonable that its Fourier diagram should also be odd, with the vectors pointing in opposite directions. Since the entire Figure represents the rightward motion of a cosine curve, we can suppose that the representation should represent some kind of progressive motion of the vectors. Clearly, that motion must be a *rotation*.

The vectors[2] (or **phasors**, to use the conveniently brief and descriptive term from electrical engineering) can be described in terms of either Cartesian or polar coordinates (see Fig. 3.7). In polar coordinates, it is defined[3] by its length or magnitude M and its phase angle φ. In Cartesian coordinates, it is defined by its x-component ('real') and y-component ('imaginary'). This is because the x-axis is labeled 'real'[4] (R_e) and the y-axis 'imaginary' (I_m). Polar coordinates are best for multiplication, Cartesian for addition or subtraction.

---

[2] In mathematics and physics, these are called 'complex numbers'.

[3] Unfortunately, there are many alternative names. The length is sometimes called 'amplitude' (in physics) or 'modulus', and the phase angle is sometimes called 'argument' or even 'amplitude' (in mathematics).

[4] Avoid confusing this 'real' – the opposite of 'imaginary' – with the other 'real' – the opposite of 'reciprocal'.

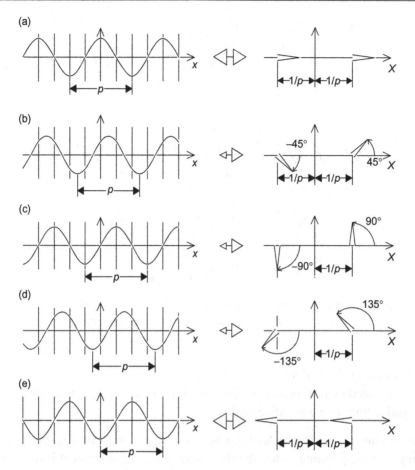

Figure 3.6: Central peak shifts and phase-angles (φ) of phasor on extreme right. **(a)** Shift = 0, $\varphi = 0°$; **(b)** Shift = $p/8$, $\varphi = 45°$; **(c)** Shift = $p/4$, $\varphi = 90°$; **(d)** Shift = $3p/8$, $\varphi = 135°$; **(e)** Shift = $p/2$, $\varphi = 180°$. (Note that the transform double arrows are *symmetric* with cosines, at (a) and (e), but *asymmetric* otherwise. See section §3.4.2 for the meaning of the asymmetry.)

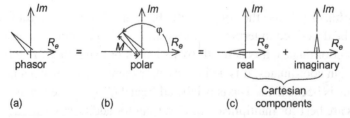

Figure 3.7: Representing a phasor in terms of either polar **(b)** or Cartesian **(c)** coordinates.

3.3.3 Multiplication of Phasors

We have seen that, in Fig. 3.6, shifts of the sinusoid correspond to rotations of the phasors. Thus, from (a) to (e), a shift of $p/2$ corresponds to a rotation of 180°, accomplished in four equal steps. But the final result, changing the phasors on the right of (a) to those on the right of (c), is equivalent to *multiplication* by −1. Thus rotation of a phasor must be equivalent to multiplying it by some unit phasor (unit, because one of these phasors is −1). This must also apply to the four intermediate steps between (a) and (e). Consider the rightmost phasor, which starts at (a) with $\varphi = 0°$, then at (b) $\varphi = 45°$, at (c) $\varphi = 90°$, at (d) $\varphi = 135°$ and finally at (b) $\varphi = 180°$. Each stage involves a rotation of 45°, equivalent to multiplication by a unit phasor with $\varphi = 45°$.

Thus multiplication of unit phasors involves simply adding their angles. But what if the phasors are *not* unit? Whatever rule applies here, must also apply when the phase-angles are zero so the phasors are just ordinary numbers whose values are the phasors' magnitudes. Then of course the magnitudes are to be multiplied. So the overall rule is: *multiply magnitudes, and add angles*.

3.3.4 Addition of Phasors

We have seen that (b) in Fig. 3.6, which is a cosine wave shifted by one-eighth of a period, corresponds to a rightmost phasor with $\varphi = 45°$. But there is another way of looking at line (b): as proportional to the *sum* of lines (a) and (c). Line (a) has a peak at $x = 0$, and line (c) has a peak at $x = p/4$, so (since the curves are otherwise identical) their sum should have a peak at $x = p/8$. So, if we calculate the sum of the phasors for (a) and (c), we should get a phasor with the same angle as that in (b) (for the proportionality constant only affects the magnitude).

This is shown in Fig. 3.8, and we see that we get the correct answer (a phasor with $\varphi = 45°$) by adding the Cartesian components at (g). Adding Cartesian components is thus the rule for phasors, just as for ordinary vectors (which add like successive journeys).

3.3.5 Phasor Conjugates

Hitherto we have concentrated on the rightmost phasor of Fig. 3.6, which was understandable since the left phasor is nearly the same. However, we should take note of the difference between them. The two phasors on the right side of Fig. 3.6 rotate, in opposite directions, by the same amount. Since, in (a), they start with the same $\varphi = 0°$, their angles will have opposite signs.

Thus the left and right phasors have the same magnitudes, but angles with opposite signs. That polar coordinate description corresponds to an equally simple description in Cartesian coordinates: they have the same real parts, but opposite imaginary parts. Phasors having that relationship are called **conjugates**, and when the phasors at $\pm X$ are conjugates, the Fourier diagram is said to show **Friedel symmetry**.

3.3.6 Phasor-Waves

We found in Fig. 3.6 many examples of curves giving two peaks. But we are interested in finding the curve that gives a Fourier series consisting of just *one* peak, a single phasor

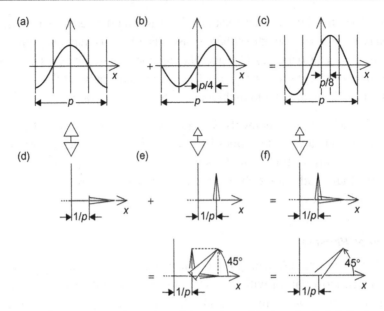

Figure 3.8: Below, phasors in Fourier series of three sinusoids (top line). **(a)** Cosine, represented in **(d)** by phasor with $\varphi = 0°$. **(b)** Sine, represented in **(e)** by phasor with $\varphi = 90°$. **(c)** Sum of (a) and (b), represented in **(f)** by the sum of (d) and (e). **(g)** and **(h)** show that, if this addition adds components, the sum is the expected phasor with $\varphi = 45°$.

with $\varphi = 0°$. We want this because all curves can be divided by 'salami-slicing' into many adjacent single peaks, so we can use this to do Fourier analysis on *any* curve.

But how can we get a single-peak Fourier diagram? The key lies in our representation of curves. We usually construct a curve from a series of peaks (or ordinary numbers) in infinitesimally close contact. But we could equally well construct it from a series of phasors. This is a trivial process if we replace the numbers by phasors with the corresponding amplitude but zero phase; so we start with this.

A cosine ((a) or (e) in Fig. 3.6) gives (on the right) two identical ($\pm$) peaks. This is shown again as (d), (e) in Fig. 3.9, where (e) shows the curve as a series of phasors with $\varphi = 0°$ (positive) or $180°$ (negative). Its Fourier diagram (i) needs to have one of the peaks removed to leave the desired single peak. To get this, we start with a sine (a), (b), whose Fourier diagram (g) has different peaks, but with $\varphi = \pm 90°$ and therefore 'imaginary'. So we need to rotate them by $90°$ through multiplication with a unit phasor with $\varphi = 90°$. This however multiplies both sides of the diagram, giving (c) where it is the *curve* that is 'imaginary'. If we now add (h) and (i), we shall eliminate the peak at $X = +1/p$, leaving only that at $X = -1/p$. On the left (f) we shall have the sum of (c) and (e). If, at each x, we add the phasors in (c) and (e),

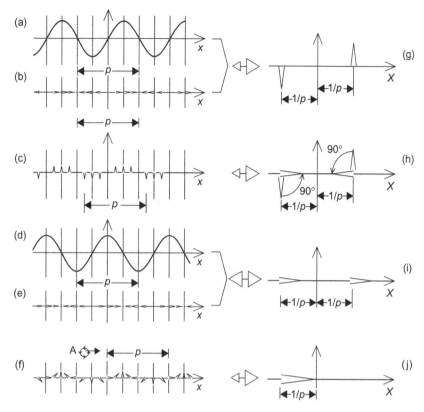

Figure 3.9: Finding the curve **(f)** whose Fourier diagram is a single peak **(j)**. **(a)**, **(b)**: sine and its Fourier diagram **(g)**. **(c)** and **(h)** are **(b)** and **(g)** multiplied by a unit phasor with $\varphi = 90°$. **(d)**, **(e)**: cosine and its Fourier diagram **(i)**. (f) = (c) + (e); (i) = (h) + (i). In (f), the 'A' symbol indicates *anti*-clockwise rotation of phasors when moving right.

we obtain the pattern shown in (f), where a unit phasor rotates uniformly anti-clockwise as we move to the right. (This addition is shown most clearly at multiples of $p/4$, where one or other of (c), (e) is zero.)

The final conclusion, line (f), (j) is an important relation between a rotating 'phasor-wave' and its Fourier diagram, a single peak. The 'phasor-wave' is also important in its connections with the sinusoids. Obviously, it can rotate either anti-clockwise or clockwise (when moving to the right along x), as in the last two lines of Fig. 3.10. These lines are identical only when the rotating phasors are horizontal ($\varphi = 0°$ or $180°$). (This is because lines (c) and (d) are (complex) conjugates; they can also be viewed as running in opposite directions.) At the point where $\varphi = 0°$, both phasor-waves also equal the cosine (shown in line (a) when the axis is positioned as marked). Thus the cosine function, at this point, is the average of (c) and (d). The interesting point is that it is *always* this average (e.g. on the origin-line marked 'sin' the phasors are ± 1, adding to zero, and $\sin 0° = 0$).

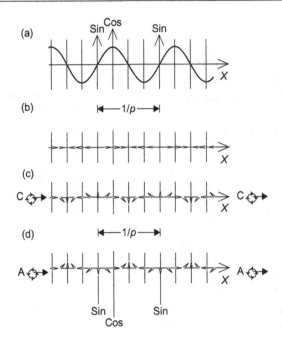

Figure 3.10: Cosine and sine shown as curves **(a)** or phasors **(b)**. Clockwise phasor-wave ($e^{-2\pi i Xp} = \cos 2\pi Xp - i.\sin 2\pi Xp$) in **(c)**. Anti-clockwise phasor-wave ($e^{2\pi i Xp} = \cos 2\pi Xp + i.\sin 2\pi Xp$) in **(d)**. (See Chapter 19, §19.4.2 in Part V for an explanation of these formulae, and those for $\cos 2\pi Xp$ and $\sin 2\pi Xp$ in terms of phasor-waves.)

So the average of lines (c) and (d) is line (a). But line (a) comprises *both* sine and cosine functions, depending only on where we choose the origin. Therefore, starting at the origin marked 'sin', the average is a sine-curve. This explains the Fourier diagrams in Fig. 3.9. The diagram in (i), two identical peaks at $\pm 1/p$, describes the phasor-waves (c) and (d) of Fig. 3.10, with the origin at 'cos'. To move the origin to 'sin', we must rotate the anti-clockwise wave by $-90°$, and the clockwise one by $+90°$; hence the anti-symmetric pair of phasors in (g) of Fig. 3.9.

3.3.7 *Fourier Series of General Functions*

We have almost completed our discussion of Fourier series. We dealt with even functions, represented as sums of cosines, in §3.2.1. From that, we found (by shifting curves) that the same procedure is applicable to sines (Fig. 3.5), but we needed a way to represent their Fourier diagram.

This was solved by the introduction of phasors in §3.3.2. So now we can represent any Fourier diagram peak as a phasor. Thus we can analyze a general periodic function (i.e. neither even nor odd). We split it into even and odd components, and analyze the former in terms of cosines, and the latter in terms of sines.

We also extended Fourier analysis to analyzing functions of phasors, and found one important function (phasor-wave) that gave the simplest possible Fourier diagram: one peak. Moreover, phasors allow us to combine our cosine and sine analyzes into one Fourier diagram. The cosine 'peak' is the 'real' component of a phasor, whose 'imaginary' component is the sine 'peak'.

3.4 Fourier Transforms

3.4.1 Fourier Series and Transforms

Despite its advantages, Fourier analysis is applicable only to periodic phenomena. However, this was useful for its original applications, because the experimental data ('curves') were very detailed, but the interesting part of their representation ('peaks') relatively simple. Thus the Fourier series was a tool for *data-compression*. This is the reason why 'curves and peaks' are so useful for removing noise. However, this method is apparently an illogical mixture, suggesting that there might also be 'all-peaks' and 'all-curves' forms of Fourier analysis.

Moreover, the 'peaks and curves' form is not strictly applicable to experimental data that, in practice, are series of numbers, i.e. 'peaks'. So actual calculations necessarily involve 'peaks and peaks'. Thus 'all-peaks' Fourier analysis, known as the 'digital Fourier-Transform' (DFT), is how we must actually *calculate* all Fourier analyses (e.g. with a computer). (We did even encounter one example of this kind of Fourier analysis in Fig. 3.4, where the two combs are 'all-peaks'). The DFT covers all practical applications of Fourier analysis, and it is explored in Chapter 4.

However, the 'all-curves' continuous FT (CFT) turns out to be the form in which we *think* about the subject. This situation is very common in applied mathematics, where the theory is expressed in conventional mathematics (e.g. the contents of Part V). This is designed primarily for comprehension and proof, and needs to make use of concepts (like infinite series or the distinction between rational and real numbers) that are abstract and remote from practicality. By contrast, the practical applications of applied mathematics usually need computers which employ algorithms whose design is the subject of *numerical analysis* (or methods). This is a special branch of mathematics whose main feature is the need to express all continuous functions in terms of finite samples, which are separate numbers of limited accuracy.

This usual dichotomy is also found in Fourier transforms, where understanding and proofs are expressed in conventional mathematics using CFTs; and this language is also used in discussions of its application to optics and image-analysis. But discussion must eventually lead to practical applications, and then we turn to the computer that necessarily employs digital DFTs.

Here we want to show how the abstract but exact CFT connects with the approximate but useful DFT, so we trace the path leading from the DFT, via the Fourier series, to the CFT. Its first half, from 'all peaks' to 'peaks and curves', is illustrated in Fig. 3.11. The DFT (a) is

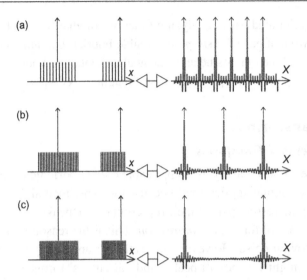

Figure 3.11: The connection between Fourier series and DFTs, with real space on the left, and Fourier space on the right. **(a)** DFT consisting of a series of adjacent peaks separated by a gap. **(b)** Real space peaks get closer, so the FT period gets wider. **(c)** Real space peaks are approaching the limit where they form a continuous curve. Correspondingly, the FT period is approaching infinity. The DFT is approaching the Fourier series. (DFTs are indicated by double-headed transform arrows with heads of the same shape, but different line-thickness; see Chapter 4, Fig. 4.8.)

periodic with a repeat defined by the number N of peaks per repeat (the size of the transform box, §4.1.1). As this number is increased, there is more detail in the left repeat, which remains the same size while its pixels (Δx) shrink to zero; so, at the limit, we have continuous real data (curves). Simultaneously, the reciprocal period enlarges, separating the overlapping rectangle FTs in (a). This is nearly the level of Fourier analysis, with curves in x but peaks in X.

Figure 3.12 shows the second half of the journey. The Fourier peaks in reciprocal space are clearly separated by ΔX, which will shrink to zero in (b) and (c). As the sampling in X becomes finer, the repeat in x becomes correspondingly bigger. Eventually, in (c), both pixel sizes (Δx and ΔX) are infinitesimal and both repeats are consequently infinite. This 'all curves' or 'continuous FT' is usually just called the FT.

However, to be quite general, the continuous FT must also handle cases where the data are peaks. The old mathematical theory of ordinary smooth curves lacked any satisfactory concept of a 'peak'. Therefore the theory was expanded to include a *continuous* function[5] that is also a *peak*, an oxymoron called the 'δ-function'. This has a precise position (and hence

---

[5] It was proposed by physicists (Heaviside, Dirac) and justified by mathematicians (especially Schwartz) in 'distribution theory'.

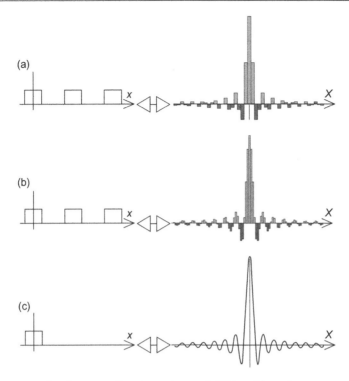

Figure 3.12: The connection between Fourier series and CFTs, with real space on the left, reciprocal space on the right. **(a)** The stage after (c) of Fig. 3.11; the pixels Δx are now infinitesimal, and the period on the right is infinite. **(b)** The pixels ΔX in X become smaller and the x-repeat correspondingly enlarges. **(c)** Both ΔX and Δx are now infinitesimal, and both sides have continuous curves. This is the 'all curves' *Continuous Fourier Transform* (CFT). (Both Fourier series and CFTs of even functions are indicated by symmetrical double-headed arrows.)

zero width) but finite area. It is the limit of a Gaussian or a rectangle, with diminishing width and increasing height, but always a curve. Thus it differs from the DFT 'peak', which is only a number (real or complex) with an associated coordinate position.

3.4.2 Fourier Transform of a Peak

Figure 3.9 showed how the Fourier series of a phasor-wave is an isolated peak, perhaps the most extreme case of a non-periodic function. But such functions, as we have seen, can give **Fourier transforms** (FTs), so we should find an isolated peak's FT. An FT is found in just the same way as a Fourier series, explained in §3.2.1: we scan the 'experimental function' with all possible 'candidate' sinusoids, measuring the CC which is then plotted at a position X that measures the spatial frequency of the 'candidate' sinusoid which gave this CC. If we are finding a Fourier series, which has the same repeat as the experimental curve, there is a limited choice of 'candidate' sinusoids. When finding an FT, however, there is no such limitation, so we have to vary the spatial frequency from minus to plus infinity. But we can

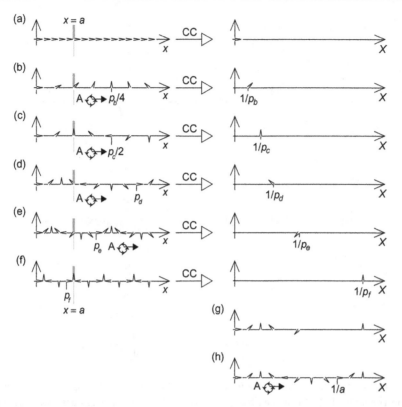

Figure 3.13: The FT of a single peak. On the left, a single peak (shaded) at $x = a$ runs down lines **(a)–(f)**. We are calculating the FT of this peak, using the 'cross-correlation sieve'. So we cross-correlate it (indicated by the 'CC arrow') with various test phasor-waves, of spatial frequency increasing from zero (a) to some maximum at (f). (Each phasor-wave's period is marked on its x-axis.) On each line, we calculate the CC of the phasor-wave with the peak at $x = a$; this CC is simply the overlap of the peak with the phasor-wave. On the right, we place on each line the CC at the spatial frequency (= reciprocal of the period) of the corresponding phasor-wave. **(g)** represents the sum of the six X-plots above it. **(h)** is the regular phasor-wave from which (g) has samples (note that it contains each phasor of (g)).

mitigate this by replacing the tedious search through cosines, followed by sines, with the more compact phasor-waves.

We now follow this procedure with an 'experimental function' that is a peak at $x = +a$ (Fig. 3.13). The figure only shows a few excerpts from the infinite search, starting at zero spatial frequency and ending at $1/p_f$. The CC is simple, since only that part of the phasor-wave coincident with the peak (i.e., at $x = a$) counts. These fragments, placed at their correct X, gives us line (g) from which we deduce the true FT (h), for positive X. It is a phasor-wave, and we can see that it extends backwards to $-\infty$.

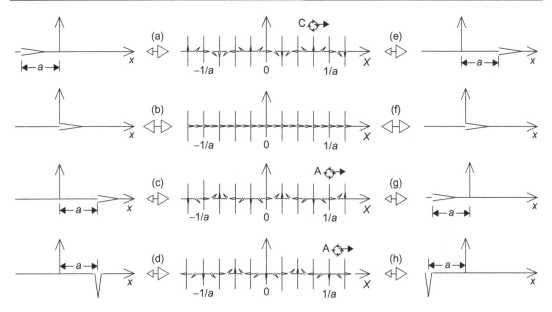

Figure 3.14: (a)–(c): FTs (middle) of isolated peaks (phasors with zero angle, left). Right: FTs of middle using same sign-choice. **(d)** is (c) multiplied by a unit phasor with angle −90°.

The full curve is copied into Fig. 3.14 (c). The rest of the line, (g), comes from Fig. 3.9 (f). We can see how the two upper lines ((a), (b), (e), (f)) of Fig. 3.14 make sense by the following argument. Start with the anti-clockwise phasor-wave between (c) and (g), and slow its rotation, increasing its period. So $1/a$ increases, meaning that a decreases, eventually becoming zero before (b) and after (f). Another way of looking at the process of slowing its rotation is to say that it makes the rotation more clockwise, cancelling the anti-clockwise rotation in the (c)–(g) line. Then the rotation becomes even more clockwise in the top line. At the same time, the movement of the peak towards the origin can be continued in the upper line, where the peaks are at the opposite side of the origin, compared with the third line (c)–(g).

The first three lines of Fig. 3.14 comprise six FT relationships, marked by double-headed transform arrows. Excluding the relatively trivial two in the second line, we have four interesting relationships (a), (e), (c) and (g), which all have asymmetric double-headed arrows. These symbolize a choice[6] we had to make when we calculated the FTs: whether the 'candidate' phasor-waves in Fig. 3.13 (and also effectively in Fig. 3.9) should be clockwise or anti-clockwise. We chose anti-clockwise for positive X (implying clockwise for negative X), and that gives us an anti-clockwise phasor-wave for the FT in Fig. 3.14 (c), where the choice is indicated by the unsymmetrical double-arrow.

---

[6] In formulae, the choice is the sign choice in $F(X) = \int f(x)e^{\pm 2\pi i x X}\,dx$, where we use the positive alternative. (See §19.5.4 in chapter 19 Part V, where e^x is written exp(x).) Some choice is enshrined in every FT program, and it is easily discovered by calculating the FT of an isolated peak at positive x.

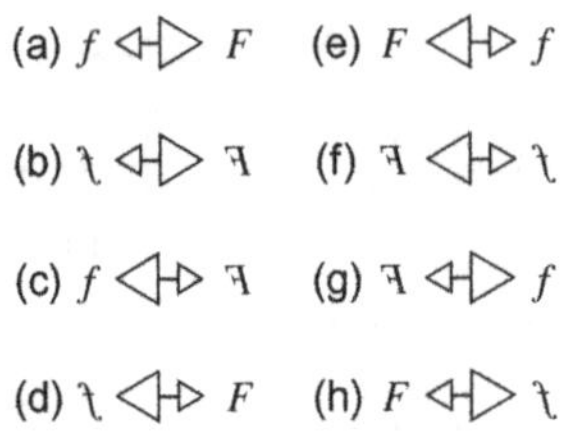

(a) $f \lhd\!\!\rhd F$ (e) $F \lhd\!\!\rhd f$

(b) $\backprime \lhd\!\!\rhd \exists$ (f) $\exists \lhd\!\!\rhd \backprime$

(c) $f \lhd\!\!\rhd \exists$ (g) $\exists \lhd\!\!\rhd f$

(d) $\backprime \lhd\!\!\rhd F$ (h) $F \lhd\!\!\rhd \backprime$

Figure 3.15: Rules for rearranging FT diagrams like those in Fig. 3.14, when using an unsymmetrical double-arrow needed when f and F are not even functions. 'Flipped' functions run the opposite way, so 'flipped' F is F^* (the complex conjugate of F; see §3.3.5).

Ordinary algebra starts with rules about how to change equations without invalidating them. We need similar rules for the visual equations in Fig. 3.14, so that we could start with (c) and deduce the other equations in the first three lines. A list of alternative valid equations is given in Fig. 3.15, where f is a function whose FT (with our convention) is F. In Fig. 3.14, f is a peak at $x = +a$, and F is an anti-clockwise phasor-wave. Thus our main relationship, (c) in Fig. 3.14, corresponds to (a) in Fig. 3.15. 'Flipping' f gives a peak at $x = -a$, and 'flipping' F gives a clockwise phasor-wave (§3.3.6). With this notation, (g) in Fig. 3.14 corresponds to (h) in Fig. 3.15; (a) in Fig. 3.14 corresponds to (b) in Fig. 3.15; and (e) in Fig. 3.14 corresponds to (g) in Fig. 3.15.

Figure 3.15 shows the eight equivalent FT diagrams; if we fit any diagram to (a), we can write down the other seven. And all eight diagrams can be deduced using just two rules, the *flip rule* and the *equation-reversal rule*. Flipping means reversing[7] either f or F or the unsymmetrical double-arrow, without changing the sequence of the items. Sequence reversal means reversing the order of these items (so f, arrows, F becomes F, arrows, f), without flipping any item. The rules are:

- *Flip rule*: we can flip any *two* of f, arrows, F. Thus an *even* number of flips keeps the relationship valid.
- *Equation-reversal rule*: we can reverse an entire equation, so the items occur in reverse order, provided we do *not* flip either f, or F, but we *do* we flip the arrows. Thus whichever of f, F was originally next to the bigger arrow, still remains next to that bigger arrow after sequence reversal. Thus equation-reversal, so that items occur in the opposite sequence, yields an invalid relationship unless the arrows are flipped. (Compare the same lines in left and right columns of Fig. 3.15.)

All these modifications generalize the first three lines of Fig. 3.14. The last line shows a different modification: as in an ordinary equation, we can multiply each term by the same constant. (This must be a *single* 'number' and *not* a function or curve.) Of particular interest

---

[7] The 'reversal' of a function around the horizontal axis is best described as its *reflection* in the vertical axis. But reversing the direction of a curve can also be viewed as rotation by 180°, the only rotation allowed in 1D.

is the case shown in the bottom line, obtained from the line above by multiplying by a 'number' which is a unit phasor with phase-angle $-90°$ (actually, this is $-i = -\sqrt{-1}$). Every phasor gets rotated clockwise by $-90°$, most obviously the peaks at the left and right. But it also affects the phasor-wave in the middle, where each phasor gets this clockwise rotation. Note, however, that the most obvious effect on the phasor-wave is a shift or translation by a quarter-period to the right. This effect of multiplication on phasor-waves leads to the most important rule for FTs, which we discuss in §3.4.4.

3.4.3 Fourier Inversion

Pixel Method
Line (c) of Fig. 3.14 shows that a peak at $x = +a$ (call it f) can be Fourier-transformed into an anti-clockwise phasor-wave (call it F). This corresponds to line (a) of Fig. 3.15, where line (h) shows that the same phasor-wave (F), after the same Fourier-transformation (small arrow left, big arrow right) yields the flipped peak, i.e. at $x = -a$. This means that, if we feed an FT program with data consisting of a single peak, take the FT it has just calculated and feed *that* into the same program, we shall get the original peak data flipped over. Apart from that trivial change, the FT program has reversed its own operation. Exact reversal (i.e. with no flipping) is called *inversion*.

Although that is just an isolated peak, it has much bigger implications. Any ordinary real (not imaginary) curve can be 'salami-sliced' into adjacent peaks, and each peak gives a phasor-wave (clockwise or anti-clockwise, according as the slice was left or right of the origin). All these phasor-waves must be added to give the overall FT. If we now take the FT of this FT (using exactly the same procedure as when finding the FT), we convert each of the constituent phasor-waves into its original peak, flipped over (left-right reversed). So, when we add all the constituents, we get the original curve, flipped over. And it is only a trivial operation to flip it back, giving the original curve. The FT has been inverted, almost entirely by the same computer program that calculated it.

This reversibility of Fourier transformation (called *inversion*) means that an FT can be modified and processed and then finally reversed to give a real structure again. This is a most valuable property of Fourier analysis, as we shall see in Parts III and IV.

Spectral Method
The method just described, which depends on adding the FTs of each weighted pixel, might be called the 'pixel method'. It is also applicable to isolated peaks, such as atoms. There is an alternative method (§3.2.1), more appropriate to the Fourier-analysis of curves, which we shall call the 'spectral method'. This uses a 'cross-correlation sieve' ('CC sieve') that compares the starting curve with a range of different phasor-waves to find which spatial frequencies match it best. It finds the CC of the function with 'test' phasor-waves of each different frequency. Thus it multiplies the function with each 'test' wave, and then adds together all the multiplication products, getting the CC appropriate to that frequency. After

surveying waves of all frequencies, we have the CC as a function of frequency the CCF. This is the FT of the starting function.

The 'CC sieve' selects just one of the component phasor-waves, by converting all except one into zero: *selective destruction*. The destruction part – turning most phasor-waves into zero – is easily done with any phasor-wave, which all have phasors at every possible angle; we get zero merely by adding all these phasors together. This is shown in Fig. 3.16, where the phasors in (a) are added base-to-base in (b), and base-to-tip in (c), giving zero in each case.

How does the same sieve that removes the unwanted phasor-waves also preserve the wanted one? That is shown where the phasors in Fig. 3.16 (d) are added base-to-tip in (e). Because the phasors are all parallel, the sum never deviates from a straight line, so it grows and grows, giving a large positive value at the end of the period.

Thus the remaining essential feature of the sieve must be selection: to straighten the wanted wave, but to leave all the others oscillating through all angles. This selective process consists of multiplication by the desired wave's *conjugate* (§3.3.5). Conjugate phasor-waves have equal repeats but rotate in opposite directions, so their phase angles are equal and opposite. If we multiply a phasor with its conjugate, these opposite phase angles add and the result is always zero phase. During the CC sieving process, a phasor gets multiplied by test phasor-waves of all sorts (i.e. of all repeats). Only when the test wave is the conjugate does the multiplication yield a product phasor-wave which is completely straight, so that it adds up

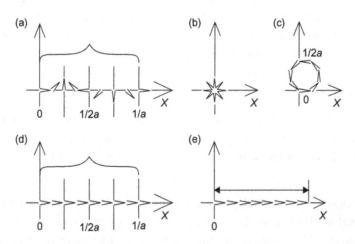

Figure 3.16: (a) Over a complete period of a phasor-wave, all the phasors are to be added together. (Here only eight phasors are drawn, but that shows how the calculation proceeds.) **(b)** The phasors are added by putting them all at a common origin. **(c)** They are added by the 'base on tip' procedure. **(d)** The same length as in (a) of a phasor 'wave' of infinite period; all the phasors are the same, with zero phase. **(e)** The sum of these phasors: unlike those in (b) or (c), they add together.

to a positive value. All other test waves give, after multiplication, a varying wave that is destroyed by summing the phasors.

Summarizing, the CC sieve works by excluding all phasor-waves except the one with precisely the same repeat as itself (i.e. same *size* repeat, but opposite *sign*). It excludes them by turning them into some wave with a fairly short repeat. The only wave to pass through the sieve gets turned by multiplication into a *constant* wave. However, each phasor-wave is 'rescued', not by itself, but by its *conjugate*.

3.4.4 Convolution Theorem

In Fig. 3.10 we generalized sines/cosines as combinations of phasor-waves. Although that figure was in reciprocal space, phasor-waves can also serve as Fourier components, like the sines and cosines in Fig. 3.1 that fit a fundamental repeat p in real space. The cosines were used in Fig. 3.3 for Fourier series, and it is to these that we must briefly return.

Phasor-waves fitting a fundamental repeat p in real space are shown in Fig. 3.17, column (B). Column (D) shows their FT, an isolated unit peak that moves rightwards from line (a) to the bottom. Associated with this movement of the peak, the rotation of the wave becomes progressively more anti-clockwise. As it is clockwise at the top, it starts by rotating more

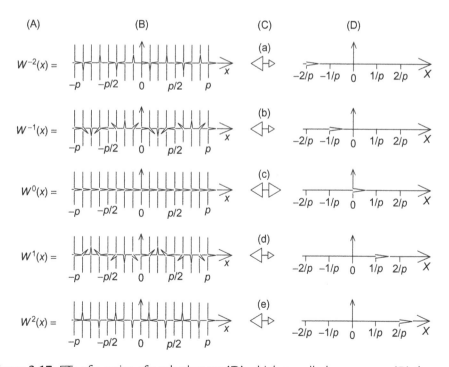

Figure 3.17: FTs of a series of peak-phasors **(D)** which are all phasor-waves **(B)** that are powers of $W(x)$ in line **(d)**. See text.

slowly until, at line (c), the rotation stops before reversing in (d), (e). Thus, when the peak is closer to the origin, the phasor-wave has a longer wavelength.

To understand further, consider the phasor-wave in column (B) on the left of (d), and note the phasor that is at $x = p/2$. This left-pointing phasor is -1, and immediately below it (in the bottom phasor-wave) is a right-pointing phasor that is $+1$, its square. Indeed, all the phasors at $x = p/2$ alternate in sign. This alternation really consists of successive 180° *rotations* in successive lines. Moreover, if we trace the phasors at any x-value, we find the same successive rotations, though by different angles at different x-values. Recall (from §3.3.3) that, when we *multiply* phasors, we *add* their angles. So successive rotations imply successive multiplications by the *same* phasor. The same phasor-wave that brings about all these changes is line (d) which is labeled $W(x)$. Thus each line is the line above it, multiplied by $W(x)$; so we can label all the lines as powers of $W(x)$, as shown on the left. Therefore we might summarize the diagram as showing that progressive shifts (translations) of the peak (on the right) correspond (in the FT on the left) to progressive multiplications of the phasor-wave by $W(x)$. In other words, all the phasor-waves are ($\pm$) powers of $W(x)$ on line (d).

All the phasor-waves could be Fourier-components of experimental curves, as in Fig. 3.3. (However, phasor-waves are complex, and a real curve would therefore require combinations that are real, as in Fig. 3.10.) The Fourier-components are shown as real peaks in Fig. 3.3. These are equivalent to the FTs of the phasor-waves shown in (D) of Fig. 3.17 (though the peaks there are shown as phasors parallel to the X-axis).

Having established the similarity between Fig. 3.17 and Fig. 3.3, we can extend it to analyzing experimental curves by finding their Fourier components. The only difference is that the components in Fig. 3.3 are cosines whereas those in Fig. 3.17 are phasor-waves; but those are more general than cosines, and allow the analysis of odd functions or general functions, as well as even ones. There is only the restriction that, to represent a 'real' experimental curve, the phasor-waves must come in pairs with Friedel symmetry.

Thus we could analyze the experimental curve into phasor-wave components, which would then be summarized as a Fourier-diagram. Next, suppose we had *two* such curves, f and g, each composed of different combinations of phasor-waves. What would happen if we multiplied them? Each curve is a weighted sum of the basic phasor-waves, so we would have to multiply the two weighted sums. This sounds a complicated operation until we recall that all the phasor-waves are ($\pm$) powers of $W(x)$ on line (d). Expressing a function in terms of these is like recording a number in digital representation[8]; so multiplying $f(x)$ and $g(x)$ is only

---

[8] Another analogy between FTs and digital numbers arises because the higher powers (positive and negative) of $W(x)$ represent fast oscillations that encode fine detail. Thus they should be paired, and then the series starts with $W°(x) = 1$; i.e. the FT resembles a number like 1.23456…which would usually be rounded to something like 1.2. Thus rounding decimals (preserving only the most important information) is analogous to omitting fine detail from an FT, equivalent to resolution loss (which also preserves only the most general features of a structure).

a matter of multiplying two digital 'numbers'. For example, suppose $f(x) = 2.W^2(x) + W(x)$ and $g(x) = 3.W^2(x) + W(x)$. Then $f(x).g(x) = [2.W^2(x) + W(x)].[3.W^2(x) + W(x)]$, which we can represent as an ordinary digital multiplication:

$$
\begin{array}{r}
21 \\
\times\ 31 \\
\hline
21 \\
63 \\
\hline
651 \\
\hline
\end{array}
$$

which means that the answer is $6.W^3(x) + 5.W^2(x) + W(x)$. This (complex) curve could be calculated and plotted with phasors, but it is simpler just to show its FT (equivalent to the Fourier-diagram). This is the same as Fig. 2.3 (e), where it was obtained as a CCF. However, as discussed in section §2.1.4, convolution is a very closely related operation. Moreover, both can be obtained by digital multiplication, giving convolution when the order of the digits is *not* reversed.

Thus the effect of *multiplying* the curves (i.e., the functions $f(x)$ and $g(x)$) is to *convolute* their FTs. Although we have discussed only Fourier series, we saw in section §3.4.1 that the FT can be viewed as a limit of a Fourier series taken to limits. So we have the convolution theorem: multiplication of functions in one space (real or reciprocal) is equivalent to convolution of their FTs in the other space.

The general convolution theorem is: if $f \leftrightarrow F$ and $g \leftrightarrow G$, then both:

$$f \star g \leftrightarrow F \times G \text{ and } f \times g \leftrightarrow F \star G \tag{3.1}$$

Indeed, the simplest summary of the convolution theorem is just:

$$\times \leftrightarrow \star \tag{3.2}$$

The convolution theorem has many important consequences, of which the leading ones are treated in the next two sections.

3.4.5 Lattice Sampling and the Convolution Theorem

For X-ray crystallography, the most important consequence of the convolution theorem is the phenomenon of *lattice sampling* of a molecule's FT. This is also relevant when crystals are studied in electron microscopy. We introduce this topic here at the 1D level, and later pursue it in Chapter 6 in 2D.

Patterns that repeat according to a 1D lattice of length p have FTs which occur only at multiples of $1/p$ (e.g. Fig. 3.3). This also follows simply from the convolution theorem. In the upper line of Fig. 3.18, (a) is a function and (b) a lattice, and convoluting the two

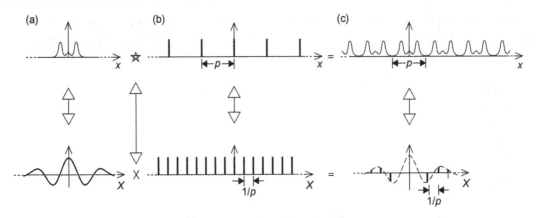

Figure 3.18: Curves (top) with their FTs (bottom). **(a)** A (limited) curve is convoluted with... **(b)** ...a comb of spacing p, giving... **(c)** ...a repeating curve. The FTs (below) show that (a) the curve's FT, times... (b) the comb's reciprocal lattice, with spacing $1/p$, gives... (c) the FT in (a) sampled at $1/p$.

gives a repeating function (c). This process resembles the formation of a crystal (c) (here only 1D) by the association of copies of its repeating structure (a). The lower line shows FTs of the upper line. The repeating structure's FT (a) gets multiplied by the lattice's FT (the reciprocal lattice of the upper line (b)). Multiplication by a lattice preserves only those values coinciding with the lattice points, so it means *sampling* of the FT under (a). This is shown under (c).

Another consequence is that a function F that has a repeat p is unaffected if convoluted with a lattice L with the same repeat. For the whole of F gets copied at each point of L and, since L has the same repeat p, all these copies are the same and get exactly superposed. So the convolution has the effect of adding together an infinite number of copies of F, i.e. of multiplying F by infinity. Now we need to ignore the scale of functions in infinite lattice convolutions (e.g. by making the weight of the lattice points of L equal to the reciprocal of the number of points). In that case, $F \star L = F$.

3.4.6 Cross-Correlation Functions and the Convolution Theorem

The convolution theorem also provides a new way to calculate correlations, since convolution is closely related to the important operation of finding the CCF; §2.1.4. There it is shown how convolution is akin to long multiplication of two numbers in digital representation.
If one of these numbers is reversed, however, the long multiplication yields the CCF instead of convolution. Thus the convolution of two functions becomes their CCF if one of the functions is reversed. Therefore $\mathrm{CCF}(f,g) = f \star g(-)$, where $g(-)$ means the function

g reversed. If $f \leftrightarrow F$ and $g \leftrightarrow G$, then $g(-) \leftrightarrow G^*$ (see Fig. 3.14 (a),(c)). So, using the convolution theorem, we get an equation equivalent to (2.5), §2.14:

$$\text{CCF}(f, g) = f \star g(-) \leftrightarrow FG^* \tag{3.3}$$

If $g = f$, the calculation yields the ACF of f, so:

$$\text{ACF}(f) = f \star f(-) \leftrightarrow FF^* = |F|^2 \tag{3.4}$$

In crystallography, the ACF is called the **Patterson function** (Patterson, 1935). This is important as it can be calculated directly from the observed X-ray intensities.

Instead of the direct calculations (such as multiplication described in §2.1.4), convolutions could be calculated by using FTs. Thus, to find the convolution of two functions f and g, we could find their FTs F and G, then multiply them to give FG, and finally find the inverse FT of FG. This route has the advantage that calculating FG involves fewer multiplications than calculating the CCF. Although it involves calculating two forward FTs and one reverse FT, there is an extraordinarily efficient way to calculate FTs (forward and reverse), the **Fast Fourier transform** (FFT); see Chapter 4. Because of that, the apparently indirect FT route is more rapid.

3.4.7 Translation Rule

Suppose we convolute a curve with a single peak, displaced from origin by a. Convolution replaces all the points of one curve by each point of the other. But the second curve (single peak) contains only one point, so convolution just *displaces* the first curve. The effect on the FT is to multiply it by FT(displaced peak) = phasor-wave, period a.

Alternatively, the translation rule can be demonstrated by supposing we multiply an FT by a phasor-wave, and then back-transform it; what do we get? Suppose we back-transform it by the CC-sieve method (§3.4.3). Consider just one Fourier-component. It gives a peak in the back-transform when the test wave matches it in spatial frequency. But it now has a shifted frequency through multiplication by a phasor-wave, so it gets shifted: it has been translated by the FT multiplication.

Note that, when a translated object's FT is multiplied by a phasor-wave, this wave has only phase variations. Its amplitude is constant, so the amplitude of the original FT is unchanged by translation. That is why we can ignore small movements of a crystal, even though they would make big changes to the phases of the diffraction pattern; for we can only record the pattern's amplitudes, not its phases; ignorance is sometimes bliss.

Finally, the first argument (that derived the translation rule from the convolution theorem) can be reversed, deriving the convolution theorem from the translation rule. This is left as an 'exercise for the reader'.

3.4.8 Choosing a Fourier Transform's Origin

Image density data, consisting entirely of real numbers, usually needs to be Fourier-transformed as a prelude for data-processing. Chapter 4 will discuss the special problems arising from digital FTs in a computer. But the computer also needs to be told our choice of the **origin**, that is, which pixel is at $x = 0$. What are the implications of this choice?

If we shift the origin, the amplitude remains the same (as we have just seen). Moreover, the phase at the FT origin must be $0°$, since the data are real, and the phasor-waves used for transformation are all constant (like (b) of Fig. 3.14) at the origin. Thus shifting the origin only affects (by the translation rule) the phases elsewhere, and the 'best' origin choice should yield the optimum phase behavior of the FT away from the origin. What is that?

The choice would be clear if the data-set happened to be an even function, since there would be a unique center and, when that is the origin, all phases are $0°$ or $180°$. Any other origin choice would create unnecessary complexity and conceal an important symmetry in the data. However, it is very unlikely that any experimental dataset could happen to be a *precisely* even function. Nevertheless it could be *approximately* even, with the deviations from evenness being negligible at low resolution. Indeed, this must be generally true, since at the lowest resolution every object approximates to an ellipsoid, which is an even function. Thus we should choose the origin at the unique center (of even-function data) or the center of the ellipsoid to which unsymmetrical data approximate.

How can we find that center in the latter case? Fig. 3.19 shows the effect of two origin positions, (b) at the center of mass and (a) at a distance $a/3$ from it. The phase in (b) is stationary at the origin (as it would be for an even function), showing that the center of mass (c.o.m.) is the best choice for the origin. The phase in (a) changes linearly at the origin, showing the effect of translation by $a/3$ from the c.o.m.

3.5 Summary of Rules

Finally, we summarize the most fundamental rules concerning the process of Fourier transformation. All the rules are summarized in Fig. 3.20.

3.5.1 The Fourier Transform and its Inversion

As explained in §3.4.3, the FT's inversion can be viewed in two ways:

(i)   The *spectral viewpoint*, in which we imagine the curve as a composition of sinusoids, elementary functions to be assayed mathematically. We use the *CC sieve* or assay, which annihilates all except the chosen 'test' sinusoid, and we apply it to all possible 'test' sinusoids.

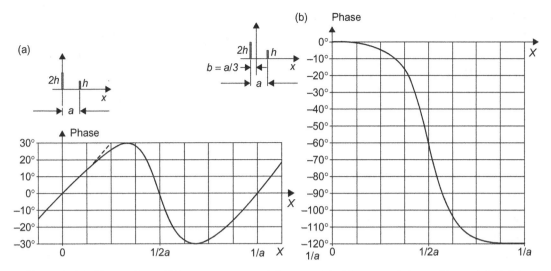

Figure 3.19: Phase plots of the same two-peak image with different origin positions. **(a)** Origin at major peak, but phase changes rapidly at origin. **(b)** Origin at center of mass, and phase is stationary at origin ($X = 0$).

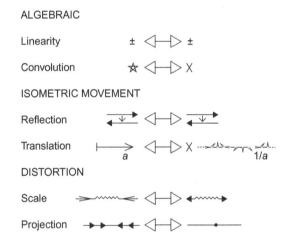

Figure 3.20: The three pairs of rules concerning FTs, explained in the text. (FT pairs are indicated by the simple double-headed arrow.)

(ii)   The *pixel viewpoint*, in which we see the curve as a series of pixels. Each weighted pixel has its own FT (a 'phasor-wave', Fig. 3.14), and we sum all the pixels' FTs to get the function's FT.

3.5.2 Algebraic Rules

If we have two functions, then adding the functions produces a combined function. If we also have the FT of each function, then adding these FTs gives the FT of the combined function.

(The same applies to subtraction.) It follows that adding a function to itself is equivalent to adding its FT to itself. Thus doubling a function doubles its FT and, in general, multiplying a function by a *number* (or even a *phasor*) is equivalent to multiplying its FT by the same number or phasor. This is shown in Fig. 3.14 (d).

But multiplying the function by another *function* leads (§3.4.4) to a new rule: multiplication of functions is equivalent to convolution of their FTs (and vice-versa), the *convolution theorem*.

3.5.3 Isometric Movement Rules

- **Translation** or shifting of the function causes its FT to be multiplied by a unit phasor-wave of period 1/(shift distance) (§3.4.7).
- **Reflection** reverses the direction of the function and also of its FT (see Fig. 3.15). (In 1D this is equivalent to rotation about 180°, a simple example of the more general **rotation rule** in 2D and 3D.)

3.5.4 Distortion Rules

Stretching a function along x causes its FT to become compressed to the same extent along X, where the 'same extent' means that (change in x) (change in X) = 1 (the **scale rule**, first met in §3.2.1).

The **projection rule** is a limiting case of the scale rule, where compression of the function becomes total. Then stretching of the FT also becomes total, leaving us with only its central part (at the origin). So the projection of a 1D function onto the origin corresponds to the value of the FT at the origin. Also, reversing real and reciprocal space, the projection of the FT onto the origin corresponds to the value the function at the origin. (This rule has more interesting consequences in 2D and 3D.)

Digital Fourier Transforms

Chapter Outline

Fourier Transforms (FTs) are important as the mathematics of wave optics in microscopes, for revealing translational symmetries in images, and for calculating Cross-correlation Functions (CCFs) between unsymmetrical images to help in their comparison and averaging. All these benefits depend ultimately on calculating FTs from data or from other FTs, and we now address this practical necessity.

Previous chapters have introduced the two fundamental types of FT that emerged after the original Fourier series were adapted to data that are not necessarily periodic. These types are the *continuous FT*, adapted for mathematical deductions, the form ordinarily implied by 'FT'; and the *digital FT* (DFT), adapted to practical calculations where the data and the results are unavoidably in numerical or digital form. However, we want the results to approximate those of the continuous FT, so we need rules that will make this possible.

Digital FTs have two special features. First: they are 'all peaks': both the original data and its FT exist only at isolated, regularly-spaced *points*, so there is no question of any continuity. Second: there is, in both data and FT, an *infinity* of points containing a strictly *limited* amount of information. This apparent paradox is explained by the way the limited information fills relatively few points that repeat indefinitely. (This set of points forms the *repeating unit*, and its length is the *period*, of the digital data or FT.) The full set of points could form an infinitely long line, or merely a circle (which avoids data-termination problems, and reminds us that the FT isn't confined to translational symmetry). If the DFT is to be completely reversible, the number of peaks per repeat must be the same for the data and its FT.

We discuss first the effects of sampling and the distribution of peaks and their heights. Then we look at the representation of periodicity in the data, and at ways to speed FT calculations.

4.1 Data Preparation

4.1.1 Effects of Sampling

Computers use arithmetic and are therefore subject to its requirement that everything must be represented as separate digits. This simple fact imposes conditions and restrictions. Obviously the image data must be digital so, if analog, the image must be scanned to give a set of numbers. These represent digitally the average image density at equally-spaced points constituting a *raster*.

We have to make two important choices when preparing our data. We must choose the distance Δx between the raster points (corresponding to the image resolution), and we must adjust the total distance d scanned in each dimension to fit the extent of the specimen. Then, combining these two choices and incorporating other considerations (see below), we must choose the desired number N of image data points in each dimension. This is called the **transform box**. (In this chapter we confine discussion to one dimension, 1D, but higher dimensions simply use the same conclusions in each dimension.)

The output is determined by these input choices, whose effects are reviewed in Fig. 4.1. We start with a continuous image (a) that we digitize by sampling with a raster (c) of spacing Δx.

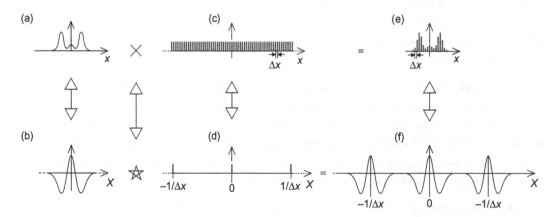

Figure 4.1: Effect of image-sampling on the FT. **(a)**, **(b)** are the image and its (hypothetical) FT. **(c)**, **(d)** are the image's sampling raster and its FT. **(e)**, **(f)** are the sampled image and its FT. (Note that (a) and (b), etc. are not an *actual* FT pair, but merely shorthand symbols.)

Sampling is equivalent to multiplying (a) with a raster function (c), and yields the sampled data (e). The effect of this sampling is revealed by taking FTs (bottom line). The FT of the continuous image (a) is a continuous FT (b). The FT of (c), the raster function (a lattice), is its reciprocal lattice (d). And the FT of multiplication is convolution yielding (f), the continuous FT that the computer should calculate from the input data. Since the continuous FT (b) has been convoluted with the reciprocal lattice (d), a periodicity of $1/\Delta x$ has been imposed on it.

But that is not all. The computer calculates the FT at equally-spaced points: it therefore imposes a sampling on the FT. The effect of this is shown in Fig. 4.2. Start with the sampled image (a) and its periodic FT (b), both the end-points of Fig. 4.1. This FT is sampled with a raster (d) of spacing ΔX, so the computer output corresponds to (f). As in the earlier diagram, we see the effect of this sampling by taking FTs. The FT of (b) is the sampled image (a). The FT of the raster lattice (d) is its reciprocal lattice (c); and the FT of multiplication is convolution. So the FT of the sampled FT (f) is the periodic function (e).

We need to be aware of the implications of Figs 4.1 and 4.2. We started with an image (a) of Fig. 4.1, and we wanted its FT (b). We thought we were giving the computer the data (a) of Fig. 4.2, but we were really giving it the data (e). Also, we thought we calculated a sampling of the FT in (b) of Fig. 4.1, but we actually got (f) of Fig. 4.2. Just as Δx sampling of the image imposes a periodicity $(1/\Delta x)$ on the FT, so also ΔX sampling of the FT imposes a periodicity $(1/\Delta X)$ on the effective image input. Computer processing makes the image, and also its FT, both evenly sampled and periodic. However, both sampling processes are under our control, so what are the implications of our decisions?

4.1.2 Digitizing Equations

We need some basic equations on which to base our discussion of parameter choice. The connection between the digitized image (a) and its digitized FT (b) is shown in Figs 4.2 and 4.6, where $p(=1/\Delta X) =$ real period and $P(=1/\Delta X) =$ reciprocal period. Since (a) is

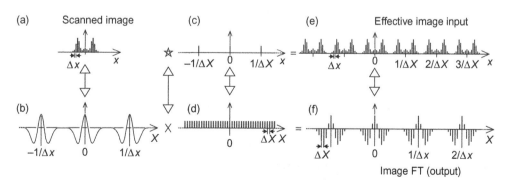

Figure 4.2: (a), (b) corresponds to (e), (f) of Fig. 4.1. **(d), (c)** are the FT's sampling raster and its FT. **(e), (f)** are the sampled image and FT, together with the periodicities they impose on each other.

periodic (period p), its convolution with a lattice (or 'comb') of period p produces no change. Therefore (by the convolution theorem) multiplication of (b) with the reciprocal lattice (period $1/p$) also produces no change. Therefore (b) exists only at multiples of $1/p$. But the points where it exists are defined as multiples of ΔX. So $1/p = \Delta X$.

The same arguments also apply to the convolution of (b) with a lattice of period P and show that $1/P = \Delta x$.

We can now deduce the basic equations for an image with N samples (transform box). The real period $p = N\Delta x$, so $1/N\Delta x = \Delta X$ or $N\Delta x \Delta X = 1 \ldots$ (i). Also $P = N\Delta X$, so $Pp = N\Delta X. N\Delta x = N^2 \Delta x \Delta X = N \ldots$ (ii). Finally, $p\Delta X = (N\Delta x)\Delta X = 1 = (N\Delta X)\Delta x = P\Delta x \ldots$ (iii).

Summarizing, we have three interconnected digitizing equations:

$$\text{(i) Pixel (or raster-sampling) equation}: \quad N\Delta x\Delta X = 1 \tag{4.1}$$

$$\text{(ii) Period equation}: \quad Pp = N \tag{4.2}$$

$$\text{(iii) Period-pixel equation}: \quad p\Delta X = 1 = P\Delta x \tag{4.3}$$

4.1.3 A Digital Fourier Transform Example

Figure 4.3 shows the Fourier-series constructed from positive cosines that have odd numbers of waves per repeat p. All waves have positive peaks at the repeat boundaries, but odd numbers of waves always have a negative peak at the center of the repeat. When all are added together, only these regular features survive, so we get alternating positive and negative peaks (f). The Fourier diagrams of course lack any even peaks from the comb, so there is a gap at $X = 0$ and at alternate pixels thereafter. The reverse diagram can be deduced similarly, giving a FT relationship that provides a good example of the three equations above. In (f), the left side has a period p and pixels $\Delta x = p/2$. On the right, the pixels are (from the period-pixel equation) $\Delta X = 1/p$, so $\Delta x \Delta X = \frac{1}{2}$, confirming the pixel equation $2\Delta x \Delta X = 1$ for $N = 2$. The period on the right of (f) is $P = 2/p$, fitting the period equation $Pp = 2$.

We now give an example of using the equations. Consider the following case.

An image photographed at a magnification of 60,000, and scanned at 7 μm, has adjacent pairs of pixels averaged. If the box size is $N = 256$, how many FT pixels equal $1/25$ Å?

First we prepare data for the equation. We recall the relevant length units: $1\,\text{mm}/1000 = 1\,\mu\text{m} = 1000\,\text{nm} = 10{,}000\,\text{Å}$. Before we can apply any equation, we need the pixel size at the specimen (Δx). As pairs of pixels were averaged, the pixel size $= 14\,\mu\text{m}$ on film $= 14/60{,}000\,\mu\text{m}$ at the specimen $= 14 \times 10{,}000/60{,}000\,\text{Å} = 2.33\,\text{Å} = \Delta x$.

Next we use an equation. Since we were given $N = 256$, and know the specimen pixel size (Δx) we can apply the pixel equation (4.1): $(256) \times (2.33) \times (\Delta X) = 1$, so each FT pixel has a width $\Delta X = 1/(256 \times 2.33\,\text{Å}) = 1/(256 \times 2.33)\,\text{Å}^{-1}$. So $1\,\text{Å}^{-1} = (256 \times 2.33)\,\Delta X$ – units or

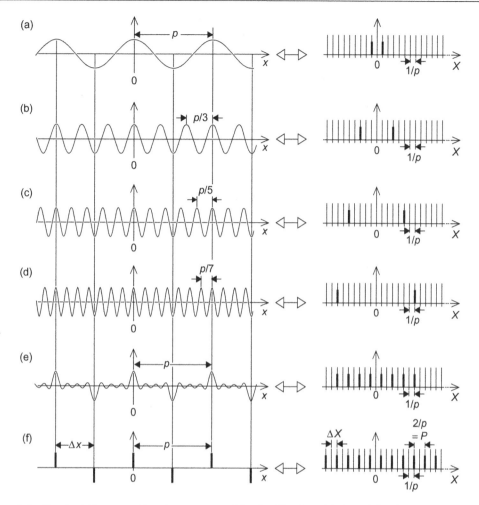

Figure 4.3: On the left, a set of cosine curves, with all *odd* submultiples of the fundamental period. **(a)**, **(b)**, **(c)** and **(d)** are positive cosine peaks of successively shorter periods. Note that all cosine peaks line up along the long vertical lines at 0, $p/3$, $p/5$, ... **(e)** Sum of **(a)**–**(d)**. **(f)** Limit of infinite sum would give sharp peaks. Left: alternating comb, period p.

FT pixels. However, we were not asked for $1\,\text{Å}^{-1}$, but for $(1/25)\,\text{Å}^{-1}$ which requires $(256 \times 2.33)/25\,\text{FT pixels} = 23.9$ pixels.

4.1.4 Resolution and Raster Size

We want the raster to preserve all available specimen detail, but exactly how small must it be?

First, we need an estimate of detail-level present after averaging. Perhaps we might find it from optical diffraction of several particles, measuring the widest diffraction spacing not caused by the *contrast transfer function* (Chapter 9, section §9.1.1). Or perhaps we just make a guess based on experience of similar specimens. Anyway, the estimate should be explicit so that we will know if the subsequent resolution limit was caused by experimental or computational problems.

From the 'spectral method' (Chapter 3, section §3.4.3), the image can be regarded as the sum of many different sinusoids, and the size of its finest detail is determined by the sinusoid with minimum wavelength λ_{min}. (Therefore the diffraction peak corresponding to the finest wave is at $D = 2/\lambda_{min}$.) To record the 'minimum wavelength' sinusoid, we need to record at least two successive turning points (i.e., an adjacent peak-trough pair, of which there are two per period). So the raster size should be exactly half the minimum wavelength. That size corresponds to the *Nyquist critical sampling frequency*. If we sample the image at that frequency, the maximum raster size to avoid loss of detail will be:

$$\Delta x_{max} \leq \lambda_{min}/2 = 1/D \tag{4.4}$$

The implications for the FT are explored in Fig. 4.4. Part (a) deals with the 'minimum wavelength' sinusoid itself. It shows (left) a cosine with minimum wavelength λ_{min}, and (right) its FT derived as in many previous diagrams (like the previous Figure). Part (b) deals with the Nyquist criterion (two successive turning points). It represents these minimal data as peaks, that form an alternating comb of period λ_{min}. This is the same as the left of (f) in Fig. 4.3, but p in that Figure becomes λ_{min} in Fig. 4.4 (b). If the Fourier-space period is P, then the period-pixel equation (4.2) tells us that $P'\Delta x = 1$. So $P \times (\lambda_{min}/2) = 1$ and $P = 2/\lambda_{min}$. This is the same as the diffraction peak spacing D: $P = D$.

Altogether, we conclude from the minimum-wavelength/Nyquist criterion that the maximum raster size that avoids loss of detail is:

$$\Delta x_{max} \leq \lambda_{min}/2 = 1/D \tag{4.5}$$

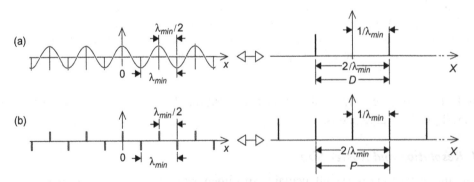

Figure 4.4: (a) The cosine wave of minimum wavelength λ_{min} is sampled at the Nyquist frequency, i.e. with a raster of spacing $\lambda_{min}/2$, shown as a series of $\pm$ rectangular peaks. Right: FT of the cosine wave: peaks at $\pm 1/\lambda_{min}$. **(b)** The Nyquist samplings are shown as peaks (left). Note its similarity to Fig. 4.3, where the FT corresponds to the right of this Figure.

4.1.5 Sampling Frequency and Aliasing

The condition just laid down is not merely desirable; its violation entails a penalty called **aliasing**. In the ideal situation (approximated by some analogue methods like optical diffraction), the continuous image curve gives just one copy of the 'true FT', which is also continuous. However, the situation changes with digital FTs. As shown in Fig. 4.1, sampling the image (a) with a raster of spacing Δx causes the 'true FT' (b) to become convoluted with the raster's FT (d) to give (f) in which copies of the true FT are repeated with a period of $(1/\Delta x)$, the reciprocal of the raster size. Varying the raster size must change the repeat period of the FT, and the consequences are explored in Fig. 4.5. With a sufficiently fine sampling raster (b), the copies of the FT are separated by a gap, letting us isolate the FT from only one period (the computer output). But using a coarser raster (c) eliminates this gap, replacing it with a small overlap; the FT gets 'worn at the edges'. Finally, a really coarse raster (d) leads to a significant overlap between the FTs, so they are 'glue-jointed together' and we can no longer isolate the true FT. This defect, aliasing, reduces the width of the reliable FT. It eliminates the sinusoids of higher spatial frequency, thereby lowering the resolution of the information.

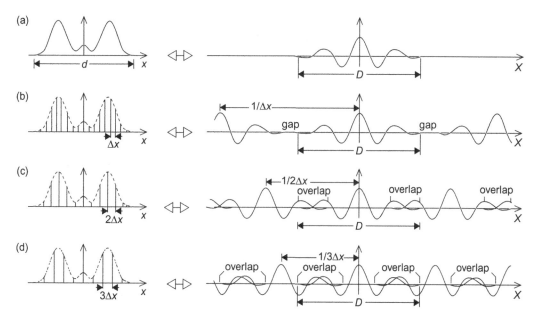

Figure 4.5: Effect of sampling raster on calculated FT: aliasing. The image has width d, and various sampling rasters (Δx etc.) in (b)–(d). Its FT has width D and a periodicity P which is the reciprocal of the image sampling raster. As the raster gets coarser (i.e., bigger), P gets smaller, becoming smaller than D in (c) and (d). **(a)** Infinitesimal raster gives isolated FT. **(b)** Fine raster Δx gives copies of FT separated by a gap. **(c)** Coarse raster $2\Delta x$ gives FT copies with no gap, but a small overlap (aliasing). **(d)** Coarser raster gives FT copies with a bigger overlap (more severe aliasing).

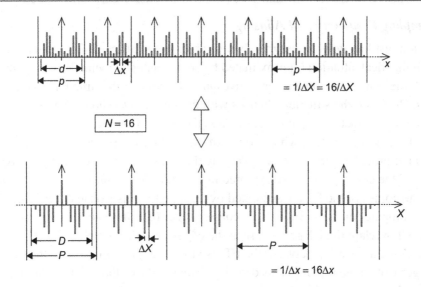

Figure 4.6: General scheme of the usual computer-calculated DFT from (e), (f) of Fig. 4.2. The digitized real-space image is above, its DFT below. The number of pixels per transform box (or repeat), N, is 16. The raster (pixel) size is Δx in real space, ΔX in reciprocal space. The real space period is p, and the reciprocal space period is P. The repeat unit's diameter is d in real space, D in reciprocal space. In practice, $(p - d)$ is zero-padding (Fig. 4.7(c)) introduced by the operator, but $(P - D)$ is the noise space ('gap' in Fig. 4.5(b)) between calculated DFTs.

The detailed implications of Fig. 4.5 are best analyzed by reference to Fig. 4.6. When the image scanning-raster is fine enough to avoid aliasing (i.e. to keep adjacent FT copies apart), the critical raster puts these copies in contact. This means that, to avoid aliasing, we must make the raster Δx_{max} small enough that $D \leq P$. From the period-pixel equation, $P.\Delta x_{max} = 1$, so $P = 1/\Delta x_{max}$. But we found above that the FT width $D = 2/\lambda_{min}$. Thus the condition $D \leq P$ implies that $2/\lambda_{min} \leq 1/\Delta x_{max}$, which implies that $\lambda_{min}/2 \geq \Delta x_{max}$. We are thus led back to the condition (4.5).

Figure 4.5 shows another feature of the DFT that needs to be considered: the spacing ΔX of *its* sampling. If that is too coarse, we may miss details of its FT or be unable to interpolate it with sufficient accuracy. We can control ΔX as shown in Fig. 4.7, where the raster sampling of the FT (right) is finer in (d) than in (b), because the transform box is bigger in (c) than in (a). It has been made bigger by including extra zeroes in the data fed into the computer (**zero-padding** or **zero-filling**).

4.1.6 Specimen Parameters for Digital Fourier Transforms

We already have three of these parameters: the specimen size (d), its resolution (λ_{min}) and its maximum raster size ($\Delta x_{max} \leq \lambda_{min}/2$). We also need the box size N, which determines

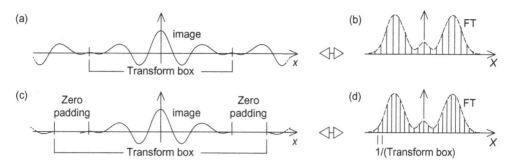

Figure 4.7: Effect of transform box size on the sampling raster of the FT. To get a suitably fine FT sampling **(d)**, we need to increase the transform box of the image by zero-padding **(c)**.

the real-space period $p = N\Delta x$ which must be bigger than the specimen size d. Therefore $N\Delta x_{\text{max}} \geq d$ or $N \geq d/\Delta x_{\text{max}}$.

That is one factor determining the box size N. The other factor is the need for speed, which favors a fast FT algorithm (the FFT, §4.4.2) that requires (i) the same number N of peaks per repeat in both real (x) and reciprocal (X) space and (ii) that N must be a number with only small prime factors, e.g. a power of 2. So we calculate $d/\Delta x_{\text{max}}$, and then take the next small-prime composite as the box size:

$$N \geq d/\Delta x_{\text{max}} \qquad (4.6)$$

Having chosen N and Δx for the specimen, the FT pixel size ΔX is set by the pixel equation $N\Delta x\Delta X = 1$: $\Delta X = 1/(N\Delta x) = 1/p$ (from the period-pixel equation). A finer sampling ΔX of the FT would shorten the distance over which we need to interpolate, improving its accuracy. If that is desirable, we can reduce ΔX by increasing N even further. That will of course increase the real-lattice period p without changing d (defined by Δx). The extra distance between p and d is filled with zeroes (zero-padding). Incidentally, just as zero-padding in real space diminishes ΔX in reciprocal space, so zero-padding in reciprocal space diminishes Δx in real space. Indeed, finer sampling in either space (real or reciprocal) decreases the overlap of (or increases the gap between) periodic copies in the other space.

4.2 Digital Fourier Transform Features

4.2.1 Amplitude Scaling Rules

In the previous section we derived the coordinate scaling rules relating periods and pixels in the original image $f(x)$ and its FT (i.e., $F(X)$). At each pixel of a DFT there is a peak, and we now consider how to scale the peaks' *amplitudes*.

The projection theorem tells us that the continuous FT's origin peak-amplitude F_0 equals the sum of all the peaks f_j in the original real function. The theorem also applies to the FT, so the reverse is valid: the origin value of the real function f_0 equals the sum of all the peaks F_j in the FT. However, some tweaking of these rules is necessary to apply them to DFTs. First, DFTs consist of *infinite periodic* series, so we replace 'the sum of *all* the peaks' by 'the sum of the peaks within one period'. Second, we shall find that scaling DFTs involves new asymmetries; we can't apply exactly the same rules to the function values (f_j) and to those of its FT (F_j), without taking account of the different scale-factors for direct and inverse FTs. In the Appendix we derive the **amplitude scaling equations**:

Projection: $$F_0 = \Sigma f \tag{4.7}$$

$$\Sigma F = N f_0 \tag{4.8}$$

Parseval: $$\Sigma |F|^2 = N \Sigma f^2 \tag{4.9}$$

The application of these rules to a simple case ($N = 2$) is shown in Fig. 4.8, where the peak amplitudes are listed in columns and arrows connect terms of the equations. A more complicated example is in Fig. 4.10 (a):

$$N = 8; \; F_0 = 8 = \Sigma f = 2 + 1 + 1 + 2 + 1 + 1$$

$$f_0 = 2; \; \Sigma F = 8 + 4 + 4 = 16 = 8 \times 2 = N f_0$$

$$\Sigma |F|^2 = 8^2 + 4^2 + 4^2 = 64 + 16 + 16 = 96$$

$$\Sigma f^2 = 2^2 + 1^2 + 1^2 + 2^2 + 1^2 + 1^2 = 4 + 1 + 1 + 4 + 1 + 1 = 12$$

$$96 = \Sigma |F|^2 = N \Sigma f^2 = 8 \times 12$$

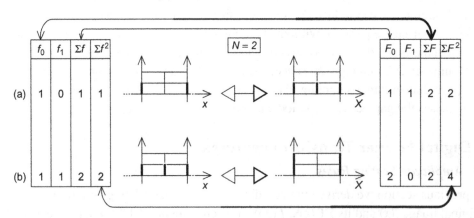

Figure 4.8: The horizontal double-headed arrows connect terms in the provisional scaling equations and establish the scaling constants for an origin peak. Where the connecting arrows are thicker, a factor of N applies. **(a)** The $N = 2$ DFT of one origin peak per period has two peaks per period. **(b)** The $N = 2$ DFT of two peaks per period. (Origin peaks are under *vertical* arrows.)

Note that the amplitudes of the FT and its inversion are not symmetrical, even where the function is even so that inversion symmetry would be expected. This is a special feature of DFTs, confined to their amplitude scaling. We indicate it in diagrams with a double-headed arrow where one arrow-head is thicker than the other.

4.2.2 Sub-Periods

In a DFT, the unit fits the period. In real space, each period p is a copy of the entire particle image. But that image may itself have translational (shift) symmetry (Chapter 10) which of course will show up in the FT. So any translational symmetry of the particle must appear in Fourier space (indeed, revealing such symmetry is the simplest application of FTs).

Translational symmetry is revealed in a DFT as **sub-periods** or **internal repeats** within the main DFT period P (that contains the whole FT of the image). As a first example, start with the simplest possible repeat: a double peak. Fig. 4.8 (a) shows that when $N = 2$, a solitary origin peak gives a DFT with two peaks per period. A sub-period within the data (b) is simply a second peak. This is actually the same as the right of (a), and (by the inversion rule) gives essentially the left of (a). The only slight difference is in the scaling of the peaks, shown as in (b) to follow the amplitude scaling rules.

The sub-peaks are shown better in Fig. 4.9, with $N = 6$. Six repeated peaks are shown on the left of (a), with a very strong reciprocal peak on the right. The inverse FT is in (b). Three peaks are shown on the left in (c), giving two strong peaks on the right (and the reciprocal FT is shown in (d)).

Note that a sub-period within the transform box, as in (a), (c) and (d), causes the FT peaks to be **zero-interlaced** (i.e., every previous pixel now alternates with a new pixel of zero content). This is biggest when the peaks in x are closest, in (a), but least (no zero-interlacing) when the x-peaks are furthest apart (b). The number of equidistant peaks must be a factor of N, so 2 and 3 are the only other possibilities, both for left and right sides. Note that, for uniform equidistant peaks, their numbers follow the rule:

$$(\text{Peaks per period on left}) \times (\text{Peaks per period on right}) = N$$

This follows from the raster sampling equation (4.1). Take, for example, (d) with a peak at $3\Delta x$: $6.\Delta X.\Delta x = 1$ so $3\Delta x = 1/2\Delta X$, implying an FT peak at $2\Delta X$, which is where it is. One consequence is that there is more diversity of sub-periods when N has more factors (and the usual choice of a power of 2 for N allows any smaller power of 2 as a factor). Another consequence is that more sub-periods generate wider zero spaces between the FT peaks. That is, internal repeats $\leftrightarrow$ zero-interlacing. This seems to fit a version of the scale rule: compression gives extra repeats, so in the FT we have stretching giving gaps.

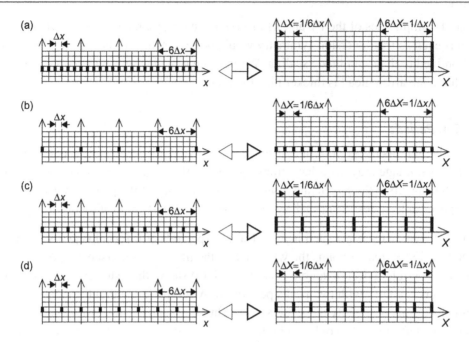

Figure 4.9: Multiple sub-periods in $N = 6$. **(a)** Six sub-periods give a single peak FT. **(b)** Inverse of (a). **(c)** Three sub-periods per period. **(d)** Two sub-periods per period; the inverse of (c).

A more complicated pattern is shown in Fig. 4.10. Consider the top line (a). Suppose we were interpreting the FT on its right, i.e. at the top-right. The FT peaks at the left are at 0 and $2\Delta X$. What real spacing corresponds to $2\Delta X$? Again, use the raster sampling equation: $8.\Delta X.\Delta x = 1$ so $2\Delta X = 1/4\Delta x$. So the reciprocal peak at $2\Delta X$ implies a real repeat of $4\Delta x$, which is true.

Another aspect of repeats and zero-interlacing is information. The box size N defines the information content of the data, which equals that of the FT (as Fourier transformation is invertible.) To get a better FT, we need finer data resolution, equivalent to higher data information, requiring a big N. Exact repeats and zero-interlacing are both features that reduce the information content of the DFT, so it is not surprising that these modes of information-reduction are FTs of each other.

Where an $N = 8$ DFT has an *exact* data repeat, and consequently *exact* zero-interlacing, then neither real- nor Fourier-space really needs $N = 8$: $N = 4$ should suffice. So how does this $N = 8$ situation relate to the corresponding $N = 4$ DFT? This is examined in (a) and (b) of Fig. 4.10, which shows repeated, or zero-interlaced, $N = 8$ patterns[1] (a) being reduced to

---

[1] Such relatively short 1D DFTs can be described in the text by using a convenient notation of Bracewell (2000). The N peaks per repeat are entered as the N peak heights which are put between ordinary curved brackets (), and it is convenient to use curly brackets { } to represent the peak heights of the DFT in reciprocal space. So Fig. 4.10 (a) could be written as (21012101) $\leftrightarrow$ {80400040}, and (b) as (2101) $\leftrightarrow$ {4202}.

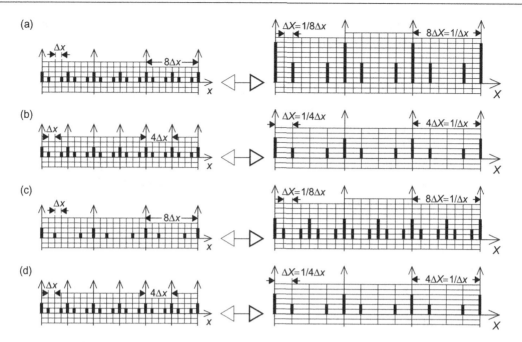

Figure 4.10: (a) Two patterns per repeat give a DFT with one such pattern zero-interlaced. **(b)** The transform box of (a) has been halved, showing the connection between zero-interlacing/sub-periods and box-bisection. **(c)** The inverse of (a). **(d)** (left) The pattern of (a) but in half its box.

$N = 4$ patterns (b). Lines (a) and (c) are $N = 8$, while line (b) is their $N = 4$ reduction. From the comparison, we see that the following rules apply:

(i) If there is an exact sub-period repeat midway within the period, we should promote it from sub-period to period. This halves the repeat and thereby halves the box-size in real space.

(ii) If there is exact zero-interlacing (every other pixel completely zero), we should omit the gaps. This halves the box-size in reciprocal space.

These rules are framed on the assumption of two sub-periods, the most important case. Of course, there could be more, leading to a DFT with zero-interlacing where every 3rd, 4th, etc. pixel is not zero.

If we had to find the DFT of data that was completely zero-interlaced, we could divide it and thereby halve the DFT calculation. We shall see later (§4.4.3) how this simplification is exploited to accelerate DFT calculations.

4.3 Digital Fourier Transform Calculations

4.3.1 Basic Calculation

It is easiest to give a simple example of this, from which general principles are apparent. For simplicity, we choose $N = 4$, the smallest N which is both a power of two and also gives a complex FT requiring phasors. We calculate the DFT by the pixel method. There are just four possible data-peaks, at 0, Δx, $2\Delta x$ and $3\Delta x$.; i.e. at $j\Delta x$ (j = 0, 1, 2, 3).

The data-peak (phase 0°) at $j\Delta x$ gives a FT which is a phasor wave (§3.4.2) whose amplitude equals the data-peak height and whose angle is 0° at the origin (X = 0) but reaches 360° when the phasor wave repeats at $1/j\Delta x$. Each data-peak generates a characteristic phasor wave. A unit data-peak (with unit amplitude and zero phase) generates one of the unit phasor waves, shown on the right of Fig. 4.11. These unit phasors constitute a framework which forms the basis for calculating the N = 4 DFT of an ordinary 'real' image (where pixels have no phase variations). We simply take the four unit density-pixels and multiply the corresponding phasor waves by each pixel's actual density.

An example is shown in Fig. 4.12, which has the same arrangement as Fig. 4.11, except that lines (a)–(c) are (on the left) individual pixels whose sum is shown in line (d). The corresponding individual phasor-waves are on the right of (a)–(c). (Note how the phasors on each axis (right side) correspond to the pixel on the left; there is no pixel at x = $2\Delta x$, so only three lines are needed instead of four.) These three lines are added together, giving the FT on the right of (d). (The addition follows vector addition, adding all the horizontal components to give the horizontal component of the sum, and the same with the vertical components.)

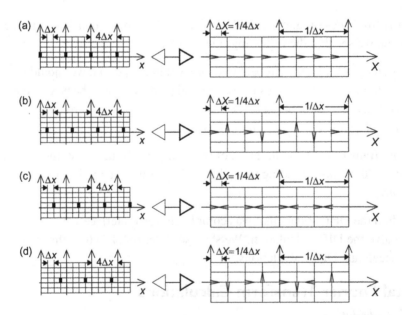

Figure 4.11: The four possible solitary unit peaks and their N = 4 DFTs. The peaks are shown on the left and, on the right, the corresponding DFT, which is a unit phasor wave. **(a)** Peak at origin (x = 0). **(b)** Peak at x = Δx. **(c)** Peak at x = $2\Delta x$. **(d)** Peak at x = $3\Delta x$.

4.3.2 Fast Fourier Transforms: Introduction

Although the calculation described above looks quite simple, a very large number of repetitions is necessary. Thus every part of the calculation must be made as simple and fast as possible. Two aspects are crucial.

(i) First, very many sines and cosines need to be calculated. This was a major problem in the earliest (pre-computer) days of crystallography, but it was easily solved in the early computer programs because of the fundamental simplicity of trigonometric functions. They simply register the vertical and horizontal projections of points around a circle. If the points are equally spaced, the same simple algebraic formula applies to all, allowing the coordinates of each point to be calculated from those of preceding points.

(ii) But even with that simplification, early crystallographic programs were very slow when very many repetitions were needed. For a box of size N, each of the N pixels of the data gives rise, in Fourier space, to a phasor wave which consists of a series of N unit phasors. To calculate the DFT for a transform box of size N, we start with the appropriate unit-phasor framework (like that for $N = 4$ in Fig. 4.11) which has N^2 phasors. Then we must multiply each line by the weight of the corresponding pixel, as in Fig. 4.12. Since each phasor in the framework must be multiplied by the appropriate pixel weight, this requires N^2 multiplications to yield modified frameworks like that shown on the right of Fig. 4.12, (a), (b) and (c). Finally, we must add together all the phasors at the same X to

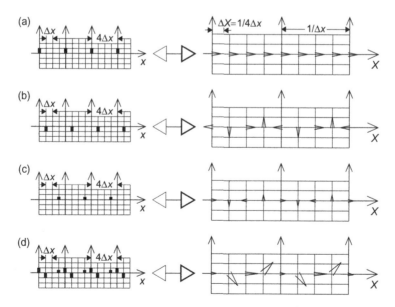

Figure 4.12: $N = 4$ DFT calculation. The diagram is set out like Fig. 4.11, with real space solitary pixels on the left and their DFTs on the right. **(a)** Pixel at origin ($x = 0$). **(b)** Pixel at $x = \Delta x$. **(c)** Pixel at $x = 2\Delta x$. **(d)** Pixel at $x = 3\Delta x$.

get the DFT. But most of the time is spent on the N^2 multiplications, and that increases quadratically. Although N^2 is trivial for $N = 4$, even quite small particles need hundreds of pixels ($N\sim100$), and crystalline sheets need many thousands ($N\sim1000$). With these large arrays, the number of multiplications becomes critical. Increasing N by 100 extends the time by 10,000, so a DFT calculation taking a second would take hours. It is this N^2 scale factor that is such a problem with large datasets.

And it is precisely this problem that is tackled by the **Fast Fourier Transform** (FFT) algorithm. Although long known, its real value became apparent only after digital electronic computers became generally available for scientists in the 1960s. Then the FFT was re-discovered by Cooley & Tukey (1965), leading to much work on its detailed implementations. There is now an extensive literature (e.g. Bracewell, 2000; Press et al., 2007; Walker, 1996; and see Brigham, 1974, for the history).

In the FFT, the number of calculations is proportional, not to N^2, but only to $N.\log_2(N)$, where $\log_2(N)$ is the logarithm to base 2. (For example, $\log_2(1024) = \log_2(2^{10}) = 10$.) Thus, for a one-dimensional image with $N = 1024$, the ordinary DFT calculation size is proportional to $1024^2 = 1048576$, but the FFT calculation size is only proportional to $1024 \times \log_2(2^{10}) = 1024 \times 10 = 10240$, which is over a hundred times faster. In general, FFT calculations take only $\log_2(N)/N$ times as long. This gain of time is of great importance for large arrays, as in two- or three-dimensional images, and is one of the main factors (along with improved computers) that makes numerical image processing a practical method in electron microscopy.

4.3.3 Fast Fourier Transforms: Essential Tricks

A brief explanation will now be given of the essential tricks of this important algorithm. (See Bracewell, 2000; Press et al., 2007 for more details.)

The basic idea is shown in Fig. 4.13. An $N = 8$ DFT (a) is 'unzipped' by separating alternate peaks into two zero-interlaced sequences, (b) and (c) (left sides); (b) has an origin peak, and can therefore be replaced directly by an $N = 4$ DFT as shown on the right; (c) is also zero-interlaced, but a Δx shift prevents the same compression being applied. In other words, (c) is a compressible sequence convoluted with a peak at $x = \Delta x$, as shown on the bottom right.

We need to correct the shift in (c) in order to extend the 'unzipping' trick. This is shown in Fig. 4.14, which starts by repeating the bottom-right of Fig. 4.13, and then takes FTs of each term. The interesting term is (f), a phasor wave that is the FT of (c). This is a simple function, so we can calculate term (d) if we can find term (e). And that is relatively easy, since (e) has an origin peak and is ready for compression to (g). We have therefore 'unzipped' and compressed the starting $N = 8$ term (a) of Fig. 4.13. Its two 'offspring' are the right of (b) in Fig. 4.13, and (g) of Fig. 4.14. Each of these new starting terms is only $N = 4$. Thus we have

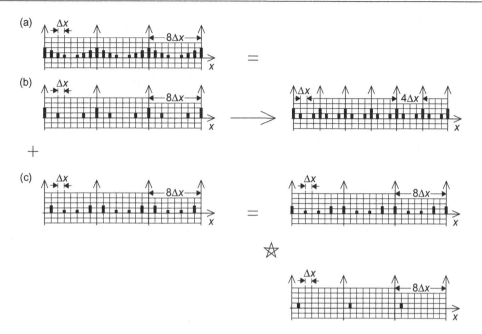

Figure 4.13: The essential core of the FFT: splitting a longer DFT (here $N = 8$) into two half-length DFTs (here $N = 4$). **(a)** An $N = 8$ DFT is divided into two parts: (b) and (c). **(b)** is zero-interlaced just like Fig. 4.10(c), and can therefore be replaced by an N 5 4 DFT, as in Fig. 4.10(b). But the other half **(c)** is shifted relative to (b). This could be reversed by convolution with a peak at $x = \Delta x$.

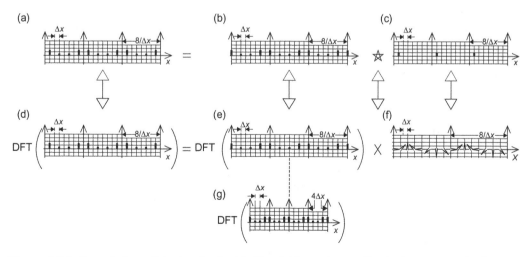

Figure 4.14: Completion of the 'unzipping' in Fig. 4.13, so the top line is the same as the bottom line in that diagram. Each term's FT is shown in the line below.

converted an original $N = 8$ term into two $N = 4$ terms of the same type. Then each of these can be halved in the same way, and so on down to $N = 1$, which is quite trivial. Therefore an original dataset with a box-size N that is a power of 2, say 2^k, can be put through the 'splitting trick' k times to reach trivial $N = 1$ terms. This is how one of the factors of N becomes converted into $\log_2(N)$.

Summarizing, there are three essential FFT tricks: unzipping into two zero-interlaced series, origin-shifting one of these and halving their transform boxes. The combination of these tricks turns one series into two, each with half the box-size. After many applications, the result is the recursive bisection of one long series into many very short ones – ultimately so short that no DFT calculation is needed. All the real work is done in the bisections, origin-shifting etc.

The speed of the FFT is really effective only with relatively large arrays (large N), making it important for image analysis. This may be one reason why FFT algorithms, though long known, were little used before the general availability of digital computers. So, besides a hardware acceleration that might appear to render any further improvement superfluous, there was an additional software bonus. But could one imagine early crystallographers hand-calculating with an FFT program? The FFT is fast but fiddly: many different numbers must be put into different boxes, and therefore moved and copied accurately. These are just the sort of operations where any mechanism, even if it were no faster than the human brain, could out-perform it in persistence, reliability and accuracy. Because of the computer's enormous speed, we tend to forget that it has other advantages as well.

4.4 Appendix

4.4.1 Amplitude Scaling Rules

As mentioned in §4.2.1, we can't apply exactly the same rules to the function values (f_j) and to those of its FT (F_j), without taking account of the different scale-factors (s, S) for direct and inverse FTs. Thus the amplitude scaling formulae[2] are $F_0 = s\Sigma f_j$ and $f_0 = S\Sigma F_j$.

These projection rules are not quite sufficient to fix the entire scaling of DFTs. For example, if either of the origin values F_0 or f_0 is zero, the scale-factors become arbitrary. We can get another rule from the function's ACF. For this, we need to connect the ACF with a function's FT. Equation (3.3) (§3.4.6) told us that $CCF(f,g) = f \odot g = f(-) \star g$, where $f(-)$ is the reversed function $f(-x)$. So $ACF(f) = CCF(f,f) = f(-x) \star f(x)$. But we saw in §3.4.2 that reversing a phasor wave produces the wave's complex conjugate. Thus, if $f(x) \leftrightarrow F(X)$, then reversing the coordinate produces the complex conjugate of the FT: $f(-x) \leftrightarrow F^*(X)$. Therefore $ACF(f) = f(-x) \star f(x) \leftrightarrow F^*(X)F(X) = |F(X)|^2$, so $ACF(f) = |F(X)|^2$. For an even function, $|F(X)|^2 = F(X)^2$. (We extend this to two-dimensional functions in Chapter 6.)

---

[2] Here the symbol Σ means 'sum of' so, for $N = 2$, $\Sigma f = f_0 + f_1$; and the same for ΣF.

If we now apply the projection theorem to the equation $\text{ACF}(f) \leftrightarrow |F(X)|^2$, we get: origin value of $\text{ACF}(f) = \Sigma|F_j|^2$. The origin value of the ACF is the sum of squares of peaks, Σf_j^2 (Fig 2.7, §2.3.1). So we get $\Sigma|F_j|^2 = t\Sigma f_j^2$, where t is another scale factor. This equation is called a *Parseval equation*.

Summarizing our three provisional scaling equations:

(i) Projection rule equations: $$F_0 = s\Sigma f$$

$$\Sigma F = S f_0$$

(ii) Parseval rule equation: $$\Sigma|F|^2 = t\Sigma f^2$$

It only remains to find the scale factors s, S and t.

The three scaling constants are determined by the 'pixel method' for getting DFTs. This implies two conditions.

(i) A unit peak in real space gives, as its FT in reciprocal space, a unit phasor wave whose period is the reciprocal of the peak's displacement.
(ii) The constants also have to be consistent with the linearity rule, whereby the DFTs of all the peaks are combined by addition or subtraction. These conditions are applied in Fig. 4.15.

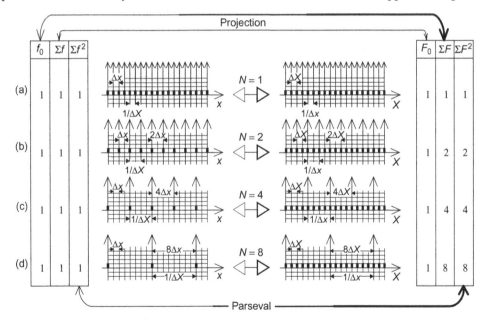

Figure 4.15: A solitary peak at the origin has $f_0 = 1$ (and all other f's are zero). N doubles successively in **(a)**, **(b)**, **(c)** and **(d)**. The horizontal double-headed arrows connect terms in the provisional scaling equations and establish the scaling constants for an origin peak. Where the lines are thicker, a factor of N applies.

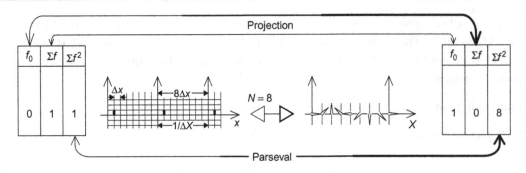

Figure 4.16: $N = 8$ DFT of a peak at Δx, so $f_0 = 0$ and $f_1 = 1$.

We start with the unit peak at every origin (Fig. 4.15). A solitary peak at one origin gives an FT that is a unit phasor wave of infinite period, and consequently of zero phase (i.e. a constant unity). So, if the transform box has N pixels, each reciprocal-space pixel has a unit peak. Thus all F's (and all F^2's) are 1. This establishes that $s = 1$, $S = N$ and $t = N$ (Fig. 4.15). We get the same result if the unit peak is displaced by one pixel, which is shown in Fig. 4.16.

As the phase is (as always for a unit peak) zero at the origin, $F_0 = 1$. $\Sigma F = 0$ because all the phasors have different directions and add to zero. But each of these eight phasors has a length or modulus $|F|$ which is 1, and so is its square; so $\Sigma |F|^2 = 8$. It can be seen that the same result will hold if the peak is displaced by other distances. So the scaling factors established for origin peaks apply to all peaks. The addition law means that combining these peaks to make an FT leaves the laws in force.

Thus the **amplitude scaling equations** are:

Projection:
$$F_0 = \Sigma f \qquad (4.10)$$

$$\Sigma F = N f_0 \qquad (4.11)$$

Parseval:
$$\Sigma |F|^2 = N \Sigma f^2 \qquad (4.12)$$

Filters

Chapter Outline

5.1 Introduction

5.1.1 Concept of a Filter

We have now developed the concepts and methodology of Fourier Transforms (FTs) to the point where we can begin to discuss their uses. Many of these fit the general description of a 'filter', a concept belonging to the field of information theory that originated in the study of long-distance communication, where information is gradually lost.

Many stages of data-recording lose information. Depth-information is lost in projections, and detail is lost in high-magnification light-microscope images. Such losses are often approximated by the effects of a **filter**. Familiarity with this class of devices, which is helped by using FTs, provides a framework for understanding instrumental limitations and their correction.

The concept originated in radio-communication. Multiple radio stations, transmitting over the same reception area, would create a confused babble unless kept separate by some form of coding, of which the earliest method used high-frequency 'carrier' waves. The listener selected

Structural Biology Using Electrons and X-Rays.
© 2011 Michael F. Moody. Published by Elsevier Ltd. All rights reserved.

a carrier by using an electronic filter to remove the unwanted carrier frequencies. After tuning with this variable 'band-pass' filter, the audio signal had to be extracted from its (now) unwanted carrier with a 'low pass' filter that rejected the carrier's much higher frequency. Similar filters removed the high-frequency hiss from the surface noise of gramophone records.

These designed filters resemble the natural filters that are unavoidable in recording data. The resolution limit of the light microscope is a filter that eliminates short distances (high spatial frequencies) through the wavelength of light. The two-dimensional character of images is a projection-filter that sums all the detail along each ray's path.

5.1.2 Operation of Simple Filters

The simplest type of filter attenuates some of the signal's frequencies. This pattern fits image-blurring, which replaces each pixel of the perfect image by a broader distribution, the point spread function (PSF). We can write this as a convolution:

$$\text{filtered image} = \text{ideal image} \star \text{PSF} \tag{5.1}$$

Applying the convolution theorem (§3.4.4), we have:

$$\text{FT(filtered image)} = \text{FT(ideal image)} \times \text{FT(PSF)} \tag{5.2}$$

The equation now applies in (spatial-) frequency space. Here the frequency-filter is FT(PSF), often called the **contrast transfer function** (CTF: see Chapter 9), which operates by *multiplying* the (frequency) signal, like the low-pass radio filter. These simple filters work by either multiplication or convolution. Usually they multiply in Fourier space (frequency-space) and convolute in real space. Thus every simple filtering operation involves a *pair* of filters (in different spaces) that are FTs of each other. We usually notice only one of the pair, often the one that multiplies in frequency space.

It is usual for natural filters to blur, and designed ones to sharpen, often to minimize a previous natural blurring (which contributes to the universal tendency for information to disappear as entropy increases). The only way to compensate for the destruction of information is to replace it by another measurement which did not destroy it ('reconstruction'). Often measurements or experiments have a variable parameter whose setting leads to destruction of a particular section of the information, so different parameter settings yield complementary information. Thus different defocus states, with different PSFs, destroy information in different parts of spatial-frequency space (§9.2.4–§9.2.5), so a combination of different images (with different defocus states) of the same object can be processed to remove defocus artifacts (§9.2.7).

When an observation is unrepeatable, information loss is irrevocable. But, although no compensating filter can reverse such information *destruction*, nevertheless a sharpening filter might *redistribute* the available information to make it more useful, e.g. by accentuating the edges of images (§5.4.3).

5.2 Blurring Filters

5.2.1 The Rectangle Filter

We begin by considering the blurring filters, which are not only natural, but also useful for reducing noise. The simplest multiplication filter is the **rectangle function** (§2.2.1), which eliminates everything outside its width. Its simplest application is in real space: selecting ('boxing') a desired particle. In reciprocal (frequency) space, the rectangle is a low-pass filter that eliminates all high-frequency signals that are probably noise.

In these examples the rectangle is a multiplication filter, but it is also useful as a convolution filter. As the entrance slit to a densitometer (in real space), or (in reciprocal space) to a spectrophotometer's photomultiplier, it averages everything that goes through it. The effect is to convolute the input signal with a rectangle function, smoothing out random fluctuations in image density or spectra (respectively).

If the rectangle is a multiplication filter in one space, it is a convolution filter in the reciprocal space, and *vice-versa*. To fully understand its effect, we must understand what is happening in the reciprocal space, so we need to find the rectangle's FT. The *projection theorem* (§3.5.4) shows that the FT's center is the total area of the rectangle – a substantial quantity, so there is a positive peak at the origin. Also the *scale rule* shows that the FT will be stretched if the rectangle is compressed, and *vice-versa*. Besides this reciprocity of width, there is also a reciprocity of shape: the sharper and more angular we make a function, the broader and smoother will be its FT. So the rectangle's infinitely sharp edges give rise to an infinitely wide FT. Nevertheless, this FT must decrease with increasing X, since its square is finite (next section).

Finally, the rectangle is an *even* function, so its FT is also even. Thus it is real and either positive or negative, only changing sign by passing through zero at a *node*. Since the FT starts positive at $X = 0$, we can get a picture of it by locating the nodes. These can be found by a simple argument (Fig. 5.1). In the top line, we take the rectangle (a), of breadth b, and convolute it with a comb-function (b) of spacing b; then all the copies of the rectangle fit exactly together, giving a uniform density (c). The FTs (bottom line) effectively constitute an equation with one unknown: the rectangle transform on the left. When this is sampled at 0, $\pm 1/b$, $\pm 2/b$, $\pm 3/b$,..., it yields a single peak at the origin. The FT must therefore annihilate the comb's peaks at $\pm 1/b$, $\pm 2/b$, $\pm 3/b$,..., so it must have nodes at these points. (d) summarizes that the rectangle's FT has an origin peak surrounded by zeroes at $\pm 1/b$, $\pm 2/b$, $\pm 3/b$... The full calculation shows that these are the only zeroes, and that they are points where the FT changes sign. So the FT oscillates evenly while slowly decaying to zero, as shown in (e) of Fig. 5.1. From (d) of Fig. 5.1, the b-width rectangle's FT has an origin peak, but is zero at all other multiples of $1/b$. This gives an outline of the FT (called the **'sinc-function**[1]), shown in detail in (e) of Fig. 5.1.

---

[1] Called the 'sinc-function' because of its equation $F = \sin(2\pi Xb)/2\pi Xb$. (See equation (19.17) in §19.5.4.)

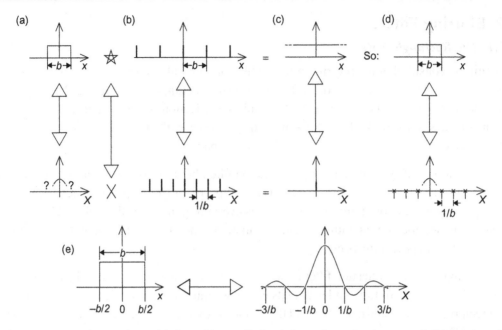

Figure 5.1: (a) Rectangle, width b, and its FT (below); but where are the zeroes? **(b)** Lattice of period b and its FT (below) which samples functions only at multiples of $1/b$. **(c)** Convolution of (a) with (b) gives a constant, whose FT (below) is an origin peak. **(d)** Therefore the FT of (a) is zero at multiples of $1/b$, except for the origin. **(e)** Rectangle-function and its calculated FT.

5.2.2 Convergence to Equality

Figure 5.1 (e) reminds us that the same filtering function is done by a *pair* of filters which are FTs of each other. One acts by multiplication in one space while its FT acts by convolution in the other space. In the usual case, like the rectangle, the two filters have very different widths and shapes. The widths (and degrees of smoothness) of a function and its FT have a reciprocal relation. The most unequal pair of filters are the sharp peak (or *delta-function* §2.2.2) and its FT, the *plateau* (or constant function §2.2.2). If the former acts by convolution, and the latter by multiplication, they form a perfectly permissive filter-pair that passes everything unchanged. But the peak acting by multiplication, with the plateau acting by convolution, form a perfectly *repressive* filter-pair that passes virtually nothing.

Could there be a pair of filters that have much the same effect, whichever of them acts by multiplication? This question can be framed differently: how similar can we get the filter and its FT? Could we even find a pair that have exactly the same shape? Such a function would be essentially unaffected by Fourier transformation; it would be an **eigenfunction** of the Fourier transform, where an 'eigenfunction' is unchanged by some operation (here Fourier transformation): a mathematical German equivalent of 'plus ça change, plus c'est la même chose'.

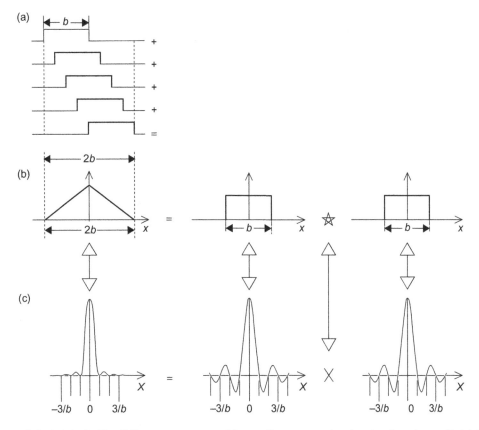

Figure 5.2: (a) As in Fig. 2.8, many superpositions of a rectangular density-function, of width b, yield a triangular density-function of width $2b$, left of **(b)**. Right: this triangle is equivalent to the self-convolution of the rectangle function. **(c)** FT of triangle (b) is the square of the sinc-function (i.e., the sinc-squared function on the left.

We start our quest by trying to smooth the rectangle. A function lacking the rectangle's sharp edges is the **triangle**, a function we encountered in section §2.3.1 as the auto-correlation function (ACF) of a rectangle. Equation (3.4) said that $\mathrm{ACF}(f) \leftrightarrow |F|^2$. Now the rectangle ($f$ in this equation) is an even function with a real FT, so equation (3A) applies: $\mathrm{ACF}(f) \leftrightarrow F^2$. Consequently, *triangle* = ACF(*rectangle*) $\leftrightarrow$ (FT(*rectangle*))^2. That is, the FT of the triangle function is the square of a real function, and therefore entirely positive but with zeroes in the same places as the corresponding rectangle. (See the sinc-squared function on the left of (c) in Fig. 5.2.) (However, we shall need to shrink the triangle in (b) to give it the same width as the rectangle; this will stretch the FT to twice the width.)

Thus, after the rectangle, the triangle is the next step: itself smoother and with its FT's oscillations both smaller than the sinc-function's and all positive. Since the remaining oscillations are presumably caused by the triangle's sharp bends, we might try a further

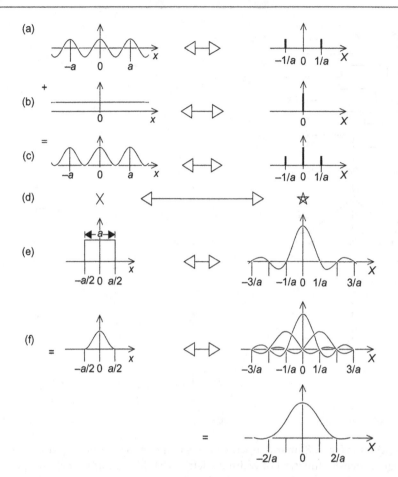

Figure 5.3: Calculating the FT of the cosine bell function. (Functions on left, their FTs on right.) **(a)** Cosine curve, plus ... **(b)** ... a constant, gives ... **(c)** ... an all-positive cosine, whose center ... **(d)** ... is then isolated by multiplication ... **(e)** ... by a rectangle, giving ... **(f)** ... the cosine bell, at bottom right. (See §3.2.1 for relative peak heights in (a), (b).)

smoothing. This next stage has the shape of a cosine's peak, down to its troughs. It is the **'cosine bell'**, shown on the left of (f) in Fig. 5.3 and of (c) in Fig. 5.4. The upper lines of Fig. 5.3 show the procedure for deducing the cosine bell's FT. We start with the cosine curve (a) of height 2 whose FT (right) is the 'twin (unit) peaks' function. We want to cut the central 'bell' out of this cosine curve. First we raise the cosine to rest on the x-axis, by adding the function (b): the constant 2, whose FT (right) is a single peak (height 2). This addition gives (c), where the left is the raised cosine curve we wanted. The FT (right) consists of three peaks, two outer of height 1 and a central one of height 2. Finally, we isolate from the left its central peak (the cosine bell) by multiplying the left with a rectangle (e) extending from $-a/2$ to $+a/2$. This multiplication convolutes the FT on the right with (e), the

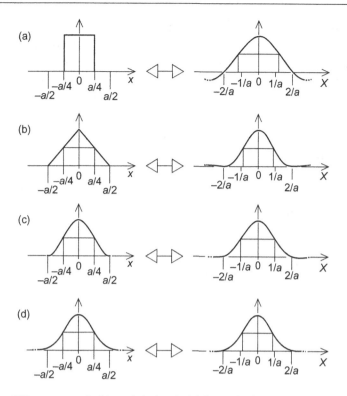

Figure 5.4: Series of filter curves (left) and their FTs (right), which become successively more similar to the functions. **(a)** Rectangle. **(b)** Triangle. **(c)** Cosine bell. **(d)** Gaussian.

rectangle's FT, a sinc-function that is zero at $X = n/a$. The right of (f) shows this convolution, whose completion is shown at the bottom right.

Line (f) of Fig. 5.3 shows that the cosine bell's FT is remarkably similar to the cosine bell itself, so we are getting very close to a curve that is its own FT. For the very last step, the equation must be demonstrated mathematically (see, e.g., Bracewell, 2000). It is a well-known function: the bell-shaped **Gaussian** distribution curve so important in statistics, and shown (with all the previous steps) in Fig. 5.4 (d). This 'eigenfunction' of Fourier-transformation is, as expected, moderately curved and rapidly declines to a low value where it asymptotically approaches the axis. But it never quite gets there, for it is impossible to get a function with its FT, without at least one of the pair having infinite width. Yet, although it is never quite zero and thus never annihilates any part of a signal (in either real or reciprocal space), the Gaussian soon gets so far below the noise level that it is (in practice) the most effective destroyer of information, compared with other functions of the same effective width[2].

---

[2] All measurable phenomena have a limited accuracy and therefore a limited information content. So any theoretical physical curve could be replaced by the same curve convoluted with some sufficiently narrow Gaussian. That would remove many of the paradoxes and problems found in 'pathological' mathematical functions.

We now examine the uses of some of the blurring filters as summarized in Fig. 5.4. The rectangle and triangle are used for the interpolation of tabulated curve-data (i.e. peak-functions; see section §5.3.1). The rectangle FT is needed for the important sampling theorem in the same section. The closely similar functions (c) and (d) are not so useful for interpolation, but are used to give a data-set a soft edge. That is, instead of sharply demarcating the required data, which would imply multiplication with a rectangle, we multiply by a **soft-edged** (or **cosine-edged**) rectangle. For this we create a function that is a rectangle, but with edges given a smooth shape. This is very similar to a rectangle convoluted with a Gaussian, whose FT is the rectangle's FT (i.e. a sinc-function) multiplied by a Gaussian – a multiplication that strongly damps the sinc-function's oscillations (see Fig. 5.5). This kind of filter reduces the ripples in real space that arise from a sharp-edged filter in reciprocal space.

5.3 Digital-to-Analog Conversion

The 'blurring' filters we have just discussed are typical of the action of natural processes or recording instruments. They remove information by reducing signals so they are closer to the noise level. But how can we measure this loss? How can we discuss the information content of a continuous analog curve? Surely it must depend on many factors, such as the accuracy of measuring each point's value, and the extent to which two nearby points can take independent values. The problem becomes clearer when we have a series of numbers obtained by sampling the curve, which we have to do in order to calculate its DFT. (The sampling of curves was discussed in Chapter 4 section §4.1.1.) But, if we want to know how much information is lost by sampling, surely the real test is to try reconstituting the curve from its sampled values. So we need a good method for digital-to-analog conversion.

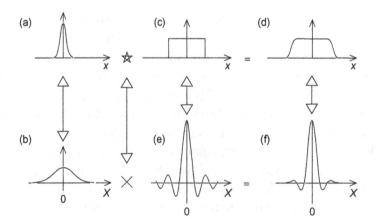

Figure 5.5: (a) Narrow Gaussian, whose FT is … **(b)** … a wide Gaussian. **(c)** A rectangle, 'softened' by convolution with (a), gives the 'softened' rectangle **(d)**. (c)'s FT **(e)** has strong oscillations, reduced by multiplication with (b), giving **(f)**, the 'softened' rectangle's FT.

5.3.1 Interpolation of Sampled Curves

The problem of reconstituting smooth curves from samples was first faced by the calculators of mathematical tables (e.g. Abramowitz & Stegun, 1964). Each mathematical function, such as a cosine, has columns of numbers which are its sampled values at regular angles. This is fine if you want the cosine of exactly 30°, but you are more likely to need some awkward angle like 30.237°. Then you will have to calculate the cosine from adjacent values like cos(30°) and cos(31°). This calculation is called **interpolation**.

The very simplest interpolation method would be to round the angle to the nearest degree, so we would replace 30.237° by 30° and use that tabulated cosine without any further calculation. This is equivalent to a 'bar chart' or block fit to a smooth curve. However, if that method is too crude, an improvement would be to draw a straight line between cos(30°) and cos(31°) and take its value at 30.237°. This method is **linear interpolation**.

Because of their simplicity, both methods can be represented as convolutions, as shown in Fig. 5.6. In (a) the original smooth curve is sampled at points Δx apart, and these sampled values are convoluted with a rectangle in (b), and a triangle in (c). This gives a bar chart on the right of (b), and a linear interpolation on the right of (c).

We examine the accuracy of these approximations, as they apply to a simple periodic curve, period a, in Fig. 5.7. The curve itself is in (a) and its peak values are shown in (b). These peaks are used as the basis for two interpolations: block in (c) and linear in (d). Their FTs are shown on the right, and the extent of the 'graph paper' indicates the extent of significant peaks in these FTs. Clearly the linear approximation (d) has a narrower extent of significant FT peaks, i.e. its FT is closer to that of the original curve (a).

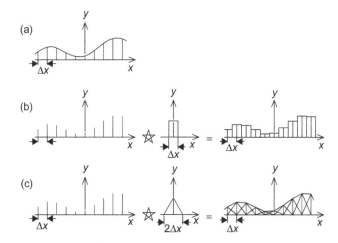

Figure 5.6: (a) Smooth curve sampled at equidistant points Δx apart. **(b)** The sampled data from (a) convoluted with a rectangle of width Δx, giving a 'block data' interpolation. **(c)** The sampled data from (a) convoluted with a triangle of width $2\Delta x$, giving a linear interpolation.

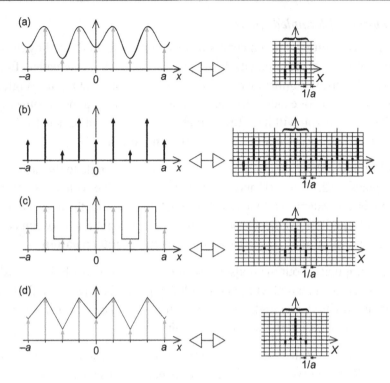

Figure 5.7: Curve **(a)** with sampling **(b)** and interpolations ((c)–(d)), from Fig. 5.6, with FTs (right) shown with 'graph paper' with squares of width $1/a$. **(c)** 'Block data' approximation. **(d)** Linear interpolation. (In the FT, right, horizontal curly brackets indicate a rectangle that would isolate the original curve's FT.)

This comparison shows the characteristics of the ideal interpolator in the Fourier domain: it should accept the inner part of the FT unmodified (a), and exclude the outer parts that are prominent in (c). But this is the description of a very simple curve or filter: the *rectangle*, but applied in *Fourier* space, unlike that in (c) (which was in *real* space).

5.3.2 Sampling Theorem

Thus the rectangle is a poor convolutor in real space, but the ideal FT filter in reciprocal space. Consequently, the rectangle's FT (the sinc-function) is the ideal convolutor in real space. When we try the sinc-function as our new convolutor, we are led to an important result: the **sampling theorem** (due to Whittaker, 1915 and to Shannon, 1949). This says something interesting about the FT of a bounded curve, i.e. one that is zero outside some given width. The bounded width obviously limits its information content, so its FT should also be somehow limited. The sampling theorem tells us this limitation: the curve can be interpolated *exactly* from a set of sampled values. The whole curve contains no more information than those values.

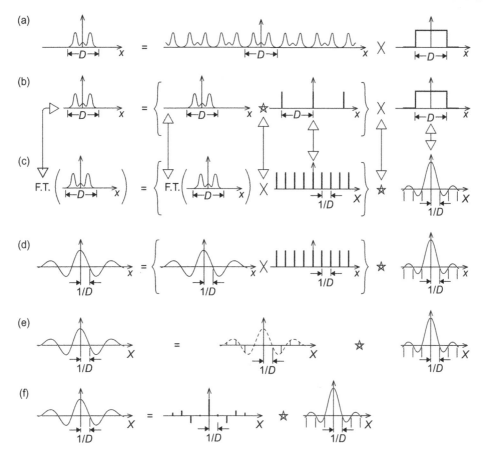

Figure 5.8: (a) Top-left is the bounded function, width D. This is converted into a periodic repetition of itself (period D), and then converted back to the original function by multiplying with a rectangle. **(b)** Spells out this identity in a little more detail, by representing the periodic repetition as the original function convoluted with a lattice. **(c)** Has the FTs of every term in (b). The lattice, period D, becomes the reciprocal lattice, period $1/D$. The rectangle becomes a sinc-function. **(d)** Introduces the bounded function's FT (do not confuse it with the sinc-function on the far right). **(e)** Multiplies the middle two terms to give the sampled FT. **(f)** States the final theorem.

How does this follow from the existence of a bounded function, width D? We sample its FT at equidistant points $1/D$ apart. This has converted the continuous FT into a list of numbers, the amplitudes and phases of peaks at the sampling points. But then we find that this list contains *all* the curve's information, because we can *exactly* reconstruct the FT by convoluting the list (i.e. its peaks) with a sinc-function which is the FT of a rectangle of width D. The proof of this crucial assertion is given in Fig. 5.8.

Thus sampled values can yield a continuous function through convolution with a sinc-function. Linear interpolation also involves convolution, but with a *finite* convolutor (the triangle). By contrast, sampling-theorem convolution uses the *infinite* sinc-function. Thus the sampling-theorem reconstruction of sampled data must also extend to infinity. (However, in practice it will soon fall below the noise-level.) The infinite size of the sinc-function has another consequence: sampling-theorem interpolation requires convolution of the *entire* dataset with the sinc-function. (This contrasts with linear interpolation, which only uses adjacent sampled values.) So sampling-theorem interpolation is a large calculation, often a high price to pay for exactness.

The sampling theorem has a theoretical problem, that the specimen width D and the extent of its FT cannot both be limited. D is obviously limited, at least in a crystal; so do we need to reconstruct an *infinite* FT? This problem is (fortunately?) removed by the presence of noise. Although the FT extends (in principle) to infinity, it soon (all too soon!) falls below the noise level and becomes undetectable.

The sampling theorem tells us how to sample an FT digitally with the greatest efficiency. It shows how the FT of a molecule of width D can be rebuilt, with complete accuracy, from its values sampled at points $1/D$ apart. These few sampled values carry all the information in the FT. So the accuracy of the samples, combined with their number, defines the information content of a FT.

Finally, the sampling theorem justifies crystallography: the smallest lattice that could accommodate such a molecule would have a repeat of D, and its reciprocal lattice would therefore have a repeat of $1/D$, the required sampling distance. (Of course, this could be viewed as an alternative route to deriving the sampling theorem.) If the molecules should become separated, e.g. by hydration, the repeat distance would increase and the FT would be over-sampled. That would provide some redundancy of information that could help to assign phases. That was first exploited with the 'minimum-wavelength principle' of Bragg & Perutz (1952), the first attempt at direct phasing of hemoglobin X-ray diffraction patterns. It has re-surfaced as the method of *solvent flattening*[3] for improving maps (i.e., images) in protein crystallography (Chapter 13, §13.4.3), and the method of 'convex sets' in image analysis (Frank, 2006).

5.4 Correcting Blurring Filters

Most filters, like those just described, tend to blur data and lose detail. While it is impossible to replace information that has been completely destroyed, many natural 'filters' only *attenuate* our data. In theory, the weakened signal only needs re-scaling. But, although attenuation reduced both the data and its immediately associated noise, new sources of noise then entered, degrading or even obliterating the reduced data. Thus it is impossible to reverse completely the action of a blurring filter; but *any* reversal, however partial, is worthwhile. So this section considers a few simple ways of doing that, starting with an obvious one.

---

[3] Also known as *density modification* or *solvent leveling/flipping*.

5.4.1 Contrast Transfer Function Correction: Wiener Filtering

We need to reverse, or at least mitigate, image blurring by the PSF. This gave us equation (5.2): FT(blurred image) = FT(ideal image) $\times$ FT(PSF).

We can replace FT(PSF) by the contrast transfer function, CTF (see §5.1.2 and Chapter 9), and then we can abbreviate the equation as $F_{obs} = F.C$, where F is FT(ideal image). Rearranging this equation to give: $F = F_{obs}/C$, we can apparently calculate F since we know F_{obs} and C. Unfortunately, C is very small or zero in some regions of frequency-space, so dividing by it requires us to *enlarge* F_{obs}; yet, in these regions, F_{obs} is mostly noise!

Thus we have serious problems doing the correction when the CTF is small, so we need to avoid this contingency by adding some constant s to the denominator; s must depend on the noise level, for dividing by a small C arises where F_{obs} is noisy. On the other hand, a very high noise level would spoil our prospects for correction, and we would lose little if we reduced the CTF's contribution by making s big. Thus s depends on the noise, relative to the signal; i.e. on the reciprocal of the signal:noise ratio. (So s is usually taken as that reciprocal.) Now this is a real quantity whereas we would add it to C which is complex, so we must first make that real by multiplying $F_{obs} = F.C$ by its complex conjugate: $(C^*) F_{obs} = F.C^2$, which gives:

$$F = (C^*) F_{obs}/(|C|^2 + s) \tag{5.3}$$

This is our best option[4] (called **Wiener filtering**) if we have only one defocused image available. However, as pointed out in 5.1.2, it is far better to combine different images of the same specimen (or its copies), taken at different focal states with different CTF's (§9.2.4). Then we can down-weight those parts where the correction has to be big, and emphasize those where it is small, i.e. where the imaging conditions are better. The Wiener filtering formula then helps to define the relative weights.

5.4.2 Blur-Correcting Filters: General Approach

A different approach would seek to compensate a blurring natural filter, with an artificial sharpening one, which we consider first. Then this approach leads us to a discussion of edge enhancement, and how that can be achieved by manipulation an image's FT. Edges involve intensity gradients, and gradients are handled by calculus; a simple introduction to which comes next. Finally, we consider the FT of a *resonance curve*, which is relevant to anomalous scattering (§13.3.4) and NMR.

So we might try to design an opposite, sharpening, filter to place after the blurring filter for compensation. If the sharpening filter has to cancel out the blurring filter, what are its required

---

[4] A more sophisticated version of Wiener filtering is given in Press et al. (2007).

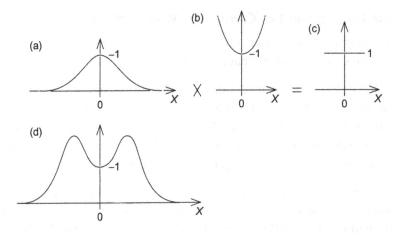

Figure 5.9: A sharpening filter as the 'antidote' to a blurring filter (Gaussian). **(a)** Gaussian filter, peak height 1, multiplied by ... **(b)** ... its reciprocal (the 'antidote'), gives ... **(c)** ... a constant filter, height 1 (no change). The improved 'antidote' **(d)** modifies (b) to avoid amplifying high-frequency noise.

characteristics? Blurring filters use convolutors that are entirely positive: they blur by *adding* adjacent portions of the image. So the opposite sharpening filters must surely *subtract* some of these adjacent image portions; they need negative terms as well as positive in their convolutors. Thus, while convolution with an *all-positive* filter blurs (lowering resolution), convolution with a *mixed-sign* filter might sharpen.

5.4.3 Sharpening Filters

The classical blurring filter is the Gaussian curve (in either multiplication or convolution). Consider its effect as a multiplication filter in frequency space. It progressively weakens higher spatial frequencies, but never completely destroys them. We might therefore attempt to find a mathematical inverse filter: something that, after multiplying a Gaussian function, gives a uniform constant (i.e. a neutral 'filter'). This would be the reciprocal of the Gaussian function, as shown in Fig. 5.9 (b). Unfortunately, the two rising arms of this filter would infinitely magnify the high-frequency noise. If there is any useful signal needing amplification, it would probably lie just outside the inner part of the FT that codes for the obvious detail. So we should just amplify that band of frequencies, as shown in (d).

What would be the effect of filters like Fig. 5.9 (d)? It is the double-hump region around the peaks that is significant, whereas the innermost part of the curve is somewhat arbitrary. This curve can be approximated by a difference of two Gaussians as in Fig. 5.10. As that diagram shows, the FT of the difference curve (g) is the negative of the same curve (h). (Its invention as a 'sharpening' filter appears to have been anticipated long ago, since this is the response

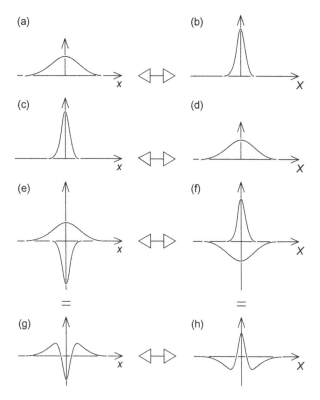

Figure 5.10: (a) A broad Gaussian, whose … **(b)** … FT is a narrow Gaussian. **(c)** The narrow Gaussian's … **(d)** … FT is the broad Gaussian. **(e)** is (a) and – (c). **(f)** is (b) and – (d). **(g)** and **(h)** are, respectively, (e) and (f) after doing the subtraction.

curve of the 'eccentric' cells in the eye of the horseshoe crab *Limulus*; see Hartline's (1959) classical studies.)

These two curves have other points of interest. First, their near-identity (apart from a multiplication constant of -1) makes them (approximate) 'eigenfunctions' of Fourier transformation, just like the Gaussian curve itself (§5.2.2). Second, they are closely related to the 'second derivative' of the Gaussian curve. This takes us into the realm of gradients that is briefly explored in the following section.

5.5 Gradients and Derivatives

5.5.1 Introductory

Gradients are important as a measure of detail in a signal. If it contains more information over the same signal length, there must be more signal alterations over a given distance. Now a bigger change over the same distance of, for example, height means a bigger gradient. So blurring, which diminishes the information content, will generally diminish the gradients of the signal curve. Gradients highlight details. Any attempt to compensate for signal loss due to blurring must therefore involve some way of increasing the signal gradient.

The mathematical treatment of gradient stems from common observation. We estimate the gradient of a road from its rise over a fixed distance, and the variation of gradient is very noticeable to anyone using their muscles for propulsion. A positive gradient is rising, a negative one falling, so a zero gradient is level. We could plot the road's *height* as a function of distance. We could also plot, underneath this curve, the road's *gradient* at the same points. (Call this the 'gradient function' for now.) A smooth road would give a smooth 'gradient function'.

When a curve is represented as an *equation* that relates height (as a function of distance), its gradient curve can be *calculated* by the operation of *differentiation* (part of the calculus) which yields the curve's *derivative* (which is our gradient function's accepted name). However, we do not need to go into this[5], but only to look at a few numerical examples.

We can portray a curve by giving its successive heights as a series of digits. Thus (following on from section §2.3.1) the sequence 12321 describes a triangle five digits long. We can alter it by convolution ($\star$), e.g. with 11, which we do (as described in section §2.1.3) by multiplication. Thus $12321 \star 11 = 12321 \times 11 = 123210 + 12321 = 135531$, which is a somewhat blurred version of the triangle. To get the opposite, i.e. more detail, we need the gradient. What would be the appropriate convolutor? A single 1 (corresponding to the 'impulse' or 'delta' function δ described in §2.2.2 and §3.4.1) gives 12321, which is just the same.

However, we can get a 'gradient convolutor' by combining an impulse (δ) with its reverse (a negative impulse) following immediately after: 1**1**, where 1 is positive but **1** is negative. This is, appropriately, the gradient (derivative) of the impulse function which, starting from zero, first rises to its peak and then falls back to its original zero. (As the impulse's derivative, this convolutor is written δ'; but it apparently has no name, so we shall call it a 'jolt'.)

Convoluting with the jolt, itself a gradient, produces other gradients. Thus the '5-triangle' gives: $12321 \times \star 1\mathbf{1} = 123210 + \mathbf{12321} = 123210 - 12321 = 111000 - 111 = 111\mathbf{111}$. (This is shown in Fig. 5.11.) The output is $+1$ throughout the rising phase and -1 throughout the falling phase, so it describes the *gradient* of the 5-triangle. So 111**111** is the 'gradient function' (called the derivative or 'differential') of the 5-triangle 12321.

5.5.2 *Fourier Transforms of Gradient Function*

We can calculate a derivative (gradient function) f' from any smooth curve f; moreover, that curve has an FT, FT(f). Its derivative f' also has an FT, FT(f'), which is closely related to

---

[5] Those interested in going further can read Chapter 19, §19.5 in Part V.

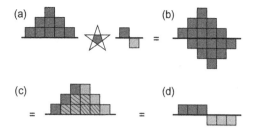

Figure 5.11: (a) The '5-triangle' is convoluted with the 'jolt' (δ') giving **(d)** the difference between two copies of opposite sign. This equals a low positive gradient, followed by a low negative gradient.

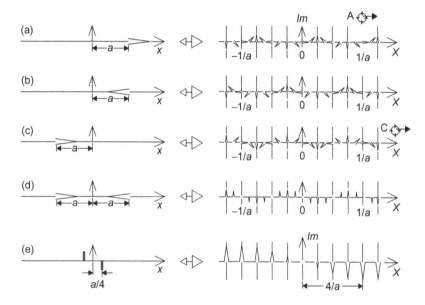

Figure 5.12: Deriving the FT of a 'jolt' (on left of (e)). Fig. 3.14 (c) is copied onto **(a)**, and Fig. 3.14 (a) is copied onto **(c)**. Multiplying (a) by a unit 180° phasor gives **(b)**, where each phasor points diametrically opposite to the corresponding phasor above. (b) plus (c) gives **(d)**, where all phasors on the right have angles 90° or 270°, and are called 'imaginary' (Im). On the left the left phasor is +1, the right one is −1. These would make a 'jolt' if they were brought extremely close together, as in **(e)**.

FT(f), as we can see by using the 'jolt' (δ') to get its derivative: $f' = f \star \delta'$, so FT(f') = FT($f \star \delta'$) = FT(f) × FT(δ'); so:

$$\text{FT}(f') = \text{FT}(\delta') \times \text{FT}(f) \tag{5.4}$$

To use this equation, we need FT(δ'), the FT of a jolt, which is derived in Fig. 5.12. The first four lines, based on Fig. 3.14, derive (d) which resembles a 'jolt' (a positive peak left of a negative peak) except for being stretched out to a width of $2a$. That is partly remedied by compression (reduction of a to $a/4$) in (e), where the right side is proportionately stretched. Already the character of the jolt's FT is apparent: it is entirely 'imaginary' (i.e., with phase-angles all $\pm 90°$), its phase-angles all $+90°$ on the left and all $270°$ (or $-90°$) on the right, and the phasor tips lie approximately on a straight line. If we compressed the left side of (e) to the limit, the phasor tips on the right side would lie on an *exact* straight line. We could write this FT as $-i(\text{constant})X$, where 'i' (the symbol for a unit phasor with angle $90°$) makes it imaginary and the '$(\text{constant})X$' makes it a straight line. (The constant turns out to be 2π with our usual FT formula.) Thus we can now write equation (5.4) as:

$$\text{FT}(f') = (-i2\pi X) \times \text{FT}(f) \tag{5.5}$$

Now we have the jolt's FT, we can apply it to a curve, the Gaussian on the left of Fig. 5.13 (a), with its (similar) FT on the right. On the left of (b) we get its derivative (gradient function): the rising left side corresponds to a rising gradient, reaching a maximum before the Gaussian starts to flatten out. When it is completely horizontal, at the axis, the gradient is zero. The right of the axis just reverses the left, as in (a).

The FT on the right of (b) is got by multiplication with the FT of the jolt, following equation (5.5). The positive part on the left of $(-i2\pi X)$ gives the left positive peak, while the negative part on the right of $(-i2\pi X)$ gives the negative peak on the right. The 'i' also makes it

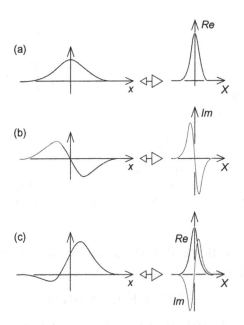

Figure 5.13: (a) Gaussian function (left) and its FT (right); see Fig. 5.4 (d). **(b)** (Left) derivative (gradient function) of (a) left, with its FT on the right. **(c)** is (a) minus (b). '*Re*' = 'real' (phase-angles all $\pm 180°$); '*Im*' = 'imaginary' (phase-angles all $\pm 90°$).

completely 'imaginary' (phase-angles all $\pm 90°$). This can also be deduced from the fact that the left of (b) is an entirely 'odd' function, so its FT must be composed entirely of sines (see section §3.3.1).

This derivative has some interesting features. The derivative (δ') is odd, so taking a derivative reverses parity. The original function (a) was even, so its derivative (b) is odd. Indeed, if we were to find the derivative of (b), we would get a curve like the 'difference of Gaussians' in Fig. 5.10 (g), (h). We noted that those two curves were like 'eigenfunctions', but so also are the two curves we found in Fig. 5.3 (b).

We have (a) minus (b) in (c), whose FT is thus the right of (a) minus the right of (b). But those right sides have no connection with each other: the former is entirely 'real' (phase-angles all $\pm 180°$), and the latter entirely 'imaginary' (phase-angles all $\pm 90°$). That is, the

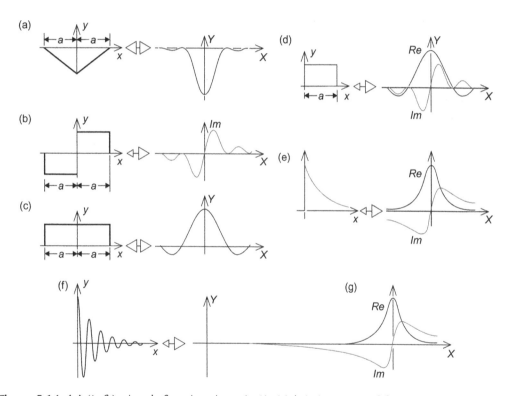

Figure 5.14: (a) (Left) triangle function times (-1); (right) sinc-squared function times (-1) (see Fig. 5.2 (b),(c)). **(b)** (Left) derivative (gradient function) of left of (a), which is a double rectangle; (right) right side of (a) times FT of 'jolt' (Fig. 5.12, (e) right). **(c)** Rectangle (left) with its FT, a sinc-function on the right (see Fig. 5.1). **(d)** Line (b) plus line (c) gives only a rectangle on the positive side of the axis; on the right, a complex FT (black: real part; grey: imaginary part). **(e)** (Left) declining exponential function; (right) complex FT that is generally similar to that in (d). The real part is a 'Lorentzian' curve. **(f)** Oscillation declining exponentially. **(g)** Its FT has a real 'Lorentzian' absorption peak, centered at the frequency of the oscillation, and an imaginary dispersion response.

two sides are composed entirely of phasors that are mutually perpendicular. So the right of (c) simply combines the right side of (a) minus that of (b). (Such a combination of pure 'real' and pure 'imaginary' is called 'complex'.)

5.5.3 Fourier Transforms of Impulse Response

Another example is shown in Fig. 5.14 (a)–(d), which shows that the FT of a rectangle entirely in the positive quadrant has a double real-imaginary FT of the same general type. Note that Fig. 5.13 (c) resembles Fig. 5.14 (d). The real-space (X) sides are essentially confined to positive X, while the reciprocal-space (X) sides have positive symmetrical real parts and anti-symmetrical imaginary parts. This last feature depends on our convention for FTs (i.e. that an isolated peak with positive x-coordinate gives an anti-clockwise phasor wave).

The declining exponential (Fig. 5.14 (e)) shares characteristics with the left sides of Fig. 5.13 (c) and Fig. 5.14 (d). All are essentially confined to the right side of the axis, where they start big and end small. Thus one expects the FT side of Fig. 5.14 (e) to share the characteristics of the FT sides of Fig. 5.13 (c) and Fig. 5.14 (d). They have even 'real' parts with a strong central peak; and odd 'imaginary' parts which are negative on the left but positive on the right. And this is indeed true of the right side of Fig. 5.14 (e).

In the bottom line, (f) and (g) show the spectral analysis of the impulse response of a system which has only one vibration frequency. The response (f) is the exponential decline in the left of (e), multiplied by a cosine wave. Consequently, its FT is the convolution of their FTs, i.e. the FT on the right of (e) shifted to the frequency of the oscillation. (There should also be a peak on the other side of the axis, but that isn't relevant to electromagnetic spectra.) The width of the absorption (real) and dispersion (imaginary) responses increases if the oscillation is more heavily damped. (See section §8.5.1.)

CHAPTER 6

Two-Dimensional FTs

Chapter Outline

6.1 Two-Dimensional Fourier Transforms Rules

We start with an account of how Fourier Transforms (FTs) are extended from one- to two-dimensions, and then survey how the basic rules need to change with dimensionality (the new rules are summarized in Fig. 6.5). Then we apply these rules to examine the far richer field of 2D structures and their FTs.

6.1.1 Transition from One-Dimensional to Two-Dimensional

Starting with a 1D density-curve along x, we have so far discussed finding its 1D FT along X. If instead we start with a 2D density-distribution (an image) in (x,y)-space, how do we get its 2D FT in (X,Y)-space? We start by slicing the image along y into thin adjacent parallel strips of data all parallel to x. Then we Fourier-transform each strip into a 1D FT along X, with a phasor at each point X. Now each strip had a y-coordinate, and so does its FT, so we now have a 'hybrid' image with real space along y but reciprocal space along X. To homogenize it, we next do a perpendicular slicing, yielding strips along y, and we Fourier-transform each strip

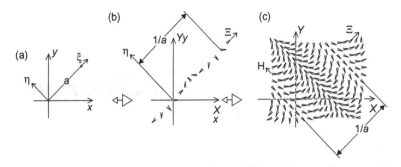

Figure 6.1: (a) Arbitrary peak in (x,y) space, with (ξ,η) axes fitting it. **(b)** 1D FT from ξ to Ξ. **(c)** Second 1D FT from η to H. This gives the 2D FT of (a) in (X,Y).

into a 1D FT along Y. Now we have a double (2D) FT in X and Y. This is not affected by the details of the 'slicing': we could do our first slicing along Y, and the second along X. Or we could choose a pair of mutually perpendicular (orthogonal) axes oriented at an angle to (x,y), like (ξ,η) in Fig. 6.1 (a), with a 1D FT along $\xi{\to}\Xi$ in (b) before the final FT along $\eta{\to}H$ in (c). In any case, we should still get the same result.

If we apply this procedure to a single pixel (Fig. 6.1), the resulting FT is a 2D phasor-wave whose lines of constant phase (*wavefronts*) are perpendicular to the line in real space (left) joining the origin to the pixel (the vector a). We would get the same result if we started by slicing parallel to a. Then the first slice would give a 1D FT of the pixel: a phasor-wave parallel to the length of the slice. The second, perpendicular, slicing would contain just one phasor per slice, so the perpendicular FT would be a constant equal to that phasor.

If the pixel were real, so too would be the phasor at the origin (on the right), so the phasor-wave would have Friedel symmetry (§3.3.5), as shown in Fig. 6.1. Now, just as we calculate 1D FTs by dividing the line-curve into peaks and adding their FTs (phasor-waves), so we divide a 2D image into pixels, each of which has an FT that is a characteristic phasor-wave, and we combine these to get the 2D FT. If each pixel is real, then each phasor wave will have Friedel symmetry. It is easy to see that two waves with Friedel symmetry must add to give a composite wave with Friedel symmetry. Consequently, if each pixel is real, then the entire 2D image's FT must have Friedel symmetry.

6.1.2 Rule Changes in Two-Dimensional Fourier Transforms

Which of the 1D FT rules, summarized in Fig. 3.20, are applicable to 2D FTs, and what extensions are needed? As we might expect, *all* are applicable in their 1D forms. In particular, the algebraic rules for addition and multiplication are unchanged. The *addition rule* (that the sum of two images yields an FT that is the sum of their individual FTs) is just as true as in 1D. The remaining 1D rules are also applicable, though with extensions. Perhaps the smallest

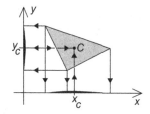

Figure 6.2: A triangle is projected onto the *x* and *y* axes (giving the compressed black density plots on these axes). From each density plot, a center of mass is found, x_C from the *x*-projection and y_C from the *y*-projection. These coordinates define the center of mass *C* of the triangle.

extensions apply to the *convolution theorem*, where the multiplication of two functions (curves) in two dimensions applies to *all* corresponding pixels in the two images, covering their entire area. As in 1D, 'zero wins' in multiplication. When the 'images' have phase variations, each point is a phasor, and phasors multiply according to the rule 'angles add, moduli multiply' (for the phase angles and the amplitudes or moduli). Naturally, convolution also gets generalized in two-dimensions: each *pixel* of one image gets replaced by the *whole* of the other image. The result is symmetrical: it doesn't matter which image we take point-by-point, and which one we copy. And, just as in 1D, the FT of multiplication is convolution.

But 2D space also brings important extensions. 1D space allows only one direction, but 2D for the first time introduces different directions. Thus operations involving some kind of uniform movement can employ it in different directions. So we have extensions concerning translation, distortion, projection and reflection. Moreover, there is something completely new: changing directions through rotation. We look at these in sequence.

The *translation rule* is easily derived from the convolution theorem, given the 2D FT of an isolated pixel (Fig. 6.1). We can translate the image to a new origin by simply convoluting the image with an isolated peak placed at the new origin. Then the peak's FT (on the right of that Figure) multiplies the image's FT. This adjustment of an image's origin can be useful. We saw in Chapter 3 (section §3.4.8) that, before calculating a one-dimensional FT, it is most convenient to adjust the origin of a curve to coincide with its center of mass. In two-dimensions we get an extension of this. We should adjust both the *x* and *y* origins of the image's coordinates to coincide with the centers of mass of the image's projection onto those axes. This is illustrated, for the image of a triangle, in Fig. 6.2. With this origin choice, we get a level phase-plot near the FT's origin.

The simple extension from one possible direction to two applies also to *distortion*. In 1D we found that stretching/compression of the curve's *x*-axis causes the FT to undergo the reciprocal distortion of the reciprocal *X*-axis. In 2D, there can be two independent distortions, with quite separate expansion or contraction factors, provided they occur at right angles.

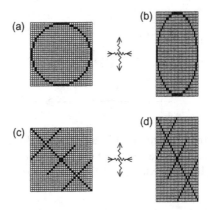

Figure 6.3: (a) Square subdivided into $30^2 = 900$ pixels bearing a pattern of a circle.
(b) The square is compressed to 2/3 its width in the x-direction, and stretched to 4/3 its height
in the y-direction, so the circle becomes an ellipse. **(c)** The same square bearing a pattern of
four straight-line segments. **(d)** The square after undergoing the same x-compression and y-stretch;
the line-segments have sheared.

The effect of two such distortions is shown in Fig. 6.3, where there is a horizontal
compression combined with a vertical stretching. A circle (a) becomes a vertical ellipsoid (b),
and this kind of horizontal/vertical effect is not seen with oblique lines (c), (d). These get bent
like a hinged lattice, and appear to show the effects of shear.

The *projection* rule undergoes an extension. If we parallel-project a 2D image we get its
projection onto a *line* perpendicular to the projection direction. The 1D FT of this line-density
will be a parallel *line-section* of the 2D FT of the image. Of course, that projected
line-density could again be projected, as described in Chapter 3, §3.5.4, giving a point
density. After the second projection we effectively get a radial projection of the entire image
onto the origin. The projected density is the density at the center of the 2D FT of the image.

1D FTs allow *reflection* in one direction, but in 2D this can be in any direction. Moreover,
in 2D we can have *two* independent reflections (but they have to be at right angles). One
reflection has the additional consequence in 2D that it changes *chirality*: e.g. changing
ordinary writing into mirror-writing. Reflections are preserved in the FT.

Some images have *rotational symmetry*: if they are rotated by some angle, they have exactly
the original appearance. (Of course, there is a unique point, the axis, that is not moved by the
rotation.) By the rotation rule, rotating the image causes the same rotation of the FT. And,
since a symmetrical image looks the same after rotation, it must again give the same FT.
Thus the rotational symmetry of the image is conveyed to its FT. We shall be making much
use of this rule in Part IV.

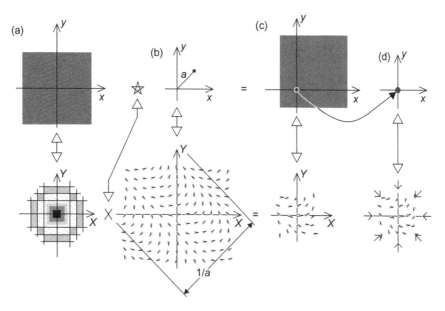

Figure 6.4: Proof of inversion theorem in 2D, using the 'CC sieve' (§3.5.1). Real-space two-dimensional images above, their FTs below. **(a)** Starting image and its FT. **(b)** (Above) a peak; (below) its FT, one of the 2D phasor waves used to multiply this FT as part of the CC sieve. That multiplication is equivalent to convoluting the starting function with a peak displaced by a ... **(c)** ... which displaces the starting function by a. **(d)** If we take that displaced function and stretch it wide in all directions, space gets filled with copies of the pixel at the origin. The corresponding FT is the complete projection of the phase-shifted image. So the 'sieved' FT gives the corresponding real-space pixel.

An image with 6-fold rotational symmetry looks the same after rotation by $360°/6 = 60°$; so it also looks the same after rotation by $60° \times 3 = 180°$. Since the same symmetry applies to its FT, that is also the same when rotated by $180°$: it has 2-fold symmetry. (This argument applies to 2-fold, 4-fold, 6-fold, 8-fold... etc. rotational symmetry.) We saw earlier that, if the original image is real, its FT must have Friedel symmetry: any two points related by $180°$ must have the same amplitude and phases, except that the phases have opposite signs. However, the FT's 2-fold symmetry implies that the phases with opposite signs are also equal, which is possible only if the phase angles are $0°$. Thus the FT must be *real*. Thus, when we deal (as usual) with real images, 2-fold symmetry implies that the FT is real.

Finally, *inversion* is demonstrated in Fig. 6.4, and the rules are summarized in Fig. 6.5.

Most of the important results can be deduced from these basic rules, but some results deserve a special place. Particularly important is the '*repeat lemma*': if any structure has an exact repeat a, then the FT exists only on a 'comb', i.e. repeating points (1D) or lines (2D) or planes (3D). This follows from the FT of the comb (Chapter 3, section 3.2.2); since a repeating structure is unaltered by convolution with the corresponding comb, its FT is

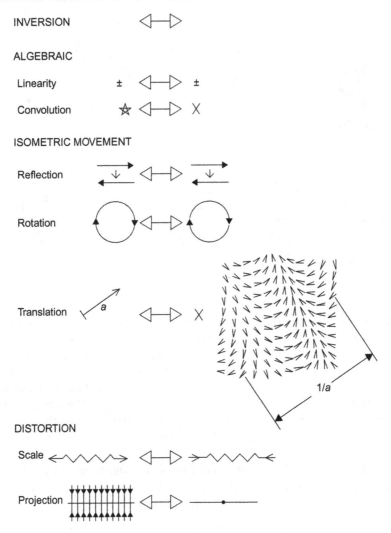

Figure 6.5: Rules for two-dimensional FTs.

unaltered by multiplication with the reciprocal comb, and therefore that FT only exists on the reciprocal comb. Associated with this is the '*crystal sampling lemma*': the FT on the 'comb' is simply the repeating unit's FT, sampled at the points, lines, etc. of the comb. This follows from: (repeating structure) = (repeat unit)★(lattice)↔FT(repeat unit) × FT(lattice), where↔ means taking FTs of each term. Now FT(repeat unit) × FT(lattice) is simply the repeating unit's FT, sampled at reciprocal lattice points.

6.1.3 Rule Changes in Three-Dimensional Fourier Transforms

Three-dimensional FTs are needed for the discussions in Chapter 13 in Part IV. However, the rule extensions from two- to three-dimensions are fewer than those from one to two.

The *addition* and *multiplication* (i.e. *convolution*) rules are the same, and the FT of an isolated pixel is still a phasor wave perpendicular to the pixel vector. However, the wave is now 3D and has *planar* wave-fronts. From this, the *translation* rule follows as in 2D.

Rotation is of course the same, except that the rotation axis is not a point but a line, which can have any direction in 3D space. Moreover, 3D space allows a combination of rotation and translation – the screw, described in Section §10.2.3. However, reflections in 3D change the handedness of amino-acids, etc. to an unnatural form; this restricts their use in symmetry operations.

Distortion is similar in 3D and 2D, except that in 3D it is described by an ellipsoid.

6.2 Points and Lines

The FT intensities (squared amplitudes) of simple cut-out patterns are easily obtained with the *optical diffractometer* (Taylor et al., 1951; see section §8.4.2), and these *optical transforms* have been useful in X-ray crystallography and optics: Lipson & Taylor (1958); Harburn et al. (1975); and Lipson et al. (1995). It is a challenging exercise to explain their salient features solely from the rules listed above, without needing any actual calculation.

6.2.1 Lines and Lattices

A long, thin, vertical line of density (Fig. 6.6 (a), left) is a 1D 'peak' stretched out indefinitely along *y*. It is equally unaffected by stretching along *y* or compression along *x*. Applying the distortion rule, its FT must be equally unaffected by *compression* along *Y* or *stretching* along *X*. So the FT must be a long, thin, *horizontal* line of density (right).

The rotation rule allows us to rotate both image and FT of (a). Such FT lines are 'infinitely narrow' and therefore cannot interfere with each other, so a pencil of lines intersecting at the origin yields an FT that is the same pencil, but rotated by 90°, as in (b).

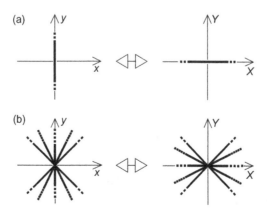

Figure 6.6: (a) Vertical line and its FT (right). **(b)** A pencil of lines passing through the origin. Right, their FT, which is the same diagram turned through 90°.

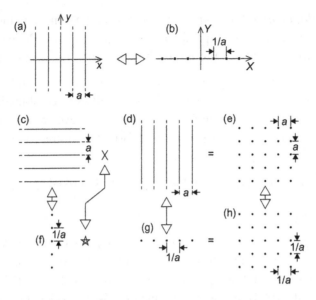

Figure 6.7: (a) The left is stretched, giving a grille of parallel lines. The FT **(b)** is correspondingly compressed, giving a 1D comb. **(c)**, **(d)** Two identical grilles of parallel lines oriented at right angles. Multiplication preserves only their common regions, which lie on a square lattice **(e)**. The corresponding FTs (**(f)**, **(g)**) are convoluted, giving a square lattice **(h)**.

Next we are interested in the FT of lattices, and approach it via sets of parallel lines, like a grille. Figure 6.7 (a) shows such a set, and its FT is shown in (b). The vertical lines in (a), like that in (a) of Fig. 6.6, must give horizontal lines in the FT, but the lines' continuity is interrupted by interference. The projection of (a) is a 1D comb whose 1D FT (a reciprocal comb – Fig. 3.4) is the 2D FT's parallel section, i.e. the X-axis. Thus we get the same horizontal comb as in Fig. 3.4(g) but its FT is different in 2D.

Two such (perpendicular) 'grilles' are shown in (c) and (d) of Fig. 6.7, where they are multiplied together to give a square lattice (e). (Multiplication of two structures – 'factor-structures' – allows zero density in one factor-structure to destroy anything in the other; zero always wins, so the only structure to survive is found where *both* of the factor-structures existed.) Below (c)–(e) we have the grilles' FTs but convoluted together; this gives us a square lattice. Thus we find that the FT of a square lattice is a reciprocal square lattice (i.e., with the reciprocal lattice-spacing).

The FT of a lattice, called its **reciprocal lattice**, is much used in analyzing symmetrical structures. Further examples can be derived from the simple square reciprocal lattice we have just found. It can be distorted into other shapes, the simplest distortion being along the directions of the grilles, giving a rectangular lattice (Fig. 6.8).

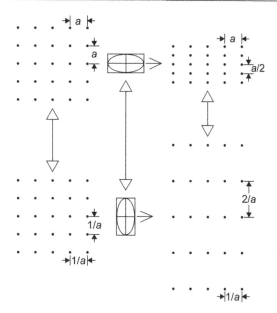

Figure 6.8: A square lattice uniformly distorted along a lattice direction. Its FT is reciprocally distorted.

We see that the two rectangular lattices on the right are FTs of each other. They are actually the same shape (rectangles twice as long as broad), but they are oriented perpendicular to each other.

The *shape* of a lattice (the shape of its unit cell) is defined by the included angle γ and the ratio of the two sides, a and b. It is fairly obvious that any desired *shape* can be produced by some appropriate uniform distortion of the square lattice. Then one can match this to the actual lattice by appropriate rotation and size change. Thus a square lattice can become any desired lattice by three appropriate operations: (i) a uniform distortion (*shape*); (ii) a rotation (*orientation*); and (iii) a change of size (*scale*). When applying these three operations to a lattice, we subject its FT to their reciprocal effects. By the distortion rule (§3.5.4), the uniform distortion causes the same distortion turned through a right-angle. The rotation has the same effect on the FT; and the size-change alters the size of the FT by the reciprocal amount. The first two operations combine together, giving the same distortion turned through a right-angle. So the *shape* of the FT, i.e. of the reciprocal lattice, is the same as that of the real lattice turned through a right-angle. This is shown in Fig. 6.9.

In this way we can get the correct *shape* of the reciprocal lattice, but we also need its correct *size*; we need the precise *scale factor*. We can usually find that from projections of the real lattice, as can be seen in the following concrete example: the *hexagonal lattice*, the most symmetrical lattice, important in the packing of identical rods, e.g. muscle filaments.

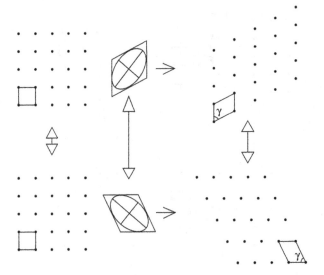

Figure 6.9: Uniform distortion of a square lattice (upper left) can give a two-dimensional lattice of any desired shape (upper right). Below are shown the FTs. These undergo the reciprocal distortion, which gives the same shape as the same distortion rotated by a right angle.

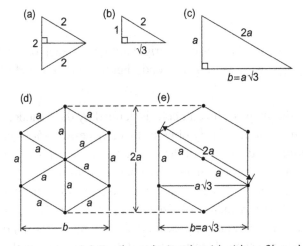

Figure 6.10: Hexagonal geometry. **(a)** Equilateral triangle with sides of length 2. **(b)** Cut in half, it gives a right-angled triangle with hypotenuse of length 2 and one side of length 1. The length of the remaining side is $\sqrt{(2^2 - 1^2)} = \sqrt{3}$. **(c)** Enlarge (b) a-fold. **(d)** Now apply this result to the hexagon, made of six equilateral triangles. We define the length of each side as a, and the hexagon's width as b, and need to know their ratio. **(e)** The hexagon contains the triangle shown in (c), which shows that $b = a\sqrt{3}$.

The basic geometry of the hexagon, shown in Fig. 6.10, is based on the *equilateral triangle*, i.e. one with equal sides and angles and hence 3-fold symmetry. Starting with a simple equilateral triangle (a), we find in (b) the ratios of its lengths which are used to find, in (c), the two widths of a hexagon ($2a$ and $b = a\sqrt{3}$) that are used in (d) and (e).

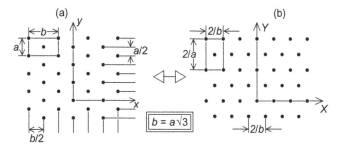

Figure 6.11: (a) Hexagonal lattice with minimum distance *a*. Its *x*-projections are $b/2$ apart, its *y*-projections are $a/2$ apart. **(b)** FT of these projections, so $X = 1/(b/2) = 2/b$, $Y = 1/(a/2) = 2/a$. (c) FT of the lattice has the shape of (a), rotated 90°, and fits the FT in (b).

Figure 6.11 shows the FT of a hexagonal lattice (its *reciprocal lattice*). We start by rotating the real lattice (a) through a right-angle to get the shape of the reciprocal lattice (b). Next we must scale the reciprocal lattice. In (a), the short vertical lines at the bottom show the projection of the lattice down *y*: a comb of spacing $b/2$. Its reciprocal comb is the *X*-section of the FT in (b): a comb of spacing $2/b$. The projection along *x* in (a) gives the second comb, short horizontal lines at the right with a spacing $a/2$. These give rise to a vertical comb with points spaced $2/a$ along the *Y*-axis of (b).

The most convenient cell for a hexagonal lattice is rectangular with an interior point (top left of (a) in Fig. 6.11). The corresponding rectangular (double) cell in reciprocal space is shown at the top left of (b). It has sides $2/a$ and $2/b$, and the *a*-side is parallel to the *a*-side in real space (and similarly with the *b*-sides). This is a convenient way to remember hexagonal lattices and their FTs. (However, each of these convenient cells is double, and they are not true reciprocals of each other.)

6.2.2 Reciprocal Lattice and Crystal Planes

Images are often calculated from crystallographic data arrayed on a reciprocal lattice. What is each reciprocal lattice point's contribution to the image? One point generates a 2D phasor-wave (Fig. 6.1), but we need pairs of symmetrically-related points with Friedel symmetry (same amplitudes and phase angles of opposite signs) to get a real image. Such a pair (top left of Fig. 6.12) generates a density-wave with wave-fronts perpendicular to the line joining the reciprocal lattice points, whose separation is the reciprocal of the wavelength (top right of Fig. 6.12).

All the density-waves have to fit together with the real lattice. Thus each set of parallel lines fitting the real lattice corresponds to a pair of reciprocal lattice points. This was found long before the advent of X-ray crystallography, since the lines joining real lattice points are important in 3D lattices. They correspond to the *plane faces* of a crystal, those planes in which the atoms or molecules are densely packed. Their angles were measured and found to be explicable in terms of the shape of a (then hypothetical) real unit cell. That cell was

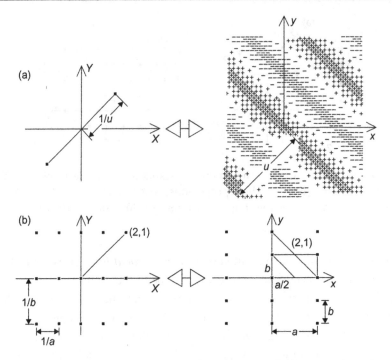

Figure 6.12: (a) Two points with the origin at their center, with the FT (right). **(b)** The points form a pair in a lattice (left) with corresponding crystal faces (right) fitting the FT above. (The rectangular ($a \times b$) unit cell lies in (x, y).)

connected with each crystal face through its three *Miller indices*, of which the basic principle is shown in Fig. 6.12 (b). The real cell is shown on the right, with its two parallel 'faces' (2,1) ('faces' are lines in two dimensions, and only two Miller indices are needed). These numbers mean that the 'faces' divide the first axis (a) in two and the second axis (b) in one: they intersect x at $a/2$, but y at b. Thus, in the Miller indices of a 'face', the number assigned to each axis is the fraction of that axis measuring the intersection of the 'face'. It is this reciprocal relationship that explains why the indices also apply to the reciprocal lattice.

6.2.3 Unit Cell Contents

A 2D repeating pattern (like Fig. 6.13) consists of an elementary pattern-motif repeated by an infinite series of translations that form a lattice. The *unit cell* is the minimum structure that can generate the whole pattern through repetition by the lattice. In Fig. 6.13 this structure corresponds to one of the big circles, plus two of the adjacent small ones. However, the cell is not unique: the parallelograms A, B or C (with different shapes or directions) would all fill the sheet after repetition by the lattice. Even when the cell's shape is specified, there are different ways of filling it, like A, A′ or B, B′ or C, C′. What *is* unique is the *area* of the unit cell and the *unit cell contents*. The one big circle and two small ones are particularly clear

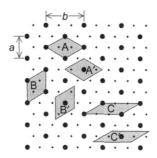

Figure 6.13: The big dots are hexagonal lattice points; the small ones are extra detail that counts as part of the repeating pattern. A, B and C are different primitive unit cells. A', B' and C' are shifted versions of those cells.

with cells A', B' or C'. However, in C the big circle is divided among the four corners, and there are four half-shares of small circles.

Since a unit cell, after repetition, fills the plane exactly, with no gaps or overlaps, the pattern is a convolution:

$$(\text{entire sheet}) = (\text{unit cell}) \star (\text{lattice}) \tag{6.1}$$

Then the convolution theorem gives us:

$$\text{FT(entire sheet)} = \text{FT(unit cell)} \times \text{FT(lattice)} = \text{FT(unit cell)} \times (\text{reciprocal lattice}) \tag{6.2}$$

This is equivalent to:

$$\text{Diffraction pattern} = \text{FT(unit cell contents)} \times (\text{reciprocal lattice}) \tag{6.3}$$

provided the 'diffraction pattern' is defined in terms of phasors (with amplitudes and phases) rather than only diffracted intensities (which are just the squared amplitudes). A useful way of looking at this equation is that the diffraction pattern (phasors) is the unit cell *sampled* at reciprocal-lattice points. This is the 2D equivalent of lattice sampling in section §3.4.5. But, in 2D, the cells A, B, C, A', B', C' are all different, so their FTs are also different. But that doesn't invalidate the equation, since all these FTs are exactly the same in those places where their FTs are sampled by the reciprocal lattice.

We close with a concrete example: the FT of the pattern[1] in Fig. 6.13 (a). This pattern is the sum of two hexagonal lattices, a larger 'myosin' lattice (with big points) and a smaller 'actin'

---

[1] This will be recognized as the arrangement of myofibrils in a cross-section of vertebrate muscle, the small circles being the thin filaments (mostly actin) and the big ones (thick filaments) mostly myosin.

lattice (with small points). Now Fig. 6.11 has already shown that a simple hexagonal lattice's FT is a perpendicular hexagonal lattice; that the rectangular body-centered cells of the real and reciprocal lattices are also perpendicular and, if the real cell is $a \times b$, the reciprocal cell is $(2/a) \times (2/b)$. In (a) of Fig. 6.16^2, the small lattice is $a \times b$ and, if all points were the same, the reciprocal lattice would be the $(2/a) \times (2/b)$ lattice of big points in (b). Correspondingly, the FT of the $A \times B$ lattice of big points in (a) is the $(2/A) \times (2/B)$ lattice of small points in (b).

If we weaken the smaller real 'actin' lattice in (a), we approach the bigger $A \times B$ 'myosin' lattice whose FT is the small $(2/A) \times (2/B)$ lattice in (b). Contrariwise, if we strengthen the 'actin' dots in (a), we approach a uniform small $a \times b$ lattice (where the 'actin' and 'myosin' dots are equally strong), whose FT is the big $(2/a) \times (2/b)$ lattice in (b). Thus muscle FTs (obtained from cross-sections in the electron microscope, or from X-ray diffraction patterns of unfixed muscle) could be used to find the relative densities of the thick and thin filaments. Huxley (1968) thereby confirmed that, in live muscle, the thin filaments are relatively weak, but rigor causes more than 40% of the mass of the thick filament to be transferred to the thin filaments (through binding of a flexible outer domain of myosin).

6.3 Polygons

Polygons are also important 2D figures. We start with 'point polygons', rings of points that are symmetrically arranged around the origin, and consider their ACFs and FTs.

6.3.1 Auto-Correlation Functions of 'Point Polygons'

Points give ACFs that consist solely of points (the faint 'construction lines' were added for clarity). We can get the ACF of a figure by convoluting it with exactly the same figure, rotated by 180°. This is shown for an equilateral triangle in Fig. 6.14, where each triangle center is indicated by a cross, and the three crosses in (b) fit the corners of the same triangle rotated by 180°. The ACFs of this and the next three symmetrical polygons (square, pentagon and hexagon) are shown in Fig. 6.15.

Because of the way they are constructed, ACFs always have 2-fold rotational symmetry. There are two ways to view the ACF (§2.3.1): the 'overlap' and 'vector' viewpoints. The former implies that the ACF is twice as large as the figure (as no overlaps are possible beyond that), that the centers have the biggest peaks (because the maximum overlap occurs there) and the simplest figures have the simplest overlaps (seen in Fig. 6.15 (b) for the square). The same conclusions follow from the 'vector' viewpoint. The center has all the numerous zero-length vectors, the width is determined by the maximum vector, and the highly symmetrical square has most of its vectors superposing.

---

2 Figure 6.16 is placed later because the relation between the two hexagonal lattices is also relevant to the FTs of two polygons, as we see below (§6.3.2).

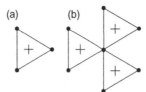

Figure 6.14: (a) Equilateral triangle, with the center marked. **(b)** The triangle (a) convoluted with an identical copy of itself, but rotated 180°; this is the triangle's ACF.

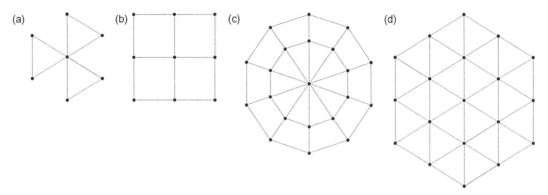

Figure 6.15: ACFs of **(a)** equilateral triangle; **(b)** square; **(c)** regular pentagon; **(d)** regular hexagon. All except (c) fit a lattice.

The pentagon uniquely gives an ACF that is plainly not part of a lattice. This fits with the fact that 5-fold rotation is the only one (out of 3-, 4-, 5- or 6-fold) that is inapplicable to a lattice (see Chapter 10). (Of course, the same applies to 7-fold and other *higher* axes.)

In 2D, the ACF is the structure convoluted with the same structure rotated by 180°. A 180° rotation changes the signs of all phasor waves in both X and Y. Consequently, the FT has changed signs for all the imaginary quantities, and is called the 'complex conjugate' (§3.3.5), written with an asterisk. So, if $f \leftrightarrow F$, then $f(-) \leftrightarrow F^*$. Therefore (as we have already seen in 1D, see section §3.4.6), the ACF $= f \star f(-) \leftrightarrow FF^* = |F|^2$.

6.3.2 Fourier Transforms of 'Point Polygons'

The square and regular hexagon have 2-fold symmetry so the FT must be entirely real (§6.1.2) if the origin is placed at the center. By contrast, the equilateral triangle lacks 2-fold symmetry, so its FT is complex, wherever the origin is put. We shall find, however, that these two polygons have closely related FTs. Both are shown on the left (a) of Fig. 6.16, which contains two real lattices: a small lattice fitting both big and small dots; and a big lattice fitting only the big dots. We have already seen (§6.2.3) that the smaller rectangular real lattice (sides a and b) gives the bigger reciprocal lattice (sides $2/a$ and $2/b$, and fitting only the big

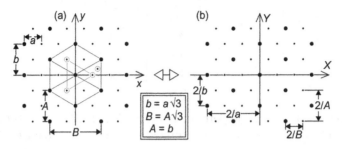

Figure 6.16: (a) The equilateral triangle's points are connected with grey lines, its ACF's points with black lines. The lattice fitting the equilateral triangle comprises all points, that fitting the triangle's ACF consists of only big points. **(b)** Both reciprocal lattices: the big points form the reciprocal lattice of *all* the points in (a); whereas *all* the points in (b) constitute *only* the reciprocal lattice of the larger lattice in (a).

dots on the right); while the bigger real lattice (left, with sides A and B) gives the smaller reciprocal lattice (right) with sides $2/A$ and $2/B$ (fitting both small and big dots on the right).

Now, in the left (real) diagram, consider the small central faint-line equilateral triangle with sides b. It has three vertices (circled), but the central big black dot is not a part of it. Therefore it has no points in common with the bigger lattice (big dots), so multiplying it with this lattice will annihilate the triangle. However, multiplying by the small lattice (big and small dots) preserves the triangle. Therefore the triangle's FT will be preserved if convoluted with the small lattice's reciprocal lattice. That reciprocal lattice is the big lattice (only big dots) on the right; this has sides $2/b$ and $2/a = 2\sqrt{3}/b$ (since $a = b/\sqrt{3}$). So the equilateral triangle, side b, gives an FT that repeats on a rectangular lattice with sides $2/b$ and $2\sqrt{3}/b$. This result is used on the right of Fig. 6.18, where the equilateral triangle has sides A and its FT repeats on a lattice with sides $2/A$ and $2\sqrt{3}/A$.

Next, in the left (real) diagram, consider the central hexagon with sides $b = A$. It has six vertices (big dots), which all fit the bigger real lattice. Therefore, by a similar argument, the hexagon's FT will repeat on the smaller reciprocal lattice which has sides $2/A$ and $2/B = 2/A\sqrt{3}$. This result is used in Fig. 6.17, and the same reciprocal lattice is repeated on the left of Fig. 6.18.

Finally, we interpret these results. First, the hexagon (Fig. 6.17). Because the origin is put at its center, its FT (right) is real and thus consists of positive or negative regions. Its FT must have positive peaks at the origin, and these will repeat according to the pattern just described. On the 'minimum wavelength' principle, there is only room for one negative region surrounding these peaks (light grey) and the intervening zeroes are shown as white rings in Fig. 6.17.

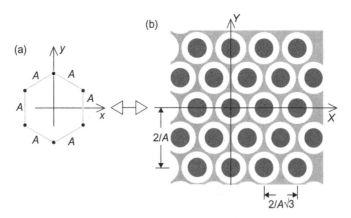

Figure 6.17: (a) Regular hexagon consisting of six points. **(b)** Schematic image of its FT. Dark shading: positive; light shading: negative. Zero lines lie within the white rings.

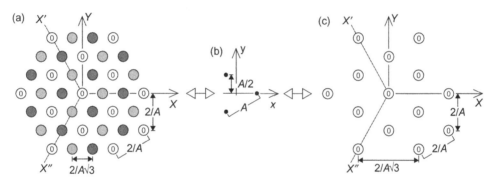

Figure 6.18: (a) FT intensities (though different shadings have different phases) of equilateral triangle **(b)**. **(c)** Only the regions of zero phase in (b)'s FT. Compare (a) with Fig. 12.9.

Last of all, the equilateral triangle (Fig. 6.18 (b),(c)). As this triangle lacks a 2-fold axis, its FT is complex and the phase can change without passing through nodes (points or lines). Nevertheless it contains 'blobs' of higher intensity, and one such must occur at the origin where the phase will be $0°$ (since the origin is at the triangle's c.o.m.) Our earlier conclusion shows that this $0°$ blob must repeat on the hexagonal lattice shown in (c). (The blobs have 0 in their centers to indicate their identity.)

However, there is a connection between these polygons. The small equilateral triangle's ACF is the hexagon that surrounds it in Fig. 6.16 (a), plus the point at the origin (see Fig. 6.14 (a)). That point makes no difference to our earlier argument showing that the hexagon fits the lattice in Fig. 6.17 (b), which is the same as the lattice in Fig. 6.18 (a). Thus the equilateral triangle's ACF gives the FT shown in Fig. 6.18 (a), except that each circle should be the same.

What is the significance of this? The intensity (i.e., square of the amplitude) of a FT is the FT of the image's ACF. The ACF of the equilateral triangle was already obtained in Fig. 6.14 (b), and it is repeated in Fig. 6.16 (a), if we ignore the differences in labelling of the 'blobs'.

So we have two important features of the small equilateral triangle's FT, shown in Fig. 6.18 (a) and (c). Its 'blobs' of *amplitude* repeat according to the finer lattice of (a), while the blobs of *zero phase* only repeat according to the larger lattice of (c). Actually, we have already incorporated into (a) the labeling of zero-phase blobs from (c). The remaining 'blobs' of (a) have been shaded in the following way. The equilateral triangle has 3-fold symmetry, and the rotation rule requires the same symmetry in the FT. Consider, then, the six apparently identical 'blobs' immediately surrounding the zero-phase 'blob' at the origin. To fit the FT's 3-fold symmetry, they must consist of two groups of three identical 'blobs'. This is indicated by the two shading levels. Relative to the origin 'blob', the left blob is arbitrarily assigned a dark level and the right one a light level. Now the strict lattice observed by the zero-phase 'blobs' (containing zeroes) requires that each such 'blob' must have a dark one on its left and a light one on its right. That leads to the pattern in (a). This is confirmed since it makes the X-axis a mirror, which is necessary as the x-axis is a mirror of the triangle on the left.

We notice that a light-grey 'blob', rotated 180° about the origin, brings us to a dark-grey 'blob'. These two 'blobs' must have Friedel symmetry, since the triangle is real. Thus the light grey and dark grey regions are complex conjugates, i.e. their phase angles have opposite signs. This fits the phase sequence 0°, 120°, 240°, … when moving to left (or right) along the X-axis. Calculation shows that, with our FT convention (§3.3.6), left is the correct direction for this phase sequence, and the FT is shown in Fig. 12.19. The regions of parallel lines correspond to the 'blobs' in Fig. 6.18(a).

6.3.3 Uniform Polygons and the Sampling Theorem

Symmetrical rings of three or six points (and also of four points, though not shown) thus give endlessly repeating 2D FTs. This should not be surprising, since they fit-lattices and since small groups of points give repeating 1D FTs. However, the rectangle-function's 1D FT does not repeat, but fades away with oscillations (Fig. 5.1(e)); filling the space between the boundary-points eliminates the repeat. Its 2D equivalents would be uniform polygons, and we would expect their 2D FTs to fade away similarly. Indeed, the case of the square is so closely related to the 1D rectangle-function that this is obvious. Therefore we shall discuss only the hexagon, another lattice-fitting uniform polygon that is an even function (relative to its center) and consequently has a real FT that can change phase only across nodal lines.

The most important application of the 1D rectangle-function's FT (the 'sinc-function') was in the sampling theorem (§5.3.2). We can generalize this theorem to 2D, using the FT of a polygon, and thereby obtain a 2D version of the sampling theorem, adapted to a particular lattice. In general, the most useful polygon is a parallelogram, but here we choose a

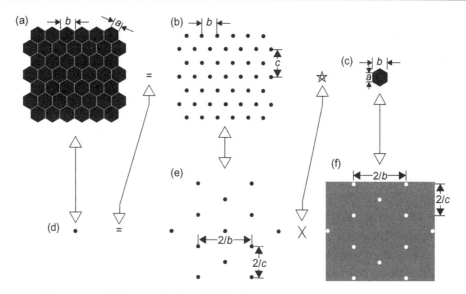

Figure 6.19: (a) Hexagons close-packed to fill the plane. **(b)** The lattice of this packing, convoluted with.. **(c)** ... one of the hexagons. **(d)** FT of uniform plane is a peak. **(e)** FT of the lattice is its reciprocal lattice. **(f)** So the FT of a hexagon has zeroes on a hexagonal reciprocal lattice, except for the origin.

hexagonal box. We want its FT, which is the interpolation function for a hexagonal lattice, and we get information about it from an extension of the argument used to find the sinc-function (§5.2.1).

The uniform hexagon (Fig. 6.19 (c)) has side-to-side width b, as in Fig. 6.10. Many such hexagons pack together, as in the top of (a), fitting the lattice at the top of (b). (There is supposedly a lattice point at the center of each hexagon in (a).) This gives us a graphical 'equation' at the top of Fig. 6.19, whose FTs (shown underneath) we need to interpret.

Starting at the left, the FT of a uniformly filled plane (a) is just a peak at the origin (d). The hexagonal lattice (b) has an FT that is another, perpendicular hexagonal lattice with relative dimensions shown in Fig. 6.16. (In Fig. 6.19 (b), the rectangular lattice is short b and long c, corresponding, respectively, to short A and long B in Fig. 6.16, where the reciprocal lattice was long $2/A$ and short $2/B = 2/(A\sqrt{3})$; so the reciprocal lattice in Fig. 6.19 (b) is long $2/b$ and short $2/c = 2/(b\sqrt{3})$.) Then the convolution of (b) and (c) in real space becomes multiplication of (e) and (f) in reciprocal space. So finally we come to the matter of interest: (c), of the uniform hexagon, whose FT is represented rather cryptically as (f). We must deduce some of its features from the equation at the bottom of Fig. 6.19. This tells us that the hexagon's FT, when multiplied by the reciprocal lattice, gives just one point at the origin. Thus the hexagon's FT annihilates all other points of the lattice, which it does by having nodes at those other points. These are shown as 'holes' in the cryptic diagram (f).

These 'holes' provide a framework on which to construct the rest of the uniform hexagon's FT (Fig. 6.20). There are three general guiding principles. (i) The minimum wavelength principle, which shows that the FT must consist of regions of dimensions no smaller than $1/b$; (ii) The hexagon, as positioned on the left, has 2-fold symmetry, so its FT is entirely real and therefore contains $\pm$ regions separated by nodal lines which include the nodal points found in Fig. 6.19 (f); (iii) The FT has the hexagon's symmetry, i.e. a 6-fold rotation axis and six mirrors. Thus the entire FT can be constructed from copies of one 30° segment.

The FT is shown on the right of Fig. 6.20. Its center is obviously positive, and the positions of its zeroes (as well as the symmetry of the hexagon itself) show that this inner positive part must be roughly circular. Outside this, there has to be a negative zone, also roughly circular (though with some hexagonal deformation). To get an insight into the further regions, we consider some of the hexagon's projections. For the x-projection, consider the decomposition of the projection into a rectangle plus a smaller triangle. The rectangle gives a sinc-function, with sign alternation and slow attenuation. The triangle gives a sinc^2-function, with zeroes but all positive and fast attenuation. The sum of these resembles the sinc-function, and generates an intensity 'spike' in the FT (though here 6-fold symmetry makes six 'spikes'). By contrast, the y-projection derives from the difference between two triangles, so its FT is entirely made from sinc^2-functions. Therefore it attenuates more quickly than the 'spikes' from the sinc-function, and these more quickly fading lines separate the six 'spikes'.

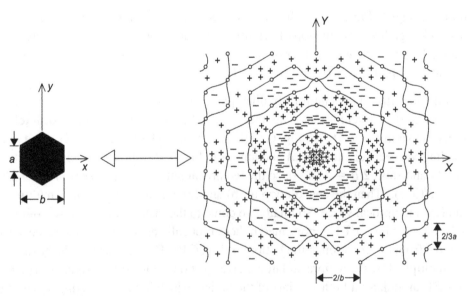

Figure 6.20: Left, a regular hexagon of uniform density, with width b; right, its FT.

This FT, like all box FTs useful in the sampling theorem, has two special features. There are nodes, that allow all the sampled values to be independent of each other, and intensity 'spikes', that carry information from sampled values far into the continuous FT.

6.4 Polar Coordinates

We started this chapter by extending 1D FTs into 2D FTs, i.e. by extending one translation to two translations. However, 1D FTs could be viewed as having (real, reciprocal) angular coordinates instead of (x,X), to represent *rotations* in 2D. Many biological structures show rotational symmetry, looking exactly the same after rotation by some fraction of a complete revolution. Such structures have an *N*-fold *rotation axis* (seen clearest when viewed down the rotation axis; Chapter 10). Such objects are periodic with *angle*, but this is best revealed if we take their FT with *polar coordinates*, using the radial distance r and the angle φ. There exists a complete mathematical system, closely analogous to Fourier analysis (indeed, largely based on it), but appropriate to images described in polar coordinates (r,φ) (Fig. 6.21). This is used to analyze the images of rotationally-symmetric and helical structures discussed in Chapters 15 and 16 in Part IV.

6.4.1 Jinc-Function

Polar coordinates can cover a 2D feature in just one dimension. A most important filter in 1D is the rectangle function (Fig. 5.1(e)), whose FT is the 'sinc-function'. Its 2D polar equivalent, a uniform aperture, has an FT called (by analogy with the sinc-function) the 'jinc-function'. Since the uniform circle (Fig. 6.22(a)) is an even function, the 'jinc-function' is entirely real, and we can find its principal features by analogy with the hexagon's FT, viewing the circle as a low-resolution approximation to a hexagon with the same width b (the distance between parallel sides). At least the central (low-resolution) parts of their FTs should be very similar. We conclude that the center of a circle's FT has a positive peak followed by a negative ring; and we might expect these to be followed by other rings of alternating signs. The comparison is shown in Fig. 6.22.

Therefore the circular aperture, used as a spatial-frequency filter, leads to the convolution of an image with Fig. 6.22 (b). That happens in optical or electron microscopes with a circular aperture, and the FT generates the 'diffraction rings' so familiar in high-magnification

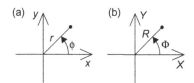

Figure 6.21: Polar coordinate system in **(a)** real and **(b)** reciprocal space.

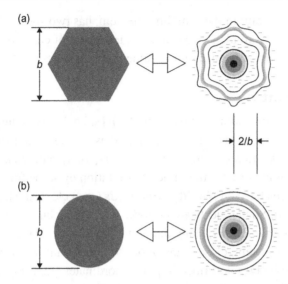

Figure 6.22: (a) Uniform hexagon and its FT, as in Fig. 6.20. **(b)** Uniform circle and its FT. Black and grey shows positive regions of the FT. Black lines indicate nodes, and minus signs show negative regions.

microscopy or telescopy. We can also view the circle's FT as the 1D FT of the projection of a uniform circle, as shown in Fig. 6.23 (a). Thus we only need to define how the circle's FT varies with radius, and Fig. 6.23 (b) shows this.

6.4.2 *Bessel Functions:* J_0

Comparing Fig. 6.23 (b) and (c), we see that, viewed as a function of r, the aperture-function (uniform circle) is the polar-coordinate equivalent of the rectangle-function. Both are uniformly filled within sharp boundaries, and their FTs resemble each other. Now the FT of the rectangle function (c) is the sinc-function, $\sin(X)/X$. The '$/X$' gradually attenuates the function, whose sine part comes from interference between the sharp edges. Both sine and '$/X$' are odd functions, so their ratio is even, which is necessary for the FT of an even function (the rectangle). By analogy with the rectangle, the circular aperture's FT might be supposed to be 'something/R', where the 'something' should be an odd function resembling $\sin(R)$. What could be the appropriate function? Actually, it was discovered nearly two centuries ago by the astronomer Bessel[3], and is therefore called the *Bessel function*.

Bessel defined a whole series of similar 'Bessel functions' (see §6.4.3, §15.2.2), of which the first two resemble most closely the familiar sines and cosines (Fig. 6.24). It is (c), the (odd) Bessel function $J_1(R)$, that resembles $\sin(x)$, so the FT of a uniform circle is $J_1(R)/R$ (hence the term 'jinc-function').

---

[3] F.W. Bessel (1784–1846) was a self-taught astronomer who first detected stellar parallax, thereby finally proving Copernicus's theory.

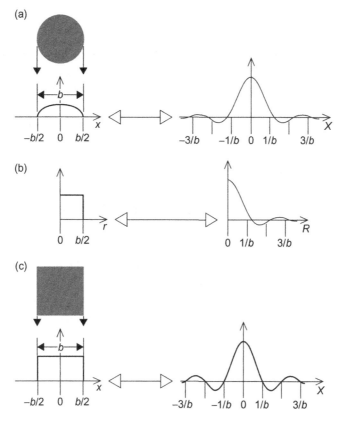

Figure 6.23: (a) Uniform circle with its projection onto the x-axis. Its FT is shown on the right. **(b)** Polar coordinate view of the FT of a circle, reduced to one dimension. The circle's density as a function of r resembles the one-dimensional rectangle function, and the right shows its FT, the 'jinc-function'. **(c)** Uniform two-dimensional square with its projection onto the x-axis. This is the rectangle function whose FT is the sinc-function (§5.2.1).

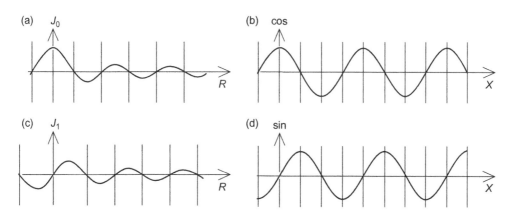

Figure 6.24: The first two Bessel functions (left), and their most similar trigonometric functions (right). **(a)** Zero order Bessel function, $J_0(R)$. **(b)** Cosine. **(c)** First order Bessel function, $J_1(R)$. **(d)** Sine.

The different Bessel functions are identified by a numerical subscript, their *order*. The first in the series, Fig. 6.24(a), the zero-order Bessel function $J_0(R)$, is an even function, like the cosine. We recall (from §3.2.1) that the cosine is the FT of a symmetrical pair of peaks. The nearest thing to this in polar coordinates is a thin circular ring, and its FT uses the zero-order Bessel function, $J_0(R)$. However, a ring is a centro-symmetrical 2D object with a real 2D FT, which is shown in Fig. 6.25 (b).

This ring can be approximated by a series of symmetrical point-polygons with an increasing number of points. Section §6.3.2 discussed a few of these (most relevantly, with four or six vertices). (See the center of the FT for six points: Fig. 6.17.) Both have a circular positive region at the FT center origin. Surrounding this is a negative region, shaped like a ring near the positive circle. Thus we can conclude that the FT of a circular ring, containing the zero-order Bessel function, will have a positive circular peak at the origin, surrounded by a negative ring. It must be real, so regions can only be positive or negative, and these must be separated by a line of zero amplitude. This is all very similar to the cosine, but converted into a function with circular symmetry. There is the big difference that the cosine repeats exactly, so all peaks have same height, whereas $J_0(R)$ has its amplitudes spread over rings whose perimeters increase proportionally with the radius R. To maintain the same total intensity, the intensity must decrease proportionally to the distance R, and the amplitude (the square root of intensity) must therefore decrease proportionally to $\sqrt{R}$.

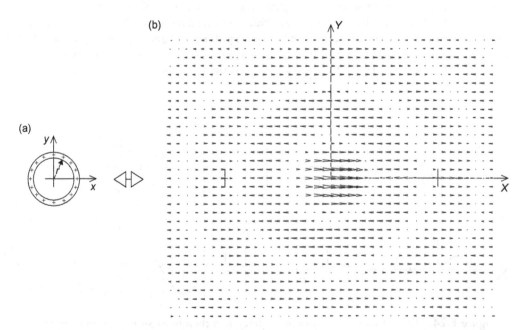

Figure 6.25: (a) Thin ring of uniform positive density, radius r. **(b)** Its FT. The mark about half-way along the X-axis indicates $X = 1/r$.

6.4.3 Bessel Functions: Higher Orders

The restriction to strict circular symmetry needs lifting if we are to apply the polar coordinate method to interesting structures. The projection of a uniform ring (Fig. 6.25) is symmetrical, approximately two peaks, whose FT is a cosine curve. The sine is the FT of an *anti*-symmetrical pair of peaks, and the corresponding feature in polar coordinates is a similar thin circular ring, but with a sinusoidally-varying density (Fig. 6.26 (a)).

What would we expect of this ring's FT (b)? The ring has a mirror on x, and so too must its FT. But on y the ring is an anti-symmetric function (reflection combined with sign-change), and so too is its FT. Next, the projections. That of the ring on y is nothing, as the positive and negative densities cancel; so the Y-section of the FT is blank (and that extends to almost cutting out an angular segment surrounding y). The projection of the ring onto x is approximately two peaks of opposite sign, the left negative and the right positive. Now the former corresponds to the left of Fig. 3.14 (a), multiplied by -1 (i.e. by a left-pointing horizontal phasor); and the latter corresponds to the left of Fig. 3.14 (c). These are respectively shown in Fig. 6.27 (a), (b), where (c) is their sum, which is a sine wave multiplied by the unit phasor with angle 90° (this is the 'imaginary' number i). So it is a sine-wave consisting of phasors with ±90°, which is similar to the X-axis in Fig. 6.26 (b).

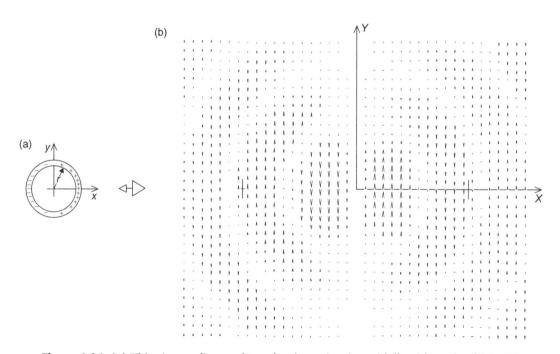

Figure 6.26: (a) Thin ring, radius r, whose density varies sinusoidally with angle. **(b)** Its FT. The mark about half-way along the X-axis indicates $X = 1/r$.

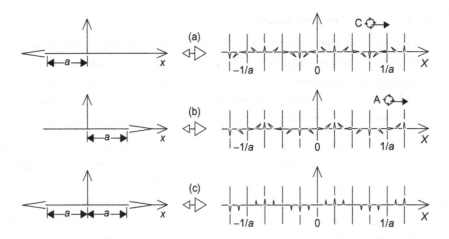

Figure 6.27: (a) FT (right) of (left) phasor -1 at $-a$; **(b)** FT (right) of (left) phasor $+1$ at $+a$; **(c)** sum of (a) + (b).

Given the FT pairs Figs 6.25 and 6.26, we could now do a very simple FT in polar coordinates (called a *Fourier–Bessel transform*). By adding, with appropriate weighting factors, the rings on the left, we can get a ring consisting only of positive density on the right (because, on the left, the positive density of the first diagram's ring can cancel the negative density of the second's). Then the FTs also add with the same weighting factors. If, in Fig. 6.26 (a), we wanted the ring's density maximum at the top, i.e. rotated anti-clockwise by 90°, we would also rotate the FT in (b) by the same angle (by the rotation rule). Moreover, we can expand or shrink the density-rings, and the FTs will undergo the reverse process (by the scale rule). Rings of different radii would allow us to use the 'salami' method for obtaining FTs, though the 'slices' would have to consist of rings.

However, if we were restricted to basing the FTs on Figs 6.25 and 6.26, our rings could only repeat after zero or one cycle. When we want a faster angular variation (more cycles per revolution), we need a higher Bessel order. We could calculate more complicated diagrams like Fig. 6.26 for $J_2(R)$, $J_3(R)$, etc., but there is a simpler way to sketch them. This is based on the pattern of nodal lines. The Bessel function, $J_0(R)$ or $J_1(R)$, which describes the FT's radial behavior, must also fit its *angular* modulation. $J_0(R)$ goes with no angular variation (Fig. 6.25 (b)), while $J_1(R)$ goes with a sinusoidal variation that, on any circle of constant radius, repeats once per revolution (or one cycle) (see Fig. 6.26 (b)). Thus the Bessel order, zero for $J_0(R)$ and one for $J_1(R)$, is also the number of angular cycles per revolution.

In Fig. 6.26 (a), there is a line (along the y-axis) where the ring's density becomes zero; similarly, the FT also becomes zero along the Y-axis. These *nodal lines* are boundaries between regions of opposite sign. There are also *circular nodal lines* which intersect the radial lines at right angles and help to demarcate FT regions of different sign. Similar circular nodes

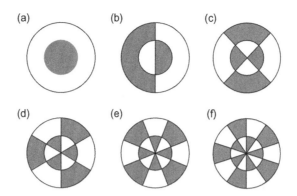

Figure 6.28: Arrangement in FTs of nodal lines and regions different sign (dark: positive, white: negative). The region closest to the center and on the right is always positive. **(a)** Zero-fold rotational symmetry (circularly symmetric): $J_0(R)$, as in Fig. 6.25 (b). **(b)** One-fold rotational symmetry: $J_1(R)$ combined with angular variation, as in Fig. 6.26 (b). **(c)** Two-fold rotational symmetry. **(d)** Three-fold rotational symmetry. **(e)** Four-fold rotational symmetry. **(f)** Five-fold rotational symmetry. (These patterns describe the normal modes of vibration (§11.3.3) of a circular drum; see (e.g.) sections 9.4 and 10.3 of Mathews & Walker, 1964.)

are found in Figs 6.25 and 6.26. They occur at zeroes of the Bessel function, $J_0(R)$ in Fig. 6.25 and $J_1(R)$ in Fig. 6.26. The nodal lines and sign-regions are illustrated in Fig. 6.28 (a) and (b). The Bessel functions have circular nodes, and the radial nodes imply an angular variation.

If we needed a ring with two sign-cycles, i.e. two positive regions and two negative ones, then the FT would also undergo two angular sign-cycles, as in Fig. 6.28 (c). That figure also shows the patterns for 3-, 4- and 5-fold angular sign-cycles. The figure also shows that, with higher-order rotational symmetries, the straight (radial) nodal lines get closer together near the center. Now the sampling theorem (section §5.3.2) says that the FT of a limited object cannot change much faster than the reciprocal of its width. But the width of FT regions shrinks towards the center, particularly when there are many nodal lines. Rings with higher symmetries give FTs with more nodal lines, so these must disappear faster towards the FT center. Thus, when the order n is bigger, the threshold value of R (i.e., where the FT becomes significant) must increase. Using higher-order Bessels, we can extend polar-coordinate FTs to higher symmetries, using circular rings with the appropriate rotational symmetry (see sections §15.2 and §16.3).

In section §6.1.1 we encountered 'hybrid' (x,Y) or (X,y) FTs which apparently lack utility. However, they can be useful in polar coordinates, where the radial part is continuous like an FT but the angular part repeats after a revolution, like a Fourier series. So intermediate stages like $(r,\varphi) \rightarrow (r,n) \rightarrow (R,n)$ can have some use, and similar hybrid transforms are used in helical FTs (Chapter 15). The existence of angular Fourier series also allows for angular convolutions and consequently also of angular CCFs and ACFs. These have an important role in single-particle analysis (Chapter 17).

Optics

The main purpose of Part II is to explain how the electron microscope is able to register high-resolution information about the smallest biological structures, and how its instrumental deficiencies (defocus and lens defects) can be corrected by computer processing. Thus the ultimate goal is the wave theory of the electron microscope's aberrations Chapter 9, but this journey is broken into three stages. (i) In Chapter 7, we look at microscopes (light and electron) using the much simpler ray theory, which nevertheless explains many imaging and aberration phenomena. We explore the basic principles of the light microscope and its lenses, partly because of their intrinsic importance, but mostly because they represent a simple analogue to the electron microscope. (ii) Chapter 8 introduces wave phenomena such as interference and diffraction, essential for Chapter 13 and for (iii) Chapter 9, which combines wave theory and microscopy. All wave phenomena (and even some ray phenomena) make use of the FT theory developed in Part I.

CHAPTER 7

Microscopy with Rays

Chapter Outline

7.1 Light Microscopy

That most important biological instrument, the light microscope, developed for nearly three centuries on the basis of simple ray-optics. The logical (if not historical) stages of its development are the pinhole camera, the single lens microscope, the compound microscope, and the same with aberration correction. These took microscopy to a level (in the mid-19th century) where the remaining problems depended on the wave nature of light.

7.1.1 Pinhole Camera

Magnification, the essential function of a microscope, can be performed by the simplest possible optical device: the pinhole camera. This creates an image, more accurate than any painting, by using just a hole (Fig. 7.1). Only one ray from each part of the object is transmitted by the pinhole, which thereby imposes a simple geometrical mapping that

Structural Biology Using Electrons and X-Rays.

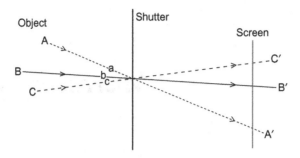

Figure 7.1: Pinhole camera image: all rays must pass through the hole in a shutter. Starting from three object points, A, B and C (or a, b, c), the rays give equally sharp images at all distances of the screen (or wall: *camera* = 'room') from the pinhole.

recreates the light distribution of the object. The pattern of the straight rays is identical on the far side of the pinhole, except for being upside-down.

The same geometry applies at all distances, so an image is formed from all points of the object visible from the pinhole, irrespective of their distance from it (infinite *field-depth*). Despite its conveniences, this loses the depth information contained in the rays leaving the object. Thus the pinhole image shows only a perspective view of the object, a kind of projection of its visible parts along the directions of the pinhole (not a simple parallel projection).

The same projection A′, B′, C′ would be given by a similar smaller object a, b, c proportionally closer to the pinhole. That is the basis of enlargement and hence of microscopy, and the reason why microscopic specimens need to be close to the first optical element (here, the pinhole).

Because of its simplicity, the pinhole has a serious problem. To get a clearer image, we need a smaller pinhole, but this transmits less light, so the theoretically perfect camera would transmit none. Since we need more light, we could try two very close pinholes; but they would give two images close together. Proceeding thus, we would enlarge the pinhole, and find that each ideal image point (e.g. a star) gets replaced by a (magnified) view of the pinhole. Thus the ideal image gets *convoluted* with an (enlarged) image of the pinhole, which is called the *point spread function* (PSF); the camera (and indeed *any* camera) acts as a convolution filter (§5.1.2). Although a big pinhole makes a brighter image, its blurring is unacceptable unless the object is extremely simple like a stellar constellation. To make images bright, we must increase the aperture; but to make them also sharp we need a miraculous object, a lens.

7.1.2 Single Lens Microscope

The light passing through a large aperture would diverge to generate an enormously wide PSF. To prevent this, the path of the light must be focused: bent by *refraction*. What is this, and how does it work?

Refraction of light is just the consequence of glass *slowing* light to around 70% of its speed in air. But explaining how this slowing bends its path required an extension of the simplest optical theory, *geometrical* or *ray optics* (originally suggested by Heron of Alexandria in the first century AD). He had supposed that light follows the *shortest* path, but this rule fails with refraction. However, only a small modification is needed to correct it. Fermat (in about 1650) changed *shortest path* to *quickest path*. Light follows the path that takes it from the starting point to the finishing point in the least time. If the transmission medium is uniform, speed is constant so distance is proportional to time, and Fermat's principle is equivalent to the 'shortest distance' rule. But the predictions can differ if light's speed varies over its path. Then this simple rule predicts the observed rules for reflection and refraction.

It also explains the function of a lens. Suppose we had an optical device that, when placed between two points O and F, arranged that *all* their connecting paths through the instrument took light *exactly the same time*. Then a light ray from O entering the instrument would not have just *one* quickest path to F. Instead, it could select to travel by any of the paths, and it would choose the one that happens to lie exactly along its own direction of travel. The result is that *all* the rays entering the instrument from O would end at F. So the instrument would bring light from O to a *focus* at F. Such an instrument is the simple convex glass lens, and Fig. 7.2 shows how we can deduce its function in stages, maintaining the principle of an instrument where all light paths are equally rapid.

Start at (a) with a point source of light O diverging to give an expanding sphere, all points of which are reached in the same time. We would get the desired 'equal time' device, if we could arrange two such point-sphere devices back-to-back as in (b). The problem is that the rays at top or bottom would have to travel further than those around the middle. So, in (c), we arrange that, although travelling a shorter distance, the middle rays take just as long because they travel further within glass (where light moves slower). This is the essential feature of the lens. Thus in (c) we have a lens focusing rays that diverged from O to a focus at F.

Light from a point source converges to a localized concentration (focus), an image of the source. This occurs at a precise distance from the lens (unlike the pinhole camera whose images are everywhere 'in focus'). What determines this distance? We started (Fig. 7.2 (c)) with the simple case where a lens has identical faces. In Fig. 7.3 (a) we split this lens into two plano-convex lenses that almost touch. This splits the lens's action into two sub-functions: the conversion of diverging rays into parallel ones; and the reverse operation of turning these back into rays converging onto the focus. That second operation is isolated in (b), and drawn as a ray diagram in (c).

Next, we transform (c) into a biconvex lens (with the same thickness variation with radius, the essential feature of a lens) which is put in (a) of Fig. 7.4. Then simple transformations (b)–(c) take us to (d), an arrangement with complete left-right symmetry. Finally, (e) through the center of the path-system adds a third 'pinhole ray', the only ray in (e) that remains when the lens

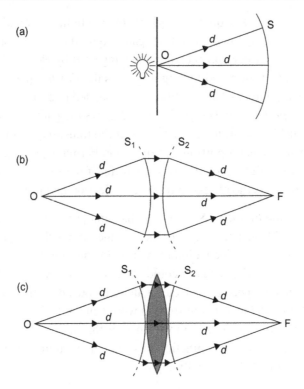

Figure 7.2: 'Equal time' principle applied to lenses. **(a)** Light rays deviating from a point O diverge in all directions, travelling equal distances *d* in straight lines at the same speed, so reaching a circular arc S of equal time. **(b)** Add the reverse of the arc S; we have two valid bundles of paths joining points to arcs. But the paths joining the arcs S_1 and S_2 would have different times. **(c)** However, a lens-shaped piece of glass slows the rays more in the center than the outer part, so all paths from O to F take light the same time.

aperture is reduced to a pinhole. We now have an essential ray-diagram of a lens, for the case where rays diverging from an object point O (left) are brought to an *equidistant* image point I (right); both object- and image-distances are 2*f*. However, (a) showed a completely unsymmetrical arrangement with an infinite 'object distance' and a minimal 'image distance' of length *f*. Thus the object and image distances (*u* and *v*, respectively) are inversely related; indeed, their reciprocals give a constant sum in the 'lens formula' ($1/u + 1/v = 1/f$: see §8.1.1 and equation (8.2)).

We can now understand the functioning of a single-lens microscope. If the specimen's distance from the lens is 2*f* (as in (e)), symmetry shows that specimen and image must have the same size. Image magnification is achieved if the specimen distance is reduced towards *f* (as in (a), reversed). If the distance is made less than *f*, the lens is too weak to focus its rays, but that is still useful if we *look* through the lens, so our eye-lens adds its focusing strength. (This is called a *virtual* image, as opposed to a *real* one.) That is how we use the

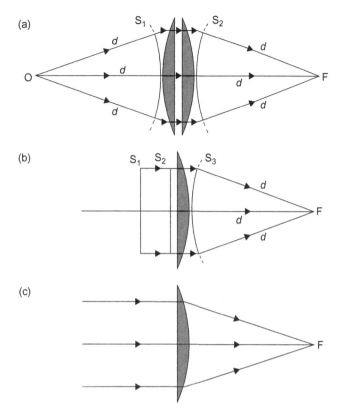

Figure 7.3: How a lens works: the continuation of Fig. 7.2. **(a)** Cut a slot in the lens parallel to its plane; the rays there must be parallel, as the system is symmetrical. **(b)** So the right half of the slotted lens turns parallel rays into rays converging onto a focus F. **(c)** This is shown with rays for a half-lens.

lens in reading-glasses, and it was how Leeuwenhoek (by 1674) used the first high-power microscope, a single lens of magnification >250.

7.1.3 Compound Microscope

Leeuwenhoek's lenses were difficult to make and use. A more convenient option was a weaker lens (easier to grind) whose strength was augmented by subsequent auxiliary lenses, rather as with the (somewhat older) telescope[1]. Here we adjust the specimen distance to between f and $2f$, giving an enlarged 'real image'. This can be magnified by another lens to give an even bigger 'virtual image' that is finally focused by the observer's eye-lens. Obviously, further similar intermediate stages can be added, remedying the purely *magnification* deficiencies of the first ('objective') lens.

---

[1]   Galileo had made such a microscope by 1614 (Harris, 1999).

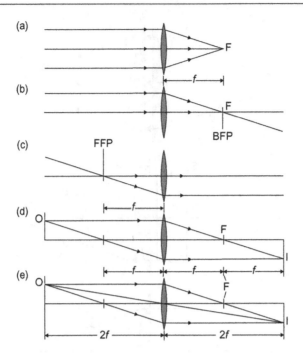

Figure 7.4: How a lens works: the continuation of Fig. 7.3. **(a)** The plano-convex half-lens can be replaced by a biconvex lens, provided the thickness changes are the same. **(b)** Two important rays isolated from (a); BFP = back focal plane. **(c)** Rotating (b) through 180° and reversing ray directions gives a valid ray diagram. FFP = front focal plane. **(d)** Combines (b) and (c). We have two rays from a paraxial object O coming to a focus (image) at I. **(e)** Finally, the 'pinhole ray' is added through the lens center. Note that the triangles around the focus are congruent, so the object and image distances at the bottom are 2f. Note the back focal plane (BFP), where a distant illumination source is focused. Rays coming to a point at the BFP originate from *all* points of the object. In ray optics, these rays merely superpose, but waves form the specimen's diffraction pattern (§9.2).

However, using many lenses to gain magnification combined and even multiplied other lens errors ('aberrations', Section §7.3). Fortunately, by varying lens parameters it was usually possible to change the sign of an error, and combining lenses of opposite sign could correct the problem. Almost all lens errors become more serious with larger apertures, so their correction allowed apertures to increase, which not only increased image brightness but was found to improve resolution. (This was later found to depend on the wave properties of light.)

The lens aperture is also important for *field depth*, which is the range of object distances for which the PSF in the image is no bigger than the detector resolution. At a given detector resolution and image magnification, field-depth varies inversely with the lens aperture (since, at zero aperture, the lens becomes a pinhole with infinite field depth). An image with less field-depth contains more depth information, which is desirable (thus the modern confocal scanning microscope was developed to increase depth information). So a smaller field-depth is yet another advantage of a large lens aperture.

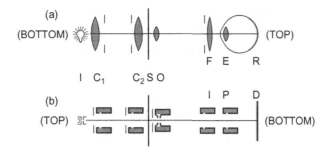

Figure 7.5: Similarities between the light microscope **(a)** and the electron microscope **(b)**. The components are, I: illumination; C_1 and C_2: condenser lenses; S: stage for adjusting specimen position; O: objective lens; F/I: field lens of eye-piece (a) or intermediate lens (b); E/P eye-lens (a) or projector lens (b); R/D: retina (a) or detector (fluorescent screen, film or electronic detector) (b).

The basic arrangement of a compound microscope is shown in the upper diagram of Fig. 7.5 (a). Light from the illumination source I is focused by a collector lens (C_1) and condenser lens (C_2) onto the specimen supported on a stage (S) where it is observed by the objective lens (O) which makes a real image viewed by a field lens (F) in the eyepiece. The final image on the retina (R) is the product of the eye-lens (E). (We shall later see the similarity with the electron microscope below.)

For the rest of this section (Light microscopy), we are interested in lens defects (called *aberrations*). This may appear to be only of specialist or historical interest, but we have an ulterior motive. The magnetic lenses of a conventional electron microscope function in a fundamentally similar way to ordinary glass lenses, which are much easier to understand. Both types of lens can have the same aberrations but, whereas those of glass lenses can be controlled by manufacture and microscope design, those of the electron microscope have so far only been minimized to a level where they can be later corrected by image-processing. For this analogy, we need to examine in some detail glass lens aberrations whose theoretical interest should have disappeared long ago (though they can still recur). As we do this, we return to considering the single microscope lens, where these aberrations remain uncorrected.

7.1.4 Chromatic Aberration and Geometrical Aberrations

This was originally the most serious aberration of the light microscope (as of the telescope). It comes in two forms. Axial or longitudinal chromatic aberration occurs when a lens's focal lens varies with wavelength (being longer for red than for blue light). Lateral chromatic aberration comes from the magnification depending on wavelength. Both problems arise from the need to use white light sources for optical microscopy; however, it can be much reduced by combining lenses made of different types of glass. In electron microscopy, nearly monochromatic sources are used, which should eliminate the effects of chromatic aberration (but see §7.3.1).

For the remainder of this section (§7.1), we concentrate on the monochromatic or *geometrical aberrations*, i.e. the inability of a practical lens to generate perfect images (even when using monochromatic light). The only geometrically perfect imaging is that produced by a pinhole camera, where the lens's effective diameter (aperture) is zero. As that is increased, *geometrical aberrations* start to appear, almost one-by-one. We shall focus on these first aberrations, which are also the simplest ones.

Glass lenses, which are stable and can be made very accurately from glasses of suitable properties, are able to avoid certain aberrations that plague electron lenses. The principal advantage of glass lenses is an accurate and permanent axial symmetry, which is unattainable with electron lenses. (However, lens misalignment can destroy this advantage.) We therefore divide the geometrical aberrations into those that arise when axial (or rotational) symmetry is present, and those caused by its lack.

Before we start this melancholy procession, there is an ambivalent 'aberration', *defocus*. It is viewed as an aberration in electron microscopy (because it is hard to correct), but not in light optics, where focusing is easy.

The rays diverging from each object point are brought together at a corresponding image point, so all the rays from a *flat* object are brought together at the *image plane*. (If the image distance varies with the ray angle, the lens – and image 'plane' – show *field curvature*.) The film or detector plane's position determines the nature of the *recorded* image. If that plane differs from the image plane, the image is *defocused*. Under-focus (a) and over-focus (b) are illustrated in Fig. 7.6.

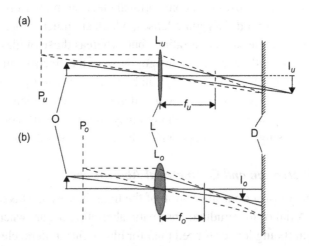

Figure 7.6: Under- and over-focused imaging. The positions of the object (O), lens (L) and detector (D) are fixed. Only the lens strength is adjusted, changing the focal length f. **(a)** A weak lens L_u with longer under-focused focal length f_u produces the object's image I_u *after* the detector, which records instead the plane P_u before the object (pre-specimen image). **(b)** A strong lens L_o with shorter over-focused focal length f_o produces the object's image I_o *before* the detector, which records instead the plane P_o after the object (post-specimen image).

7.1.5 Rotationally Symmetric Aberrations

Aberrations that are *rotationally symmetric* depend only on the angle between the ray and the optic axis. That angle determines the illuminated aperture of the lens. A simple lens is ideal at a very small aperture angle but, as this is increased, aberrations appear in increasing number. We need dip only superficially into this complicated optical pathology: the first aberration sets the resolution limit by spoiling the image, and later defects only make confusion worse confounded.

The first rotationally-symmetric aberration to appear is *spherical aberration*, so-called because it arises from the spherical shape of glass lens surfaces. (Essentially the same defect occurs in electron lenses.) Its effects are shown in Fig. 7.7.

Rays entering at wider apertures (higher rays in (a)) undergo more bending by the lens, so that they come to a shorter focus A than the true focus B, at a focal length *f*. (When rays at larger apertures are bent more, the spherical aberration is termed *positive*; this is true for both glass and electron lenses.) Thus there is no single crossing-point to give a sharp image; rather, the different crossing-points combine to give a 'circle of least confusion', the smallest PSF (PSFm in (b)). By reversing (b), we see that, if a specimen is at the same distance from the lens as the smallest PSF, the objective lens produces the most parallel rays for later focusing. This position (O in (d)) lies closer to the lens than the paraxial focus B, i.e. the object is slightly *under*-focused (see Fig. 7.6).

Thus the effects of positive spherical aberration are mitigated by a slight under-focus (relative to the focus when spherical aberration is absent). This is called *Scherzer focus* (Scherzer, 1949) and is used in two senses. Usually it means an optimal focal state, but occasionally it measures the extent of under-defocus, which might be described as 'so many Scherzer'.

A perfect lens, illuminated by parallel rays, brings them all to the same focus. There is no change in focal length *f* with the aperture *a* at which the rays strike the lens, as in Fig. 7.7 (e), on the right. In that diagram, the angle α is proportional to *a* when *a* is small. This also fits with the fact that, when *a* becomes negative (below the horizontal axis), so too does α. Thus α must always be an odd function of *a*, so it could depend on *a* or on a^3 (etc.). If it depends only on *a*, the lens is perfect[2]. So the first defect (spherical aberration) involves a term in a^3.

The size of that term measures the lens's *spherical aberration coefficient*, abbreviated and symbolized as C_S. In light optics, spherical aberration is ordinarily corrected by the manufacturer. For it is possible, with appropriate combinations of lenses of different glasses with different curvatures, to make lenses with *negative* spherical aberration. These can be combined with ordinary lenses (where that aberration is positive) to give a corrected instrument.

---

[2] Strictly, of course, $a = \tan\alpha$ so $\alpha = \arctan(a)$, which involves a^3 even without spherical aberration. However, the a^3 term is much bigger when there is spherical aberration.

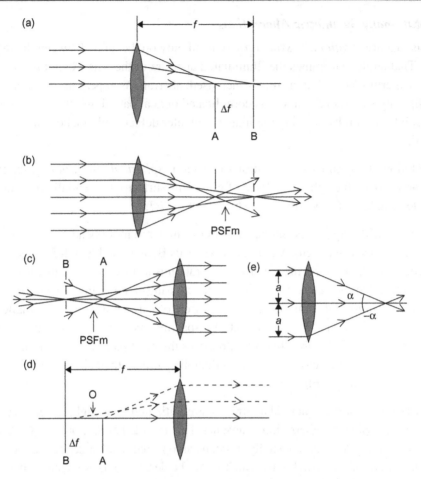

Figure 7.7: Effects of positive spherical aberration. **(a)** Rays passing through the lens's periphery come to a focus (at A) that is Δf shorter than the focus of paraxial rays (at B). **(b)** The smallest average 'focus' (i.e., the plane with PSFm, the smallest PSF) lies between A and B, corresponding to a shorter 'focal length' than the paraxial focal length. **(c)** Diagram (b) reversed. If we also reverse the ray directions, a point source will give (nearly) parallel rays. **(d)** Suppose those rays are imaged by other lenses on the right, so the best image is obtained when the rays are most parallel. That will happen if we place specimen or object (O) at the 'smallest PSF' point in (a). **(e)** The lens aperture a determines the ray angle α.

7.1.6 Lack of Rotational Symmetry

Lack of optical rotational symmetry is itself an aberration. However, the problem is mildest when the lens, while aberration-free within each axial plane, only has a slightly different focal length in different planes. Then we have *astigmatism*, the defect that the lens's strength (or its reciprocal, the focal length) varies as the ray-plane is rotated about the optic axis, changing the angle φ. The rotationally-symmetric and astigmatic ray diagrams are shown in (a) and (b) of Fig. 7.8.

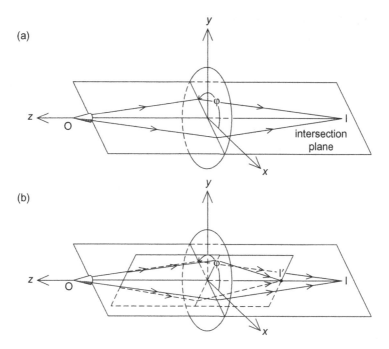

Figure 7.8: (a) A point object O sends a cone of rays into the central lens, which focuses it to a point image I. The cone is intersected by an axial plane (angle φ) on which the cone is represented by two lines. These give the ray-diagram, an adequate representation if the lens is rotationally symmetric (as here). **(b)** But here the lens is astigmatic, and the previous intersection plane is now the plane where I is furthest from the lens. It is closest at I′ on the (smaller) perpendicular plane. Images at I and I′ are perpendicular lines between which the 'mean focus' is a uniform circle.

Consider the details of this variation. When the ray-plane is rotated about the lens axis, the focal length must reach a maximum at some angle φ. As the ray-plane at $(\varphi + 180°)$ is the same as that at φ, any effect on the focal length must repeat after a 180° rotation. (If it repeats more frequently than this, the variation is not merely one of focus.) So the maximum focal length occurs on the planes at φ and $(\varphi + 180°)$, and the minimum focal length occurs half-way between the maxima, i.e. on the planes at $(\varphi + 90°)$ and $(\varphi + 270°)$, as shown in Fig. 7.8 (b). At the maxima and minima, a point source is not brought to a point image, but to two perpendicular line-images – the first produced by the stronger direction of the lens, the second by the weaker direction. The best compromise or *mean focus*, midway between these lines, is a circular disc.

In glass lenses astigmatism is under the control of the manufacturer, so it is eliminated as far as possible, with one important exception. Spectacle lenses have to correct the optical defects of the eye, and one of its most common defects is astigmatism. Since astigmatism is a directional defocus, two identical astigmatic lenses compensate each other when their strong-focusing directions are oriented at right angles. This is the principle underlying astigmatism correction in spectacles, and also in the electron microscope.

We have now followed the improvements of the light microscope up to the level it achieved in the later 19th century. All obvious problems had been solved, yet its resolution remained stuck at a level not much smaller than a bacterium.

In 1873 Abbe showed that the resolution of the light microscope is limited by the wavelength of light. To get much higher resolutions, it would be necessary to replace light by some other radiation of much shorter wavelength. The first suitable radiation, electrons (discovered by J.J. Thomson in 1897) were used in the first magnetic lens electron microscope (built by Knoll and Ruska in 1931–2). However, the availability of satisfactory commercial instruments was delayed by the war period until nearly 1950.

7.2 Electron Microscopy

Before turning to the aberrations of electron lenses, readers unfamiliar with the electron microscope may find that a short introduction to the microscope's essential parts and basic preparation techniques makes the later sections more comprehensible. There are many books on the electron microscope (Glaeser et al., 2007; Reimer, 1989; Spence, 1981–2003;) and on practical electron microscopy (Chapman, 1986; Dykstra & Renss, 2003; Sommerville & Scheer, 1987).

We start by describing the electron microscope, following the stages of its development as they accompanied the development of high-resolution work.

7.2.1 The Basic Electron Microscope

We start by describing the basic electron microscope that became widely-used in the 1950s and 1960s. This contains the essentials of the modern instrument, and the following section notes subsequent improvements.

Several differences were necessitated by the change from light to electrons. First, electrons need an evacuated space in which to move, so their entire path must lie in a high vacuum, maintained by (originally) oil diffusion pumps backed by rotary mechanical pumps, plus (more recently) ion pumps.

The illumination system is an electron source, in which electrons from a wire are accelerated to high speed in the vacuum. To get them out of the wire, it is connected to a tungsten filament which is heated as in a lamp, so that the hot electrons can 'evaporate' from its surface. Since the microscope's resolution depends on a small electron wavelength, and since that is inversely proportional to its momentum, the microscope needs very fast electrons, accelerated by a large electric potential. Thus the electron source is maintained at a negative potential around 100 kV, so the electrons are accelerated by moving into the main microscope through a hole in the anode (at ground potential). However, the 'evaporated' electron cloud must be concentrated into a fast, narrow beam ready to be focused by the lenses. Electrons are negatively charged particles, and a hole in a negatively-charged disc focuses them. So the filament is surrounded by a negatively-charged cup (Wehnelt assembly), at the bottom of which is a hole. This focuses the emitted electrons as they emerge a short distance above the anode, a metal plate at ground potential (relatively positively-charged). By the time they have reached the anode and pass through its central hole, the electrons have been accelerated to nearly half the speed of light (for 100 kV electrons, and even more at higher voltages). (So the electron microscope is one of the few biological instruments where relativity is important.)

The electron beam is now ready to be focused by lenses. Electrons also experience forces when *moving* relative to magnetic fields, and a typical magnetic lens is shown in Fig. 7.9. Electrons are forced into a helical path as they pass down the bore of the pole-pieces, where the magnetic field is strongly concentrated. This concentration focuses the electrons, the focal length depending on the field-strength (and hence on the lens-current), the bore and the gap between the pole-pieces.

At very high speeds, electrons need powerful magnets to focus them; hence the large lenses and also a relatively long optical path (making the microscope so big). The lenses, with their

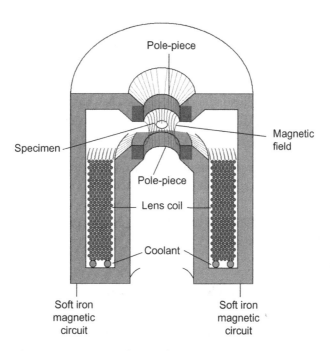

Figure 7.9: Magnetic electron microscope lens (schematic). An electromagnet generates a magnetic field, concentrated between adjustable pole-pieces.

coils and iron shielding, are very heavy, hindering mechanical alignment, which consequently is used only in the basic alignment of the microscope. Fine focusing adjustments are controlled with the lens current, which must also be very stable. (Other – very weak – magnetic lenses or coils are added to control astigmatism and beam deflections or tilts.)

The electron microscope is a compound microscope in which the arrangement of the main lenses follows the same pattern as in the light microscope Fig. 7.5. (However, the light microscope points upwards towards the eye, whereas an electron microscope points downwards onto a screen viewed from above.)

The illumination is controlled by two condenser lenses. These can provide a wide-spread beam for preliminary specimen surveys, or a concentrated one for seeing small specimens at high magnification without damaging adjacent regions. Specimens, which must be extremely thin to be transparent to electrons, are attached to an even thinner support film of evaporated carbon, held on a fine metal mesh. Each such 'grid', equivalent to a light-microscope slide, is held in a specimen-holder that is transferred to the stage. This moves the grid within the microscope, under operator control, and it must be very stable to avoid image blurring at high magnifications.

As the aberrations of the objective lens are magnified by all subsequent lenses, this lens determines the microscope's resolution. Its main aberration, astigmatism, is controlled by coils that make a very weak cylindrical lens whose strength and direction can be adjusted. (The objective lens also acts as a final condenser lens.) The specimen is immersed in the magnetic field, being inserted (through an air-lock) either from above or from the side. The intermediate lens allows a choice between the image and the diffraction pattern at the *back focal plane* (BFP) of the objective lens (just as, in the light microscope, we remove the eyepiece to see the BFP when adjusting a phase-ring). The projector lens forms the final image on the fluorescent screen (during operation of the microscope) or (for permanent record) on film or, more usually, an electronic detector.

The entire microscope column is under high vacuum, so there are vacuum-locks for entry and removal of the specimen and the film (pre-evacuated to remove water vapor and stored in the main vacuum ready for use). The microscope's dependence on magnetic fields makes it sensitive to unshielded electric cables or motors, etc.; and its very high magnification enlarges any extraneous mechanical vibrations.

7.2.2 Improvements on the Basic Electron Microscope

When this basic electron microscope became generally available, it brought in a rich harvest of discoveries that delayed awareness of its limitations. The instrument's limitations were accentuated by its casual use like a light microscope where the observer has leisure for survey, careful observation, and instrumental adjustments before photography.

Perhaps the first apparent problem of the electron microscope was the build-up of contamination, a deposition of hydrocarbon products (deriving from diffusion-pump oil) onto the most irradiated (i.e. observed) parts of the specimen. The first counter-measure was an anti-contamination device which surrounds, as far as possible, the specimen by very cold surfaces onto which hydrocarbons preferentially condense. This device preceded the more radical solution of improving the microscope vacuum, but (because of later developments) both measures are now needed.

A deeper awareness of the effects of irradiation on the finest detail was delayed by the use of negative staining (§7.2.3), which the beam rapidly cooks into a mineral casing that is then relatively damage-resistant. Destruction of any detail depending on the specimen's survival occurs before the eye perceives it; but Williams & Fisher (1970) showed that extra detail was revealed by minimal beam exposure. This is absolutely essential when examining unstained specimens, and special procedures are now used for the initial survey, for focusing and astigmatism correction (on an adjacent specimen region), and (using 'low dose' equipment) for switching the beam onto the precious specimen region only while every electron is contributing to the detector image.

The biggest change in the basic microscope followed the final development of satisfactory techniques for preparing cryo-specimens (Dubochet et al., 1982, 1988; see §7.2.3).

First, the specimen's observation requires its maintenance at below 170 K, to avoid ice sublimation, and below 110 K, to prevent ice-crystal growth (Glaeser et al., 2007, p. 125). Also, at these low temperatures, the stage must be stable against vibration from coolant boiling, or drift from thermal contraction/expansion (Heide, 1982).

Second, cryo-specimens have very low contrast which is enhanced by phase-contrast (§9.1). This requires an extremely coherent electron beam, i.e. one with the smallest possible source of electrons ('spatial coherence') and with the least possible energy-spread ('temporal coherence'). Both are improved by changing the electron source. Traditional tungsten filaments at 2800 K emit electrons with a substantial thermal energy spread. To reduce that, we must extract electrons at a lower temperature and therefore need to use a bigger field gradient. This is increased by using a small, pointed source attached to the filament: a sharpened rod of LaB_6 or CeB_6, or (much better) a single tungsten crystal in a field-emission gun (FEG or Schottky source). Besides being cooler (only just above 1000 K), these sources produce far greater coherence, because the effective source size is smaller. The relative energy spread is also decreased by using a higher accelerating voltage (up to 300 kV). (This has the further advantage of increasing the penetration of the electrons.)

Third, good images of cryo-specimens reward the use of 3D reconstruction techniques (§12.4), which usually depend on images from tilted specimens. Therefore the low-temperature stage should, in addition, allow the measurement of tilt-angles ('goniometer stage'), and also tilting of the specimen with neither horizontal nor vertical movement of the region under observation ('eucentric stage'). Moreover, the objective lens needs a sufficiently large pole-piece gap to permit the necessary tilt-angles.

*   *   *   *

These are the main improvements commonly found in modern high-resolution microscopes. However, major problems still remain. In increasing importance, these are, first, inelastically-scattered electrons (§7.4.1) leaving the specimen; these blur the image because of the chromatic aberration of conventional circular magnetic lenses (§7.2.1). Second, there is beam-induced specimen motion and charging. But, most important of all, there remains the third problem: radiation damage by electrons.

Various pieces of equipment are being tried for dealing with these problems. The first is tackled from two directions: with energy-selective optics that help remove inelastically-scattered electrons (§7.4.1), particularly important with thick specimens; and with multipole lenses with a very low or zero chromatic-aberration coefficient (Kabius et al., 2009). Specimen movement should be minimized by using a small focused beam scanned across the specimen ('spot-scan': Bullough & Henderson, 1987; Downing & Glaeser, 1986); and charging by devising conducting supports (e.g. carbon) for specimens. Finally, some reduction in radiation damage may result from having the specimen at the lowest possible temperature, that of liquid helium (around 4 K). This is achieved in experimental superconducting lenses that also have strong and very stable magnetic fields (Dietrich et al., 1977).

7.2.3 Electron Microscope Preparation Techniques

Biological specimens are the most difficult to study, being composed almost exclusively of light elements yielding only small visibility, and having weak structures that are vulnerable to damage by almost every step in microscopy. Their strongest visible signals are phase changes, of which light microscopes made poor use before phase-contrast and interference microscopy were developed in the 20th century. Instead, an elaborate histological technique had been perfected in the 19th century, so most images showed only the dyes or metals deposited onto specific cell structures.

These difficulties (plus the effects of dehydration in the microscope vacuum) were repeated with electron microscopy before the 1950s. Then great progress was made in adapting histological techniques for electron microscopy; but, although satisfactory fixation, embedding and sectioning became possible, there were no adequate substitutes for the numerous specific histological stains generated by the dyestuff industry. Metal staining, although mostly unspecific, revealed cellular fine structures (especially membranes) of great interest in cell biology, although resolution was no better than 3–5 nm.

Somewhat better resolution was obtainable with isolated particles, once they were made visible. The oldest technique was metal shadowing, following freeze-drying; then Hall (1955), Huxley (1956), and Brenner & Horne (1959) introduced 'negative staining' (the application of a previous light-microscopic method for seeing bacteria). Small particles, attached to a grid, are immersed in a thin layer of heavy-atom stain solution which, after drying, surrounds them with dried stain. The higher resolution of these images (up to 2 nm) was an early stimulus to the improvement of microscope design and practice, and to the development of image-processing.

So far, electron microscopy had been following in the footsteps of light microscopy, without yet reaching the stage of phase-sensitive detection. To repeat those developments, electron microscopy needed methods not only for phase-detection but also for specimen preservation in a high vacuum.

Deep freezing does this, while also reducing beam damage. This was originally suggested some time ago (Keller & O'Connor, 1958; Leisegang, 1954), but frozen water invariably gives ice crystals whose growth disrupts and destroys delicate specimens, and whose presence interferes with good images. The answer is to freeze so quickly that the water molecules have no time to organize themselves into crystals, but yield 'vitreous' (glassy) ice. However, that required *extremely* rapid cooling; even plunging into liquid nitrogen was insufficient. After several trials (e.g. Glaeser & Taylor, 1978; Taylor and Glaeser, 1974, 1976), Dubochet et al. (1982) developed a satisfactory procedure. The grid, held in forceps attached to the apparatus, falls until its speed is at least 1 meter/sec, and then enters a small pot of cryogen just before the grid comes to a stop. The best cryogen is liquid ethane, a compromise between diatomic nitrogen, which immediately gasifies into an insulating layer, and heavier molecules, which will not sublime from the frozen specimen in the microscope. (See the review by Dubochet et al., 1988.) Nevertheless, these specimens are very beam-sensitive (electron doses have to be less than $5 \, e/\text{Å}^2$ to preserve high-resolution detail[3]), so low-dose imaging is essential.

The final challenge was the development of phase-detection electron optics. Phase plates have so far proved impractical, so the best way is to use defocusing (and other imaging defects) to produce electron interference patterns that yield the desired information through image-processing.

7.3 Electron Lens Aberrations

Except for the single crucial advantage of employing a wavelength short enough for atomic resolution, electron microscopes have almost every possible disadvantage relative to light

---

[3] When using film with such weak electron beams, an important parameter is the magnification. Too little loses resolution but too much could reduce exposures to the film's fog-level. The best compromise seems to be optical densities >0.2 and magnifications of around 30,000–40,000.

microscopes. In particular, the aberrations of electron lenses are bigger and more difficult to correct. (Perhaps this is because the real 'electron lenses' are the ultra-smooth magnetic fields that can only be controlled from the periphery, whereas glass lenses are shaped by grinding over their entire surface.) We consider the aberrations in the same order in which we discussed those of glass lenses (§7.1.4).

7.3.1 Chromatic Aberration

Chromatic aberration (measured by the coefficient C_c, whose dimensions are length) is the inability of a lens to bring to the same focus light of different wavelengths. Since an electron's wavelength is determined by its speed, chromatic aberration in an electron lens is the inability to bring to the same focus electrons of different speeds (i.e. energies). Faster electrons are less strongly focused and the defect can only be palliated by trying to ensure, as far as possible, that all electrons have exactly the same speed. However, we should distinguish between the situation before, and after, the specimen.

Differing speeds of the electrons *reaching* the specimen are an unmitigated disadvantage which microscope makers try to minimize. The hotter the filament, the faster the average electron speed, and a higher average speed also broadens the *distribution* of speeds around the average. A 'monochromatic' electron beam needs the narrowest possible speed distribution, relative to the accelerating voltage. This is produced when the filament has the lowest possible temperature (with a field-emission gun) and the accelerating voltage is higher (§7.2.2).

By contrast, differing speeds of the electrons *leaving* the specimen contain information about its structure, and the chromatic aberration of the subsequent lenses helps (a little) to reveal this information. We consider this later (§7.4.1). However, chromatic aberration also blurs the image, but removing it is theoretically impossible for rotationally symmetric, static lenses (Scherzer, 1936), and it remained one of the hardest tasks facing microscope designers (Freitag et al., 2005), but now solved by Kabius et al. (2009).

7.3.2 Geometrical Aberrations

These aberrations are essentially the same as the corresponding aberrations of glass lenses (§7.1.4), though far more severe. The simplest rotationally-symmetric aberration, *defocus*, is controlled by adjusting the objective lens current. But there are two complications: it is difficult to determine the defocus during operation, without increasing the exposure and consequent radiation damage; and attaining an exactly focused image is not just difficult but undesirable, since it would eliminate a major source of phase-contrast information (see Chapter 9). Thus a moderate (though unknown) defocus is both inevitable and desirable.

Similar observations apply to the simplest rotationally asymmetric aberration, *astigmatism*. Unavoidable imperfections in electron lenses (or charging of the specimen) leave a small astigmatism that can be partly corrected in the way that spectacles correct for visual astigmatism. The 'spectacles' used in electron microscopes add a special lens (the *stigmator*) with *only* astigmatism that can be adjusted electrically for strength and direction. But, as with defocus, exact correction during microscopy is impractical but unnecessary, since any residual astigmatism can be compensated by image-processing which extracts (a little) phase-contrast information in the process.

Spherical aberration is the least significant aberration at moderate resolution (i.e., around 0.5 nm). Unlike defocus or astigmatism, spherical aberration cannot be corrected by the microscope operator. The only control lies in choosing an objective lens with a small spherical aberration coefficient (C_S). The units of C_S are length, and the smaller it is, the better. C_S is smallest when the focal length is shortest (the same applies to the chromatic aberration coefficient). The lowest C_S is about ½ mm, but that is possible only with an objective lens that imposes a restricted tilt angle. An objective lens allowing the widest tilt angles has a C_S about 6 mm. A good compromise for C_S is around 2 mm.

Spherical aberration correction is possible in light optics because of the existence of glass lenses with negative C_S. But Scherzer (1936) showed that electric or magnetic lenses can have only *positive* spherical aberration, when they are rotationally symmetrical and static. Thus ordinary electron microscope lenses all have positive spherical aberration, removing any possibility of correcting spherical aberration in an ordinary electron microscope. However, by using non-circular lenses, spherical aberration *can* be corrected, and such lenses are used in materials science. Presumably they will eventually be introduced into biological electron microscopes, but the modern microscope (combined with image-processing) could already have a resolving power comparable with that of X-ray patterns of good protein crystals, if geometrical optics were the only consideration.

But the overall performance of electron lenses is far worse than that of glass lenses; the modern electron microscope has not yet attained the same ratio of resolution relative to wavelength as the light microscope in the 17th century. Then there is the added problem of radiation damage that shifts the crucial limitation from the microscope to the specimen. So we leave the topic of electrons in lenses and turn to their more complicated experiences when passing through the specimen.

7.4 Contrast Mechanisms

The function of an image is to reveal the structural information carried by electrons – information they acquired during their passage through the specimen. But extracting that information is a later consideration (§9.1). First, we ask how the electrons get it: what happens to them inside the specimen?

7.4.1 *Elastic and Inelastic Scattering*

Electrons, being electrically charged, interact strongly with electrical charges in the specimen. Although the outer atomic electrons shield the central positive nuclear charge, this shield disappears progressively as an electron penetrates deeper inside the atom. Atoms are close-packed and nuclei are extremely small, so most places in a specimen lie somewhere inside an atom but outside its nucleus. Thus an electron penetrating the specimen is most likely to experience a partial positive charge, enough to deflect it – a possible source of information about the specimen, which we shall shortly examine in detail.

However, information at the atomic level comes at a high cost. A fast electron has enormous energy, and its charge allows it to transfer some of this energy to electrons or nuclei in the specimen. The resulting changes, *specimen damage*, depend on the energy transferred. Transfers affecting only the atoms' outer electrons involve the least energy. Ordinary chemical changes require only a few electron volts[4] (eV), whereas the typical transfer energy is about 20 eV. Thus *bond damage* is the most common type of change, especially at lower accelerating voltages, and hydrogen release is the most common type of bond damage in biological specimens. But specimen electrons can be not only displaced, but also ejected, leaving the damaged region charged.

The seriousness of any collision depends on the speed and proximity of the projectile. A fast electron in a head-on collision with a nucleus can eject it from the specimen (*knock-on damage*). In car collisions, lighter vehicles suffer more; similarly, it is easier to eject lighter nuclei, which also have fewer electrons holding them in place. Protons (hydrogen) may be ejected by electrons with only 5 KeV (and hydrogen atoms by much less – 2 eV via bond damage). However, knock-on damage starts to make a major contribution only in million-volt microscopes. (Electron scattering and contrast mechanisms are discussed by Henderson, 1992.) Apart from the energy of individual electrons, the overall energy flux can also be important: at high beam intensities, the specimen can be locally heated.

After emerging from a damaging collision, an electron is said to be *inelastically scattered*. Even if it has not been absorbed or undergone a big deflection, its energy loss can have much the same effect through the chromatic aberration of the electron lens (§7.3.1). Thus the image point that should receive it loses the electron, which reaches other parts of the image, generating a background illumination that lowers contrast. However, the smallest energy losses leave the electron merely unable to contribute to the phase contrast detail of the image; and that also reduces contrast, unless they are removed by energy-loss filtering (§7.2.2). (However, if the chromatic-aberration of the lens can be corrected, then these inelastic electrons form very good phase-contrast images, as seen in Kabius et al., 2009.) All these processes contribute to an image that shows intensity variations which are not critically dependent on the focus: *amplitude contrast*. It can be broadly divided into direct loss (*absorption contrast*) or loss at an aperture (*aperture contrast*). If the microscope is equipped with a special electron energy-loss filter, the remaining inelastically-scattered electrons

---

[4]   An electron volt is a little less than a hundred kilo-Joules (around twenty kilo-calories) per mole.

are made to contribute to what might be termed *energy-filter contrast*. Amplitude contrast is generally bigger at low resolution, so it was very important in early microscopy but has become progressively less so, especially after the development of cryo-microscopy.

At the other extreme, an electron may lose no energy to the specimen and yet still extract some information about it through undergoing a small deflection (*elastically scattered*). This can be exploited to give phase-contrast (or refraction contrast) images. Losing no energy to the specimen, the electron cannot damage it. This would obviously be an ideal way to study specimens, but there is no way to eliminate the due proportion of inelastic scattering. In cryo-specimens, the unwanted inelastically-scattered electrons are at least three times as numerous as the useful elastically-scattered ones; see Toyoshima & Unwin (1988), Henderson (1992) or Smith & Langmore (1992).

Electron scattering depends on the accelerating voltage used. The electrostatic interaction energy becomes proportionately smaller when the electron's kinetic energy increases; so the same attractive force causes less deflection with faster electrons. Smaller deflections allow the electron to travel further within the specimen, so high-voltage electrons can penetrate thicker specimens (or pass through the same specimen with less attenuation). (This is one reason for the higher accelerating voltages of modern microscopes.) However, although specimen damage diminishes with increasing voltage, so also does scattering. Indeed, the ratio of these quantities changes little with voltage.

Electron deflection is increased when the attractive force is greater, i.e. when the nuclear charge is bigger, so it increases with the atom's atomic number. Wide deflections (needed for amplitude contrast) require large forces, so the electrons need to get close to a highly-charged nucleus. Therefore heavy atoms produce more amplitude contrast, a feature which was essential for obtaining adequate images in earlier electron microscopy.

These considerations are relevant to the specimen preparation techniques reviewed above. Earlier methods (metal shadowing or staining, then negative staining) used metal-contrasted specimens which scattered electrons inelastically, giving amplitude contrast. But unstained cryo-preserved specimens contain little of any element heavier than calcium so, as specimens are immersed in ice of nearly the same density, amplitude contrast is very low. This demotes amplitude contrast in favor of contrast from elastically-scattered electrons (next section). The correlation between inelastic/elastic electrons and particle/wave properties is explored further in the next chapter (§8.2.1).

7.4.2 Phase Contrast: General Considerations

Elastically-scattered electrons undergo such small deviations[5] that they are rarely lost from the image, but instead contribute to it. But what is the source of image contrast? A *perfect*

---

[5] However, most inelastically-scattered electrons are scattered through even smaller angles.

image would be simply an enlargement of the rays just leaving the specimen, so it would have zero contrast. Thus an electron microscope *should not* produce a perfect image, so it is advantageous that it *cannot* produce one: its usefulness depends on imaging defects. After exploiting them, we try to remove their pernicious effects by image-processing. It is all an exercise in having one's cake and eating it.

Elastically-scattered electrons must be processed so that the subtle specimen information they carry gets turned into intensity variations: *phase contrast*. The most practical way to reveal this is also the simplest. The early high-resolution microscopists noticed the curious phenomena associated with focusing. When the image is *exactly* focused, it has minimal contrast which immediately increases on defocusing. But that is asymmetrical, and only *under*-focusing improves contrast, though it does so at the expense of resolution. So the search for image clarity (focusing a faint green image in a race against specimen degradation) tempted the operator into excessive under-focus, whose crude effects were only apparent in the developed film.

These near-focus appearances depend on wave properties (see §9.2). But a more intuitive way of understanding this, using ray optics, considers phase contrast as the consequence of a slight bending or deflection of the transmitted electrons' direction ('refraction contrast').

7.4.3 Refraction (Phase) Contrast Through Defocus

We see refraction contrast in everyday transparent objects. Patterned bathroom windows that combine illumination with privacy have thickness variations that bend the light, distorting what is seen through them. Sunlight through such a window illuminates a curtain nearly uniformly, but reveals the thickness pattern strikingly on a distant wall. A similar effect is seen when focusing an irregular transparent sheet in a projector. Such 'refraction images' are changed by adjusting the projector's focus. The *post*-sheet images disappear in exact focus, but continuing the lens movement reveals new *pre*-sheet 'refraction images'.

Similar effects are produced by a 'transparent' electron microscope specimens. A frozen-hydrated (cryo-) specimen consists of ice (density $0.92\,\mathrm{g/cm^3}$) containing slightly denser structures (protein, average density $1.33\,\mathrm{g/cm^3}$ or nucleic acids, $1.84\,\mathrm{g/cm^3}$). These contain more of the heavier atoms, exposing electrons more often to positive charge. A region of positive charge acts like a convex lens (§7.1.2 and Fig. 8.4), bending the electrons' paths. The small, slightly denser particle at the bottom center of Fig. 7.10 has a slightly bigger refractive index, so it acts as a weak lens, slowly focusing the electrons after the specimen. This generates the post-specimen 'refraction image' on the right. If the rays are traced straight back, they appear to come from a region on the left that is slightly darker: the pre-specimen 'refraction image'.

Note that the *actual* ray-pattern before the specimen is a set of near-parallel continuous lines. The ray-pattern drawn in broken lines before the specimen is the pattern *that would need to*

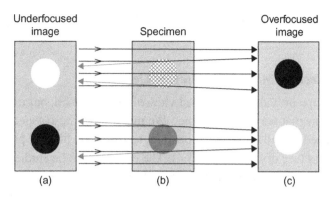

Figure 7.10: The middle frame is two transparent specimens (upper: less dense; lower: more dense). The two outer frames are de-focused images (left, under-focused, and right, over-focused). At the specimen level, only the right diagram exists, as the near-field pattern. The specimen's light circle has both a slightly lower refractive index and absorbance, the dark circle slightly higher ones. The arrowed lines show electron ray paths. (Black lines: actual paths, grey lines: extrapolated lines.)

have existed, in order to produce the post-specimen ray distribution. But any optical system using the rays on the right of the specimen must inevitably produce the same images that one would get from the broken-line ray-pattern. These two patterns are rendered visible simply by strengthening (*over-focusing*) or weakening (*under-focusing*)[6] the objective lens (Fig. 7.6).

If there is no absorption (no amplitude contrast), the pre-specimen 'refraction image' is the *exact reverse* of the post-specimen 'refraction image' (compare (a) and (c) of Fig. 7.10). That is, refraction contrast is *reversed* when we exchange under- and over-focus. Defocus-reversal produces *contrast-reversal*, provided the specimen is completely transparent, with no amplitude contrast[7].

However, suppose that the dark circle at the bottom of the middle diagram of Fig. 7.10 were not only more refractile, but also more absorbent. Then its amplitude image would be a dark circle, rather like the under-focused image, but opposite to the over-focused image. The upper, lighter, circle at the top of the middle diagram of Fig. 7.10 is less refractile, so its defocused image is just the opposite of that of the more refractile circle.

A random (i.e., complicated) specimen consists of various small regions of greater density (which act like the more refractile circle), mixed with small regions of lesser density (which act like the less refractile circle). Thus the random structure will be revealed by under- or

---

[6] *Under*-focus means having a lens that is too *weak* (so it *under*-performs); *over*-focus is the opposite.

[7] In §9.2.4 we find that this 'contrast reversal rule' also applies with waves, thereby avoiding the paradoxical effect of the large-scale features of an image (which follow the rules of ray optics) undergoing a contrast-reversal that is lacking for finer details.

over-focus, but disappear when precisely focused. Also, its size ('granularity') will be roughly proportional to the defocus.

7.4.4 Analyzing Electron Microscope Imaging Defects

An electron microscope provides an enlarged view of the specimen, but imperfectly, introducing distortions that we need to correct. We might view these aspects as a filter, whose characteristics we need to find in order to construct a correction filter, if possible, or at least a reconstruction process (see discussion in §5.4). The errors might be studied like those when sounds are imperfectly transmitted in a concert-hall. In that case, we could test the hall by getting it to transmit test-sounds, whose simplicity makes them sensitive to transmission errors, and therefore diagnostic of them. Three diagnostic sounds are commonly used for such test. A *sine-wave* or continuous note (like singing in a bathroom) reveals prominent resonances. An *impulse* or sharp sound produces an after-sound, like the echo of a gun-shot; this is called the *impulse response*. (These two methods correspond to alternative ways of collecting NMR spectra §1.2.3.) Finally, *white noise* (like the hiss made by blowing across the opening of an empty bottle) acquires a characteristic musical note.

Corresponding to these sounds are three types of diagnostic electron microscope specimen. The *sine-wave* corresponds to a thin, transparent, sinusoidal diffraction-grating varying in refractility (a *sinusoidal phase object*). The *impulse* corresponds to an isolated small particle. And *white noise* corresponds to a transparent specimen with a completely random density distribution. A good approximation to this is a thin film of evaporated carbon, which has random thickness and acts as a random phase object. It, and also the sinusoidal phase object, are best explored further using wave theory; see Chapter 9, §9.2.2.

7.4.5 Impulse Response

The 2D impulse, a small particle, would give a PSF nearly invisible above the noisy image of the supporting film. It is better to use a 1D impulse, a *step-function* (which suddenly rises from zero to a constant plateau). This sudden shift that usually stays at the new level requires a bigger energy input and therefore gives a bigger signal. In practice, electron microscopy uses a small hole in a thin carbon film. A hole's edge is a refraction (or phase) step-function from vacuum to the carbon film. The response to a 1D step-function is the convolution of the 2D PSF with a long rectangle-function. Its effect can also be found as in Fig. 7.11.

Simple ray optics can predict the main features of the 'post-specimen' near-field pattern at the bottom of (a). That also appears as the over-focused image (b). By following the rays from it upwards, through the in-focus image, we see that the under-focused image must have reversed contrast. (The same is seen with sinusoidal specimens.) The effect of this on the image of a hole is shown in Fig. 7.12, where (a), (b) and (c) respectively show the under-, correctly- and over-focused images. (The first and last were taken from Fig. 7.11.) Note that,

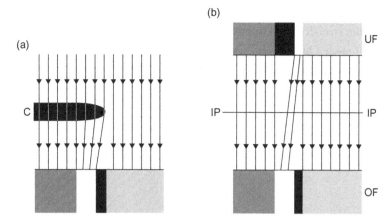

Figure 7.11: **(a)** Edge of a hole in a carbon film (C); below, its near-field image. **(b)** Region of image plane (IP) showing under-focused (UF) and over-focused (OF) images.

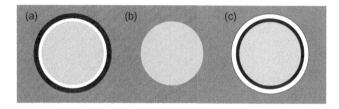

Figure 7.12: Application of Fig. 7.11 predicting the images near the edge of a refractile carbon film. **(a)** Underfocused. **(b)** Correctly focused. **(c)** Overfocused.

Figure 7.13: As Fig. 7.12, but in the presence of astigmatism. The top and bottom edges are under-focused, the left and right sides are over-focused.

because the film has weak absorption (shown in (c)), the under-focused image has increased contrast (whereas the over-focused image has an artifactual black ring that reduces contrast). A focused image of such a hole reveals any lens astigmatism, since the edges in one direction would show over-focused fringes, while those in the perpendicular direction show under-focused fringes: Fig. 7.13.

Waves

Chapter Outline

Here we concentrate on wave topics relevant to image-formation in the electron microscope, as a foundation for the following chapter, which discusses that essential topic mostly in a qualitative, intuitive way. This brief survey can be supplemented by the many excellent accounts of optics, especially with light: Born & Wolf (2002); Cowley (1975); Goodman (1968–2005); Lipson et al. (1995).

8.1 Wave Properties

8.1.1 What are Waves?

Waves comprise an extremely diverse set of phenomena whose common link is a state of oscillation spreading to adjacent regions. Their similarity allows the fundamental properties to be illustrated in a simple situation. Suppose that a uniform train of parallel shallow waves crosses the flat surface of a pond. The waves move a floating leaf up and down without displacing it along the surface. Neither the leaf nor the water undergoes net movement; both are left where they started, and only the vertical movement traverses the pond. The number of up–down movement cycles completed in a second is the wave's *frequency* ν (Greek 'nu', measured in Hertz, Hz). If the surface has many leaves, all show the same movement, but not in exact synchrony. The different movement cycles differ in their *phase*, and the phase

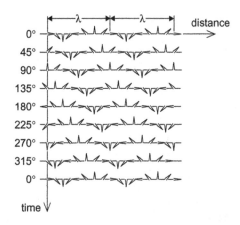

Figure 8.1: A wave train (wavelength λ) moves to the right, as time increases from top to bottom. At each moment of time (indicated by the phase angle on the left), 17 points are selected where the vibration is followed as a phasor which rotates anti-clockwise. A typical point is that on the extreme left (on the time axis) where the phase angle is marked. The 0° phasor starts at the top left and 'moves' (gets transferred) to the right at later times. Note that the phase rotates *anti-clockwise* along the time axis, but *clockwise* along the distance axis. (This depends on the convention.)

difference propagates at constant speed, the *phase velocity* (or simply *velocity*) V (m/sec). At a distance, one notices a series of parallel waves crossing the surface, the distance between them being the *wavelength* λ (meters). During one cycle-time, the wave moves λ meters and, in one second, there occur ν cycle-times, so the wave moves ($\lambda\nu$) meters in a second. Thus:

$$\text{wave velocity}\, V \,=\, \nu\lambda. \tag{8.1}$$

In a simple wave, each particle vibrates sinusoidally with a frequency ν. Phase is measured as an angle, so each particle's phase angle is increasing at ω radians[1] /second: $\omega = 2\pi\nu$. If the motion were represented as a phasor (§3.3.2), it would rotate *anti*-clockwise at ω radians/ second (or ν revolutions/second)[2]. A fixed observer would register the wave's vibrations as increasing phase. Thus, in a snapshot of the wave, phase increases in the direction towards the source (the direction in which we see more recent parts of the signal). So we conclude that phase must *de*crease in the direction *away* from the source, that is, in the direction in which the wave is moving (see Fig. 8.1.)

At the boundary between two media, vibrations in the first medium generate vibrations in the second. These have to follow the generating vibrations and vibrate at the same frequency; so

---

[1] One radian is the angle – approximately 57° – that cuts off a length of circumference equal to the radius r. A complete revolution corresponds to a full circumference, of length $2\pi r$, so it has an angle of $2\pi r/r = 2\pi$ radians.

[2] This is, of course, an arbitrary convention; it would be equally possible to suppose that it rotated clockwise. The usual convention, that anti-clockwise rotations use positive angles, may be connected with the fact that a small anti-clockwise rotation *raises* the point at (1,0), so both coordinates are positive.

the ratio: frequency = (velocity)/(wavelength) must remain constant. When light passes from air to glass, it travels slower in the new medium, so the ratio is kept constant by reducing the wavelength in the same proportion as the velocity. Examine the implications when a wave meets a small glass block. The part traveling through glass has a shorter wavelength, and therefore executes more vibrations, relative to the part that travels through air. Thus waves emerging from the glass are slightly older than adjacent waves that were always in air; so, in Fig. 8.2, they have a phase of 0° when the rest of the wave's phase has advanced to 60°. The emerging wave's phase is relatively *retarded*.

Consider the situation in Fig. 8.3, where a spherical wave-front diverging from O meets a curved medium boundary – the surface of a light-optical lens. Because light moves slower in glass, the wave-fronts are compressed within it, so that the emerging wave-fronts are retarded, as they were from the glass block. However, they are retarded unequally: more so near the axis[3], so the curved lens surface flattens or even inverts the spherical wave-fronts.

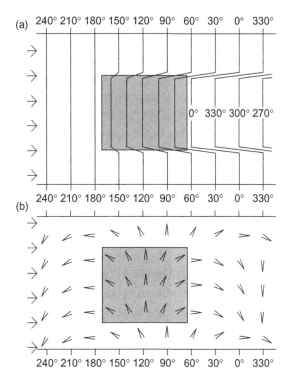

Figure 8.2: A wave is moving from left to right, and part of it passes through a small square slab of higher refractive index. The same situation is shown in two representations. **(a)** Wave fronts with phase angles. **(b)** Phasors oriented appropriately. (Compare with Fig. 8.1.)

---

[3] The 'large circle approximation', Appendix §8.5.2, shows that the relative shift x necessary to flatten it, at a given aperture a, is $x \approx a^2/d$, where d is the wave-front's diameter.

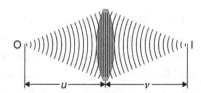

Figure 8.3: Wave picture of the process in Fig. 7.4 (e). The diverging spherical waves from O are converted into plane waves (inside the lens); the reverse process on the right makes the waves come to a focus at I.

In Fig. 8.3 it flattens them completely inside the lens, inverting them outside and allowing the symmetrical right to reverse the process, forming spherical wave-fronts that converge to an image I.

This is the wave equivalent of Fig. 7.3 (a) and Fig. 7.4 (e), but here we view the process in terms not of rays but of curved wave-fronts. The curved lens surfaces change the curvature of any incident wave-front. Indeed, each surface adds a given curvature, dependent on its own curvature and refractive index. So, on the left lens surface, it adds a flattening curvature. It adds that same curvature again on the right surface, generating the curved wave-front that converges to I. So any wave-front impinging on the lens emerges with the same overall change in its curvature. The curvature κ of a wave-front is defined as the reciprocal of its radius; so the curvatures are $1/u$ on the left and $1/v$ on the right. The lens applies a constant total curvature change $\kappa = 1/u + 1/v$. What is the constant κ? If the left wave-front were flat, the curvature $1/u$ would be zero, while v would equal the focal length f. Therefore $\kappa = 0 + 1/f = 1/f$, so the original equation becomes the reciprocal 'lens formula' (§7.1.2).

$$1/f = 1/u + 1/v \qquad\qquad (8.2)$$

8.1.2 Introducing Diffraction

Now we meet the fundamental differences between waves and rays. Unlike rays, waves are not confined to straight-line paths, but are able to spread their paths and bend near sharp edges. This allows a superposition of direct and bent waves, and the consequences constitute another difference. When several *rays* are brought together (e.g. by a lens) they simply add. But *waves* can also *subtract*, giving a local destruction of energy called *interference*.

The detailed mechanism is as follows. A sound wave (for example) originates with some vibrating particle and consists of vibrations spreading outwards to neighboring particles of the medium. Each particle is made to vibrate as a consequence of its neighbors' vibrations; and *its* vibration, in turn, helps to cause those neighboring vibrations. Thus *each particle* acts like a weak copy of the original vibrating particle, and is itself the source of a new wavelet (a *Huygens' wavelet*), as proposed by Christian Huygens in 1678. This is how waves bend round edges, their first feature.

The second difference arises if the superposed waves are vibrating in step (i.e. if they are *coherent* because they arise from a common source). Then the waves add together like phasors so, if they differ in phase, they can subtract. (If the superposed waves vibrate independently, i.e. *incoherently*, they add chaotically, like different conversations, giving an average that corresponds to adding their intensities, as with rays.)

The overall process is *diffraction*. It becomes important when the wavelength is comparable to the dimensions of important features of the optical path. If the wavelength is relatively small, we see only shadows. Thus light's very small wavelength is partly why it acts as rays in ordinary circumstances. But also ordinary light comes from many independent sources and therefore lacks coherence, unlike laser light (§ 8.4.2).

The importance of wavelength and coherence in distinguishing rays (or particles) from waves became fundamentally important after the development of quantum mechanics.

8.2 The Quantum Electron

8.2.1 Electrons as Waves

It was originally inconceivable that the electron (of characteristic mass and charge, leaving clear cloud-chamber tracks), could be anything but a particle. But, in 1927, de Broglie's earlier suggestion that it also has wave properties was confirmed by Davisson & Germer and by Thomson & Reid.

Consider, from both viewpoints, an electron passing by a nucleus (Fig. 8.4). The particle picture (a) shows the electron passing the nucleus rather like a comet passing the sun, its path bent by the attractive force; (b) shows how this bent path, interpreted as a bent wave, implies that

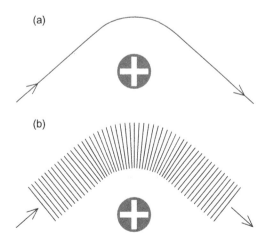

Figure 8.4: (a) Electron ray-path bent by a positively-charged nucleus. **(b)** The same process involving an electron wave (schematic).

Table 8.1: Electron wavelengths

MV	kV	λ(Å)	λ(nm)
0.08	80	0.0417	0.00417
0.1	100	0.037	0.0037
0.12	120	0.0335	0.00335
0.2	200	0.0251	0.0025
0.3	300	0.0197	0.002

Energy is proportional to V, and wavelength is proportional to 1/(momentum). Now momentum is proportional to speed which is proportional to $\sqrt{}$(energy); so λ is proportional to $1/\sqrt{}V$, where V is in volts. Spence (1981–2003), p. 30, gives the proportionality constant, approximately 1¼. So, at low voltages, $\lambda(nm) = (1.22639)/\sqrt{}V$. However, at high voltages (including those used in electron microscopes) a relativistic correction[4] is needed. Note that, because of the square root term, the wavelength varies less than the voltage.

the electron's wavelength must be shorter when it is in a more positive region, i.e. when it is traveling faster. (De Broglie had originally suggested that its wavelength varied inversely with its momentum.) Thus, when electrons are sufficiently fast, their wavelength can be extremely small, making them useful for microscopy. The exact variation is shown in Table 8.1.

Table 8.1 shows that electron wavelengths are far smaller than those of light (more than 400 nm). Indeed, the electrons in a typical microscope have a wavelength around 1/30 Å or 1/300 nm, which is nearly 50 times smaller than the length of a C–C bond (a satisfactory wavelength for X-ray crystallography). Thus a microscope's electron wavelengths are almost 50 times smaller than the size needed to resolve macromolecule structures. However, this is not a situation of over-kill. It makes the necessary resolution achievable when using aperture angles small enough to muffle the gross aberrations of electron lenses.

8.2.2 Waves Versus Particles

From the perspective of microscopy, light is usually waves but occasionally particles, whereas electrons are usually particles but occasionally waves. Depending on its design details, the electron microscope can emphasize the wave or particle properties. Early microscopes, with their relatively large tungsten filament emitters, their lack of ultra-stable voltages and currents

---

[4] This is because the rest mass of the electron is about ½ MeV, so a typical operating potential of 100 kV (100,000 volts) gives each electron 0.1 MeV energy of motion which is 20% its rest mass. This further increases its momentum, so the relativistic correction reduces the wavelength. The correct formula (e.g. Baker & Henderson, 2001) is $\lambda(nm) = (1.22639)/\sqrt{}(V + 0.000000978V^2)$. This is almost equivalent to replacing $\sqrt{}V$ by $1000\sqrt{}(v(1 + v))$, where v = accelerating voltage in MV. The difference is a factor of $1/\sqrt{}(1 + v) \approx 1 - v/2$ if v is small. So, for the usual electron microscope voltages, we decrease the wavelength by $V/20000\%$, where V is in volts. For example, at 100 kV, $V = 100,000$ volts giving $100,000/20,000 = 5\%$ shorter wavelength.

and their relatively low resolution, gave images in which the wave properties were not very obvious. All modern microscopes have general improvements (like voltage and current stability, or high resolution) that incidentally help to reveal wave properties (besides high-coherence illumination systems that intentionally optimize them).

The shift from particle- to wave-images accompanies a shift in operating practice. Earlier specimen preparation techniques relied exclusively on the amplitude contrast of heavy metals. Now these have been replaced by cryo-preparations in which phase contrast supplies most of the structural detail. We saw in section §7.4.1 that amplitude contrast is conveyed by *inelastically scattered* electrons that have lost energy, and phase contrast by *elastically scattered* electrons that underwent no energy loss. This division correlates with the dual nature of electrons as particles and waves. If an electron, in its path towards the detector, leaves information allowing it to be localized, then it loses its wave properties. For it is a 'paradox' of quantum mechanics that wave and particle properties, although coexisting, are mutually incompatible: they cannot be expressed together by the same particle at the same time. (See the discussion of the famous two-slit experiment by Feynman et al., 1965.) All electrons must behave as particles when they are registered by a detector. They are however permitted to show wave properties at earlier stages, provided the actual conditions make it impossible *in principle* to localize them as particles. But an inelastically scattered electron leaves an atomic trace of its path, permitting its localization as a particle and thereby disqualifying it from simultaneously exhibiting wave properties. Phase contrast requires these wave properties, and is therefore excluded for that electron. So phase contrast is correlated with low radiation damage, but there is no way to get 100% phase contrast and avoid a substantial proportion of radiation damage.

8.3 Fresnel Diffraction

8.3.1 Two-Slit Diffraction

We have seen in §8.1.2 the basic conditions for diffraction. An initial coherent illumination (i.e. all waves with well-correlated phases) must be separated into different beams which, after following different paths, must be brought together so that the waves can interfere. In their different paths, the waves' relative phases changed, leading to interference or diffraction patterns when together again.

Interference requires a minimum of two coherent beams from the same source, and the simplest source is (in 1D) a slit (or, in 2D, a pinhole) so the simplest possible experiment must be a two-slit diffraction experiment: that of Young[5], Fig. 8.5 (a). What does the screen DD show?

---

[5] The polymath Thomas Young (1773–1829) devised this and other experiments (1800–1804) that disproved the corpuscular theory of Newton, whose century-old authority nevertheless prevailed until the wave theory was accepted in France through the later work of Fresnel (1788–1827).

Ray optics predicts an image consisting of two points of light (b), but we find (c), with two differences. (i) Even one pinhole gives, not a point image, but one that is enlarged because the pinhole passes a Huygens wavelet that expands until it reaches the screen. (ii) Each pinhole gives a similar enlarged region and, if the two pinholes are very close, the two regions interfere to give parallel equidistant bands (c). Those bands, actually sinusoidal fringes, constitute the interference or diffraction pattern.

The wavelets from the two slits overlap and interfere, generating in space a fan-pattern that gives the fringes on the screen: Fig. 8.6 (a). Diagram (b) shows how doubling the distance between the slits shrinks the fan-pattern to half its previous size. Why is this? If Fig. 8.6 (a) were doubled in size, the distance between the slits would double to the size in Fig. 8.6 (b). However, uniform enlargement still leaves all angles unchanged, so the fan-angle would be the same as in (a), which is double that in (b). The difference in (b) is that the *wavelength* of

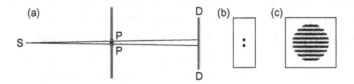

Figure 8.5: (a) Source S illuminating two pinholes P and P′; a screen DD is placed at the right. **(b)** Image expected on screen according to ray optics. **(c)** Image actually seen with light, if the pinholes are very small and separated by a small distance. Diffraction enlarges the pinhole images, which are crossed by horizontal fringes (perpendicular to the line joining the holes).

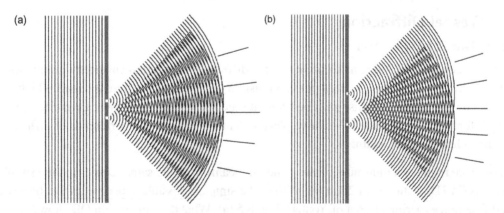

Figure 8.6: (a) An infinitely distant source of coherent light illuminates two close slits in a screen. These generate Huygens wavelets, which add and thereby interfere, producing a fan of bands. Diverging lines mark the positions of bright bands. **(b)** As in (a), but the distance between the slits is doubled, causing the fan of interference bands to diverge at only half the angle in (a). The diverging lines (far right) are copied from (a) in order to show that the bands are twice as close.

light has remained the same as in (a). All these considerations are explained by supposing that the fan angle (α) is proportional to:

$$(\text{wavelength})/(\text{slit distance PP}') \tag{8.3}$$

The two-slit diffraction pattern (in which the slit widths are much smaller than their spacing) is thus defined by just two adjustable distances: the wavelength and the slit spacing. Uniform enlargement or shrinkage does not change the _ratio_ of these distances, which therefore defines the _character_ of the diffraction pattern (though its _size_ is of course affected). Usually, a section of the pattern is recorded on a screen parallel to the wave-fronts. Moving this screen gives a range of patterns (see Fig. 8.7). Very close to the slits they act as independent radiation sources; a little further, they overlap to give a diffraction pattern which, yet further, changes as well as enlarging with distance. These are the _Fresnel diffraction_ patterns (screen position 1). Although its enlargement remains, the pattern's character gradually stabilizes, becoming a constant function of scattering angle. This is the _Fraunhofer diffraction_ pattern (position 2).

Diffraction patterns depend on path-lengths compared with wavelengths. At infinite distance, path-lengths no longer depend on the object's distance: only the _direction_ of the scattered ray (relative to the orientation of the specimen) matters. The scattering pattern is thus a function

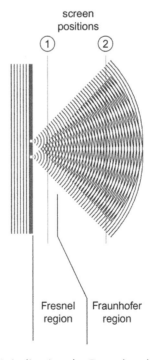

Figure 8.7: As Fig. 8.6 (a), indicating the Fresnel and Fraunhofer regions of the diffraction pattern (see text).

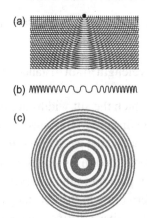

Figure 8.8: Scattering by a point: views including the optic axis (**(a)**, **(b)**) or cutting it **(c)**.

of angle (in 2D). This is the region of Fraunhofer diffraction which we study with X-ray diffraction (§8.4.3). However, at relatively short distances, the distance from the specimen *does* matter, so the more complicated Fresnel diffraction patterns depend also on distance.

However, Fresnel patterns are important in electron microscopy. If the object is a particle surrounded by a uniformly illuminated field, the scattered wavelets interfere with the uniform background (Fig. 8.8 (a)), generating a Fresnel pattern ('chirp[6]'), (c), where the spacing of successive circles decreases as in Newton's rings; both are contours of the top of a sphere. As in Section §7.4.5, Fresnel diffraction fringes are much stronger at an edge. As an edge is a line of points, its Fresnel diffraction is the convolution of a line with a pattern (including phases) like that in Fig. 8.8 (c). The result is a low resolution change at the boundary, rather like (b) starting at the middle.

8.4 Fraunhofer Diffraction

8.4.1 Fraunhofer Means Fourier Transforms

Removal of the distance-dependence makes Fraunhofer diffraction practically useful and theoretically simple. As is shown in any book on X-ray diffraction[7], the Fraunhofer pattern of transmitted coherent waves corresponds to their Fourier Transforms (FT). Here we only demonstrate this for certain cases, of which a particularly simple one is the FT of two peaks (i.e. point-sources), which is a cosine wave. Figure 8.9 shows the geometry. Two wave-trains (a) reinforce first along the centre-line between the sources, and then again at an angle α. Indeed, the intensity is a cosine function of α.

---

[6] The 'chirp' (Mertz, 1965) is a signal type used in radar because its rapid frequency rise gives sharp auto-correlations and hence precise localization. For the same reason, some animal calls (especially those used for echo-location) also scan across the frequency spectrum.

[7] See, for example, Chapter 3 of Giacovazzo et al., 1992; or Chapters 2–4 of Woolfson, 1997.

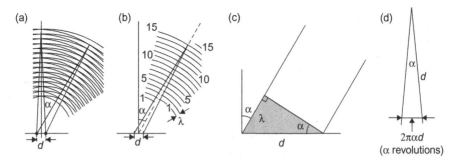

Figure 8.9: (a) Distant part of the Young's fringe experiment, with two scattering atoms.
(b) Wavefronts numbered, and distances and angles defined. **(c)** Triangle at base of (b) much
enlarged. **(d)** Shaded triangle in (c) rotated so that the vertex containing α is at the top. Moreover,
α (here in revolutions) is made small.

Parts of each wave-train, and the marked wave-fronts are numbered in (b). The numbers show
that they are displaced by exactly one wavelength, so they add together in phase; the left path
is exactly one wavelength λ longer than the right path. The base is shown enlarged in (c) to
reveal the geometry of the path difference. The source separation d, the wavelength λ and the
angle α are related by a right-angled triangle (shaded). Since α is extremely small, we can
use a convenient approximation in (d), where the triangle is turned round so that the sides
containing α (very small) are vertical. Here the base (λ) is a straight line very nearly the same
as the small circular arc at a radius d. If α is in revolutions, that small arc is a fraction α of
the complete perimeter ($2\pi d$) of the circle, so it is $2\pi\alpha d = \lambda$, giving $\alpha = \lambda/2\pi d$ (when α is in
revolutions), or:

$$\alpha = \lambda/d \qquad\qquad (8.4)$$

when α is in radians, measured as arc length/radius.

8.4.2 Optical Diffraction

The simple two- (or multi-) slit diffraction experiment is limited to thin flat specimens and (in
our formulae so far) to small angles. Before we drop both limitations in the next section, we
first discuss an important application.

The original optical diffractometer (Taylor et al., 1951), a development of the 'fly's
eye' (Bragg & Stokes, 1945), was used for the rapid analog calculation of FTs in X-ray
crystallography. Rays diverging from a point monochromatized light source entered the first
lens, which generated a plane wavefront (Fig. 8.10). That passed through the pattern or mask,
which created superposed plane wavefronts. Although they need a great distance to separate,
a lens converts the infinitely-distant diffraction pattern into points of light at the focal plane.

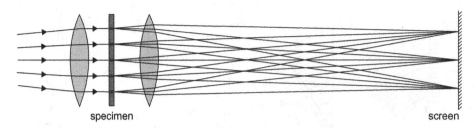

Figure 8.10: Focusing Fraunhofer diffraction patterns so that angle distributions become distributions in space. The specimen (or mask) generates the diffraction pattern.

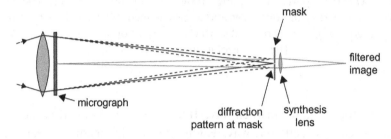

Figure 8.11: Getting and using analog (continuous) FTs: the optical diffractometer. A single long-focus lens produces converging light that generates the diffraction pattern near the middle. Usually that is recorded, but here it is followed by a mask and other equipment for optical filtering (see text).

The diffractometer was widely used in the pre-computer days for crystallography (Lipson & Taylor, 1958; Taylor & Lipson, 1964) before being adapted for analog FTs of electron micrographs (Klug & Berger, 1964). Digitization and digital Fourier Transforms (DFTs) are not only far easier, but they supply FT phases while avoiding artifacts due to thickness variations in the film emulsion. Thus, although the diffractometer is still used for fast preliminary surveys, all critical FTs are re-calculated by computer. The modern diffractometer, adapted for speed rather than precision, is shown in Fig. 8.11. The light source (not shown) is a laser, and a single diffraction lens focuses the diffraction pattern adequately (since phases are unimportant). Figure 8.11 shows additional equipment for *optical filtering*, where the diffraction pattern is modified by a mask (e.g. to select diffraction spots). Then the modified light is focused by a *synthesis lens* that generates a *filtered image* at the far right. (If one switched off the laser and put a diffuse fluorescent light just to the left of the micrograph, and then removed the mask, the synthesis lens would produce an image of the micrograph at the far right.) This analog optical filtering is now performed inside a computer; but the equipment is of theoretical interest in connection with image-formation in a microscope (§9.1).

The optical diffraction pattern or *optical transform* consists of the FT intensity, approximately the square of the wave-vector's amplitude (with the phases lost, as in X-ray diffraction). As with DFTs, the scale is important.

The diffractometer can be calibrated by using, as one specimen or mask, a set of equidistant parallel lines of known spacing on a transparent base. Its diffraction pattern will be a set of equidistant points with the reciprocal spacing. That gives us the conversion factor c = (spacing of lines on specimen) × (spacing of points in diffraction pattern). We divide c by any distances in diffraction patterns to get the corresponding distances in the specimen. Alternatively, $c = \lambda f$, which is in cm² if centimeters are used for both the focal length and the wavelength of the light used (e.g. helium-neon light of wavelength 632.8 nm, $\lambda = 6.328 \times 10^{-5}$ cm). This calculation is useful when planning a diffraction experiment.

8.4.3 General Fraunhofer Diffraction: Reflection/Ewald Sphere

Our previous introductory treatment of Fraunhofer diffraction was simplified by considering only two-dimensional specimens and diffraction through small angles. We now remove these simplifications to obtain a method for calculating the diffraction patterns of three-dimensional structures at small or large scattering angles. However, we *develop* this method from the familiar flat multi-slit specimens considered above.

Figure 8.9 (c) is copied to (a) of Fig. 8.12 and transformed into the similar triangle (b)=(c). That is copied to Fig. 8.13 (d), a triangle forming the basis of the diffraction process in (a) above it. Comparable triangles in (e) and (f) form the bases of diffraction processes in (b) and (c), respectively. Thus the diffraction processes in (a)–(c) explain the simultaneous emergence of three diffracted rays (angles α, β, γ) from the same grating (spacing d) when it is illuminated by a perpendicular beam of light, wavelength λ.

Note that the triangles (d)–(f) of Fig. 8.13, which determine the angles α–γ, are all right-angled triangles with the same hypotenuse d. These key triangles are shrunk by λd and rotated to give (a)–(c) of Fig. 8.14. Here they are turned to combine the diffraction angles α, β, γ in the same diagram (d), where they all form part of a circle of radius $1/\lambda$, a section of the 'reflection sphere' or 'Ewald sphere', discussed further in Chapter 13.

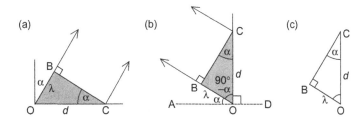

Figure 8.12: (a) Wave-front picture of diffraction from Fig. 8.9 (c). The path difference BO is one wavelength, λ. **(b)** The shaded triangle from (a) was turned round while preserving the lettering of its vertices and sides. **(c)** The triangle in (b) extracted ready for Fig. 8.13 (d).

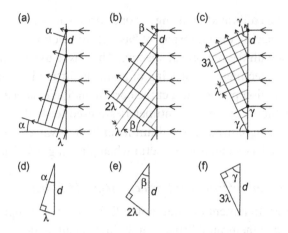

Figure 8.13: Diffraction from the line of slits (separation d) with radiation of wavelength λ. Rays are diffracted at three different angles (measured as deviations from the original ray directions). A path difference of λ gives diffraction at angle α in **(a)**, one of 2λ gives diffraction at angle β in **(b)**, and similarly 3λ gives the angle γ in **(c)**. The small bottom-right triangles in (a)–(c) are then extracted to give the triangles shown in **(d)–(f)**.

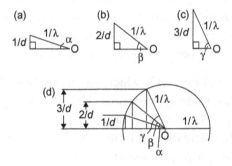

Figure 8.14: (a) is the triangle (d) of Fig. 8.13, rescaled and rotated. (The corresponding change applies to triangles **(b)** and **(c)**.) **(d)** Shows the triangles (a)–(c) assembled together within a circle.

8.5 Appendix

8.5.1 *Resonance links Scattering with Spectroscopy*

In our discussion of wave-scattering, we ignored the possibility that the molecules (etc.) transmitting the wave might also absorb it as, for example, when light passes through a colored glass or solution. Absorption involves energy transfer from a wave-oscillation to an oscillating atom or molecule, and is similar to the classical process of transferring vibration energy to a building or bridge through resonance. Resonance is a very widespread phenomenon and is relevant to any technique using spectroscopy (like NMR) and also to anomalous scattering in X-ray diffraction (§13.3.4); see Giacovazzo et al. (1992); Woolfson (1997).

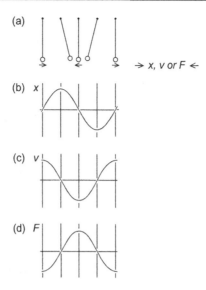

Figure 8.15: Pendulum's oscillations **(a)**, with x pointing right. Below, plots of position **(b)** (positive, i.e. rightward x, up); **(c)** velocity (rightwards pointing up); **(d)** frictional force (rightwards pointing up).

Light is an oscillating electric field that acts on the outer electrons, distorting them slightly. A static electric field produces a permanent distortion pulling the electrons towards the positive potential. The same would occur, at any instant, with a very slowly-oscillating electric field; so the distortion would follow the oscillations (i.e. they would both be in phase). This is what we have been assuming hitherto. However, if the electrons had a natural frequency of oscillation which matched that of the wave, then the wave's oscillations would drive those of the electrons, so the wave would lose energy and we would get absorption. They would also have a changed phase-relationship. (Thus we consider the case of a very *rapidly*-oscillating electric field.)

This situation of *resonance* is most easily seen in a simpler system: a pendulum (Fig. 8.15). Here the position (b) and the speed (c) have the same oscillation frequency, but a different phase. The pendulum's oscillation (like all oscillations) is eventually dissipated (damped) as it loses energy to friction. This frictional force is proportional to the speed (at least, it must be zero while the pendulum bob is stationary); but it is of course opposed to the direction of movement. Thus it will follow curve (d). So the damping force lags 90° behind the oscillator from which it draws energy.

But suppose that there were an external source of oscillatory motion, and that this source were always 90° ahead of the pendulum's position. Then the pendulum would lag 90° behind the external source; so, by analogy with the previous case, the pendulum would 'damp' the source. By drawing energy from the source, the pendulum would act as an 'absorber'. So we see that, at resonance, the absorber lags 90° behind the driving oscillations.

We have covered the case where the driving oscillation has an extremely low frequency (oscillator lag = 0°), and when it has the resonance frequency (oscillator lag = 90°). What about the case where it has an even higher frequency? An oscillating particle, like a mass attached to a spring, exerts two separate forces in response to an external oscillation. There is the force of the spring, and the inertial force of the mass. That inertial force depends on how fast the mass is forced to move. Thus static displacements are opposed only by the spring. But, at the other extreme, very high-frequency driving oscillations are opposed only by the mass. In this case, the mass will lie at one extreme when the force is at one maximum, and then move to the other extreme by the time the force is at the opposite maximum. So it will lag (or lead?) by 180°.

These phase-changes have two important effects. First, they cause the scattered waves to change their phase relative to the driving oscillation. Second, these scattered waves add to the driving oscillation (which has the same *frequency*) but they add with different *phase* (under these different circumstances). This addition causes the overall oscillation to change its phase while it is travelling through the medium containing the *driven* oscillation. And that phase-change is effectively a speed-change, as we saw in §8.1.1.

As mentioned in Chapter 7 (§7.4.4), resonance may be studied in two ways, by scanning across a frequency range and measuring the frequency-response directly; or by supplying an impulse, recording the response (which is a damped oscillation) and calculating its FT. That FT, discussed in Chapter 5 (§5.5.3), gives us the resonance curves directly. (Thus, in assessing a violin's quality, the traditional method of playing every note is replaced by Fourier-transforming the sound of a light hammer-tap on the bridge.)

The FT method works because an impulse is the sum of very many sinusoids, each of which elicits its characteristic response. With small impulses, a system responds linearly, i.e. the response to two signals is the sum of the responses to each (etc.). Therefore the response to all these sinusoids is the same as the impulse-response. Now the FT of the impulse is a sum of many sinusoids, sorted by frequency; so the FT of the impulse-response will be the sum of many frequency-responses, sorted by frequency. And that is what we require.

8.5.2 Large Circle Approximation[8]

In Fig. 8.16, the y-axis is tangent to a circle so as to touch it at the origin. Parallel to the y-axis (but close to it) is a chord of length y that meets the circle at P. In the shaded right-angle triangle:

$$r^2 = y^2 + (r - x)^2 = r^2 + y^2 + x^2 - 2rx \tag{8.5}$$

---

[8] Although very useful, this equation seems to lack a standard name and form, and thus is frequently re-deduced, something quite redundant for a formula of such antiquity. For it must surely have been familiar to Newton (necessary for his work on the interference rings between convex lenses, 1675), and also to Huygens (necessary for his work on the pendulum, 1673).

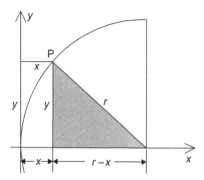

Figure 8.16

The r^2 terms cancel from both sides, leaving $2rx = y^2 + x^2 \approx y^2$, as $x^2 < < y^2$. Therefore:

$$x \approx y^2/2r, \text{or} \, x \approx y^2/d \qquad (8.6)$$

where the circle diameter $d = 2r$.

This is fairly easy to remember. (i) The circle's deviation from the tangent increases quadratically, i.e. as y^2 or the square of its length. (ii) The deviation is smaller for a larger circle, so the proportionality constant is 1/diameter. (It is only necessary to remember that the constant is diameter, not radius.) The equation can be remembered as deviation = (chord length)^2/diameter. We shall call this useful formula the *large circle approximation*.

CHAPTER 9

Wave Imaging

Chapter Outline

9.1 Overview of Wave Imaging[1]

9.1.1 Double Diffraction Imaging

Although wave-imaging in the microscope is inherently complicated, the special conditions for electron cryo-microscopy allow us to make simplifications. We start with the very simplest treatment of image defects. Rayleigh[2] (1896) was interested in telescope images of isolated stars (where diffraction effects are important for each star image), and where each point-image becomes a blur called the **point spread function** (PSF). (The PSF depends on both the design and focussing of the instrument.) Thus the defective image is a *convolution* of

---

[1] Useful books include Heidenreich, (1964), Reimer (1989–97) and Spence (1981–2003), and of course Glaeser et al. (2007). A good book about optics using FTs is Goodman (1968–2005).

[2] J.W. Strutt, third Baron Rayleigh (1842–1919), was perhaps the most fortunate of scientists, working in a private laboratory on his estates when not replacing Maxwell as Cavendish professor in Cambridge, or traveling to receive the Nobel prize (for discovering argon).

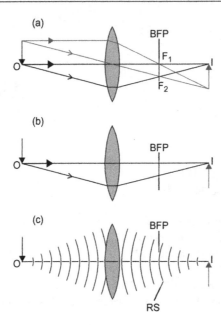

Figure 9.1: (a) Ray diagram of image formation, as in Fig. 7.4. Parallel rays come to foci (F_1,F_2) at the back focal plane (BFP). **(b)** An axial specimen point gives an axial image point. **(c)** The wave picture of (b). At the BFP the converging spherical wave-front ('reference sphere' RS) registers the object's diffraction pattern.

the ideal image with the PSF: defective image = ideal image $\star$ PSF, where the image and the PSF are measured in units of brightness.

Now the microscope image of an extended object can also be conceptually divided into tiny pixels, each giving its own copy of the PSF. If all pixels were independent sources of light, like stars (i.e. if the illumination were incoherent), then each of these copies would add its brightness to the image. Then the same convolution equation would apply, and we could get a simpler relation by taking the Fourier Transforms (FT) of each term: FT(defective image) = FT(ideal image) $\times$ FT(PSF). This equation considers a microscope's defects concentrated in a single function, FT(PSF), which is called the **contrast transfer function** (CTF). Thus:

$$\text{FT(defective image)} = \text{FT(ideal image)} \times \text{(CTF)} \qquad (9.1)$$

(We shall call this the 'CTF equation'.) So far, this is simply the Fourier version of the previous convolution imaging theory. But it is made real by the physical existence of the two Fourier transformations.

The first occurs when each set of parallel rays from the object gets focused at the back focal plane (BFP) (Fig. 9.1 (a)), so the BFP shows the FT of the object, or (more correctly) of the light[3]

---

[3] Because ordinary lenses are easier to understand, we often refer to the radiation as 'light', but the same arguments apply to electrons.

leaving it. This is the first Fourier transformation. The second FT occurs when the rays at the BFP go on to form an image[4] (I in Fig. 9.1 (a)), a process that involves further diffraction. Thus, with perfect imaging, the image plane shows the FT of FT(object wave); and we know that two successive FTs regenerate the original function. (More precisely, there is a net inversion, and the image is indeed inverted.)

The CTF equation provides a new viewpoint on image defects. These are seen, not as a PSF applying in the image plane, i.e. in real space, but as the corresponding reciprocal-space filter (CTF). But where does this reciprocal-space filter apply? It modifies FT(ideal image) which actually exists as a diffraction pattern at the BFP; so we can suppose that the lens's defects are applied there, in the form of the CTF, and they are only later revealed in the image. This is the essential idea behind the wave theory of aberrations.

So the CTF equation (9.1) would have many advantages, if only it were applicable. But it applies to *incoherent* images, that is, images of independently illuminated points like stars. But their independence prevents the existence of phase effects, whereas it is precisely such effects that are the source of our information. Phase effects are maximized by using highly coherent illumination, i.e. beams where relative phases are not lost or dissipated. When coherent beams add, they add *phasors* where the phase angles are quite as important as the amplitudes. Thus the original equation becomes:

$$\text{defective image(phasors)} = \text{ideal image(phasors)} \star \text{PSF(phasors)} \qquad (9.2)$$

where the different copies of the PSF are added as phasors (i.e. adding real and imaginary parts separately: see §3.3.4). But we can't observe the image *phasors*; we see only the image *intensity*, which is the square of the net phasor amplitude.

Is the equation therefore useless because it deals with unobservable quantities? Fortunately, two simplifications come to our aid. The first derives from the nature of the specimen. To see the finest detail, we need to use cryo-specimens that have very little amplitude contrast. Ignoring that is our first simplification.

Phase contrast is thus essential for visibility, and the only way to get it is by defocus. Even so, the phase contrast is very weak. Of course, this weakness is an experimental difficulty, but it is also our second useful simplification. A weak phase-contrast means that the component containing all our information about the specimen is in the presence of a large uniform background. This linearizes the changes (as with holography or finding heavy-atom positions in crystallography: §13.3.2–§13.3.3).

Thus we use our experimental difficulties to solve our problems. We are forced to use specimens of which perfect images would lack all contrast; and our images of them are also spoiled by lens aberrations. However, these imperfections provide us with the required image contrast. Unfortunately, this contrast is very weak; but that lets us use a simple convolution formulation for correcting the image-spoiling by lens aberrations.

---

[4] This diffraction theory of the microscope was first proposed in 1872 by Abbe.

This opportunistic 'linearized' method of image-correction is the subject of this chapter, illustrated first by the most important correction, defocus. It introduces two important concepts, to which we now turn: the CTF being localized at the 'reference sphere', where it connects with lens aberrations; and the signal being conveyed mostly by *weak* phase-contrast.

9.2 Defocus

9.2.1 The (Flattened) Reference Sphere

We have seen how images result from two successive FTs, first at the BFP, and second at the image plane. Figure 9.1 (b) shows how a point on the axis generates rays that are converging at the BFP. The ray directions are normals to wavefronts, which are shown in (c), showing that the back focal plane is *not* a plane but part of a sphere.

However, if the imaging is imperfect, the shape of the wave-front at the BFP cannot spherical, converging onto the image plane: it must either be non-spherical or (if spherical) not converging to the image plane. So it must deviate from that perfect image-converging sphere, which we take as a standard or reference: the **reference sphere** (RS, Fig. 9.1(c)). The actual wave-front's deviation from this sphere, after conversion into a phase-shift γ, gives us the CTF. From this point of view, all the microscope's optical deficiencies can be seen as acting on the reference sphere at the BFP. They act by applying, at different parts of the sphere, phase-changes called the **aberration phase-shift** γ.[5]

But discussing and representing phase advances and retardations is difficult enough, without having to imagine them occurring relative to a sphere. We would prefer a 'reference *plane*', so we imagine the sphere temporarily flattened for making the γ phase-shifts. Therefore we introduce an imaginary concave 'flattening lens' to be followed (after the CTF multiplication) by an imaginary convex 'bending lens' that restores the wave-front's convergent shape. These imaginary additions are shown in Fig. 9.2. Besides using them to obtain a conceptual 'flattened reference sphere', we shall also introduce the CTF in the form of another imaginary 'γ-lens' that produces the aberration phase shift γ. This, and the 'flattening' and 'bending' lenses, constitute an imaginary apparatus that gives exactly the same optical effects as defocusing the lens on the left. Its justification is making these effects clearer.

It should be clearly understood that the 'flattened reference sphere' (FRS) is *not* a wave-front, but an imaginary geometric construction like a circle of longitude. (That's why it always remains flat, despite all the aberrations.) On a wave-front, the phase must (by definition) remain constant, but phase usually varies over the FRS. There are two distinct sources to this variation: the specimen (which determines the phases of the FT); and the lens aberrations (which are supposed to contribute an *additional* 'aberration phase shift' γ to the FT). This

---

[5] There are various notations for the aberration phase shift. We have so far call it γ, but it is also denoted by χ. Some people use the equation $\gamma = 2\pi\chi$, implying that χ is in revolutions and γ is in radians. Here we avoid using χ, because of its similarity to the reciprocal-space coordinate X, and we state the units of γ when they matter.

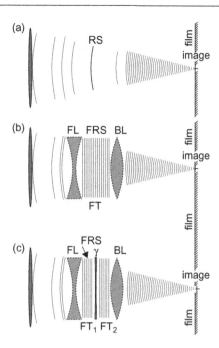

Figure 9.2: **(a)** Waves as on the right of Fig. 9.1 (c), with a curved reference sphere (RS) at the back focal plane. **(b)** Conceptually flattened reference sphere (FRS) between two additional *imaginary* lenses, 'flattening' (FL) and 'bending' (BL). **(c)** Added *imaginary* 'γ-lens' introduces the CTF as a phase-shift γ which changes the FT$_1$ to FT$_2$.

contribution actually comes from defects or defocus in the main lens, but we are representing these effects as an imaginary separate lens, the 'γ lens' (an imaginary objectification of lens-aberration, rather like a wicked spirit).

To understand imaging defects, we need to consider three parts of the optical path. First, the specimen imprints changes on the transmitted wave. Second (and most important), the aberration phase shift γ acts on the reference sphere (or, preferably, the flattened reference sphere). This combines with the specimen's changes to give a modified FT. Finally, this FT gets transformed to give the image, so the last region is the image plane.

We therefore start by considering the specimen.

9.2.2 Thin Sinusoidal Specimen-Components

There are two kinds of image contrast, amplitude- and phase-contrast, and these are used by both types of high-resolution biological specimen. (The older negative-staining specimens produce both phase- and amplitude-contrast, while the newer cryo-specimens produce almost only phase contrast.) Suppose that the complicated structure of an actual specimen has been expressed as a Fourier sum of simple 1D sinusoidal specimen-components[6] with different

---

[6] As the specimen is two-dimensional, the sinusoids will resemble the plane wave in Fig. 6.1 (c).

spatial frequencies (see Fourier series in §3.2). Each specimen is expressed as the sum of such simple specimen-components, but any specimen-component will be a combination of two basic types: amplitude- or phase-contrast, corresponding to the two basic types of image-contrast. Thus there are amplitude-contrast specimen-components and phase-contrast specimen-components. If we study the imaging of these two basic types, we can understand the imaging of any weak specimen. (To understand cryo-specimen images, we mostly need to understand the imaging of phase-contrast specimen-components; amplitude amounts to only around 7% of the total, but its imaging is not ignored: see p. 176 and 179–180.)

These basic types are shown in Fig. 9.3, with amplitude-contrast above and phase-contrast below. In each case, a uniform plane wave enters the sinusoidal specimen from the left (a) and

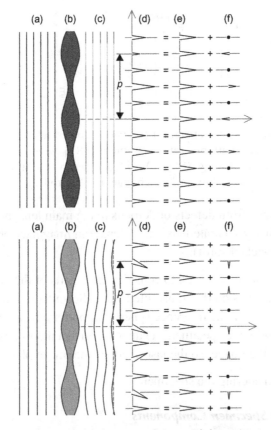

Figure 9.3: Sources of image-contrast: basic types of sinusoidal specimen-component. Upper, pure amplitude contrast, lower, pure phase contrast. (The upper specimen's thickness variations aren't intended to cause any phase changes.) In each case, sinusoidal thickness variations (period p) of the specimen-component **(b)** get translated into sinusoidally varying wave-fronts **(d)**, which are decomposed into a constant wave **(e)** (the same for upper and lower diagrams) and a varying wave **(f)**. Upper: amplitude variations generate 0°/180° varying wave. Lower: phase variations generate ±90° varying wave.

emerges changed on the right, as what we shall call the 'specimen wave'. This is drawn as wave-fronts in (c) and as phasors in (d). In each diagram, the phasor-wave (d) is decomposed into a uniform wave (e) added to a fluctuating phasor-wave (f) that carries all the image information. The top diagram shows amplitude-contrast, where only the amplitude of the transmitted wave (d) changes. This is represented as a uniform wave (e) plus a small varying wave (f) with the same phase (0°/180°). The bottom diagram shows phase-contrast, where only the phase of the transmitted wave changes. So the transmitted wave is represented as a uniform wave (e) plus a small varying wave (f), but here its phase is ±90°.

The microscope lens creates the FT of the constant wave (e) and small varying waves (f). These FTs are shown in Fig. 9.4. The constant wave's FT (a) is a peak at the origin, with the same phase as the constant wave. The varying wave with 0°/180° phases (b) give peaks at $\pm 1/p$ which have the same phase (0°) as the varying wave at the origin. The ±90° varying wave (c) also gives peaks at $\pm 1/p$ which have the same phase (90°) as the varying wave at the origin. These FTs appear at the BFP, i.e. at the reference sphere and will be shown at the 'flattened reference sphere'.

9.2.3 Double-Diffraction Perfect Imaging

Figure 9.3 showed the two types of specimen wave, and Fig. 9.5 shows how they get processed by a single-lens microscope when it is exactly in focus. A glance at this figure will show that, in each case, the image-wave is an exact copy of the specimen-wave (apart from magnification). However, the details are spelled out below to help with the corresponding figures (§9.2.4) that portray out-of-focus imaging.

Parts (a)–(d) of Fig. 9.3 are copied to parts (a) and (b) of Fig. 9.5, where parts (c) show the FT of the specimen wave (b), produced by the lens[7]. The components of these FTs were

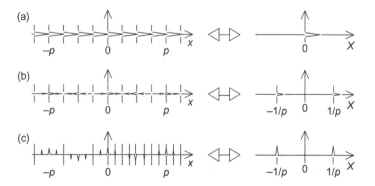

Figure 9.4: FTs of the output waves from Fig. 9.3. **(a)** Constant wave; **(b)** 0°/180° varying wave; **(c)** ±90° varying wave.

---

[7] For clarity, this lens, which would lie between (b) and (c), has been omitted from Fig. 9.5.

shown in Fig. 9.4. Thus the amplitude wave (upper (b) in Fig. 9.5) was upper (d) in Fig. 9.3 where it consisted of a constant wave (e) and a 0°/180° varying wave (f). Figure 9.4 showed that the constant wave's FT is an origin peak, while the 0°/180° varying wave's FT is a pair of 0° peaks either side of the origin. These three components are shown in Fig. 9.5, upper (c). Similarly, the phase wave (lower (b) in Fig. 9.5) was lower (d) in Fig. 9.3 where it consisted of a constant wave (e) and a ±90° varying wave (f). Figure 9.4 showed the constant wave's

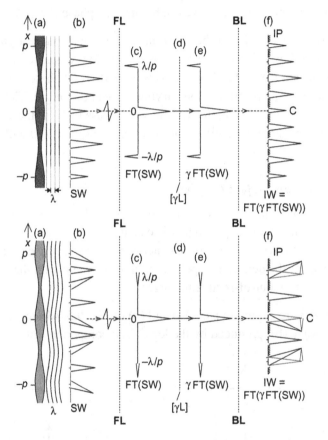

Figure 9.5: Perfect imaging of specimens, amplitude (above) and phase (below). Only three sections of the optical path are represented: the specimen region **(a)** and **(b)**; the section around the back focal plane with the 'flattened reference sphere' (including the 'flattening' and 'bending lenses', FL and BL), where the main action occurs; and the image **(f)**. (Thus the microscope lens is also omitted before FL.) Lines **(a)** and **(b)** correspond to parts (a)–(d) of Fig. 9.3. Line **(c)** shows the FT of the 'specimen wave' (SW); this is taken from the right of Fig. 9.4, but the scale refers to the aperture angle α, so the peaks at $X = \pm 1/p$ have $\alpha = \pm \lambda/p$, where λ is the electron wavelength. (See equation (8.4) in section §8.4.1.) Line **(d)** should have the 'γ lens', but none is present with perfect imaging. So the 'modified FT' in line **(e)** is the same as the original FT in line (c). Its FT is shown at the image plane IP, line **(f)**, where we have the perfect image, a copy of the specimen wave in line (b). (The phase image is somewhat enlarged.)

FT (as we have noted), while the $\pm 90°$ varying wave's FT is a pair of 90° peaks either side of the origin. These three components are shown in Fig. 9.5, lower (c).

In Fig. 9.5, position (d), where the 'γ-lens' should be placed, is vacant so the FT is copied unchanged from (c) to (e), which is therefore the FT of the specimen wave in (b). Consequently, the FT of (e) is just the specimen wave (though enlarged). Now the specimen wave in the top diagram has amplitude (and therefore intensity) variations because electrons are removed. However, the specimen wave in the bottom diagram has a constant amplitude (and therefore constant intensity) and so also does its (enlarged) image in (f). All the specimen information is encoded as invisible phase variations.

Thus, at exact focus, only amplitude-contrast specimens show any contrast, and a pure phase-specimen (e.g. a cryo-specimen) shows nothing. So we need to reveal the phase variations in the specimen wave by changing the focus, i.e. by changing the lens strength, which can be represented as adding a small (convex or concave) 'γ-lens' at (d).

9.2.4 Defocused Imaging

Now we introduce, after the first FT, a 'γ-lens' which defocuses the image. However, before discussing its effects, we need to understand clearly the effect of *any* lens on a plane wave. Figure 9.6 shows the effect of a convex lens. On the left are vertical wave-fronts where the wave has constant phase, 180° on the far left, with the angles reducing (i.e. the phasors rotating

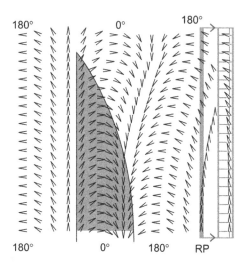

Figure 9.6: Effect of a convex lens on plane waves. At the left, plane wave-fronts move towards lens which bends the wave-fronts by retarding waves below (along the axis) more than those at the periphery (above). If we intersect the curved wave-fronts with a flat 'reference plane' (RP) we get a phase variation shown in the box at the extreme right.

clockwise) as we move rightwards, fitting the convention in section §8.1.1. Inside the lens, the wave moves slower so its wavelength is shortened and its wave-fronts are retarded, shifting left of the top wave-fronts (outside the lens). This leftward shift increases where the lens is thicker, producing curved wave-fronts converging towards the right. If we now set up an imaginary 'reference plane' (RP) on the right, it will have phase-variations over its flat surface. If we choose a moment when the phase is 90° at the lens centre, it will be nearly 180° outside the lens (at the top). Thus, within the 'reference plane' and starting at the lens centre, the phase-angle *increases* (the phasor rotates *anti-clockwise*) towards the periphery. This is characteristic of convex (focusing) lenses, and of course the opposite happens with *concave* (diverging) lenses.

The same applies to the weak 'γ-lenses' which act on the specimen's FT. Figure 9.7 shows both types of lens acting at (d) on the FT of a weak sinusoidal phase object. In the upper

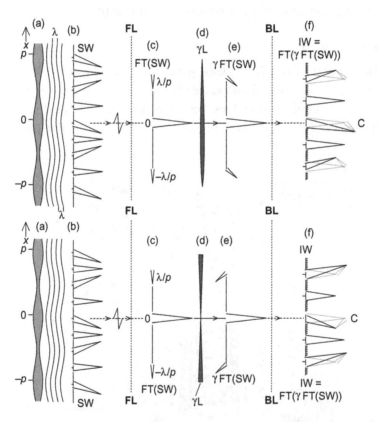

Figure 9.7: The effect of a slight defocus of the system in Fig. 9.5, where **(a)**–**(c)** are the same as in this figure. **(d)** is the 'γ-lens', and (e) is the FT of the specimen wave after phase-modulation by the 'γ-lens'. In the upper diagram, its rotation is anti-clockwise after a convex 'γ-lens' provides extra focusing, hence over-focused. In the lower diagram, its rotation is clockwise after a concave 'γ-lens' corresponding to under-focusing.

diagram the 'γ-lens' (γL) is convex, in the lower diagram it is concave. Since only *relative* rotations of phasors affect the image, we are assuming that, at (c), the first FT's big phasor on the axis is unchanged, so only the peripheral diffraction spots at $\alpha = \pm\lambda/p$ can rotate. The convex 'γ-lens' rotates these spots *anti*-clockwise, so the modified FT in Fig. 9.7 upper (e) has had its peripheral phasors rotated anti-clockwise. The concave 'γ-lens' in the lower diagram rotates these same phasors clockwise in (e).

The effect of these small rotations is seen in the images in (f), the FTs of the γ-modified diffraction pattern in (e). Each image curve is shown as a series of phasors, of which the most important is on the axis, next to C. The final image phasor, in black, is the sum of two light-grey phasors. Of these, the bigger is constant and horizontal; it derives from the constant phasor in the FT at (c) and (e). The smaller grey phasor next to C is essentially a copy of the corresponding peripheral phasor in (e).

The two images are very similar: if you turn the upper image upside-down and shift it vertically, you get the lower image. The essential difference is contained in the vertical shift. Whereas the upper image has next to C the longest black phasor, the lower image has the shortest black phasor there. That means that the phase contrast revealed by over-focus (upper diagram) is opposite to that with under-focus (lower diagram).

On the axis, over-focus (upper) gives a bright line whereas under-focus (lower) gives a dark one. This is the wave explanation of contrast-reversal, discussed from a ray viewpoint in §7.4.2. Moreover, the point C on the axis is where amplitude contrast gives a dark line in the image (top diagram in Fig. 9.5). So amplitude contrast is enhanced by phase contrast with under-focus, but weakened by over-focus. This was suggested by ray-optical arguments in §7.4.3.

This shows the underlying principles of how defocus changes contrast. But we need to know how the contrast varies quantitatively with defocus and spatial frequency, and we now turn to these questions.

9.2.5 How Defocus Affects Amplitude/Phase Contrast

Onto the incident wave phasor, the specimen adds a scattered wave phasor, but amplitude and phase specimens do this differently. The scattered wave phasor of an amplitude specimen is *parallel* to the incident wave phasor, but that of a phase specimen is *perpendicular*. These relative phases undergo further changes when the microscope is defocused, adding an aberration phase-shift γ which rotates the scattered wave phasor. The consequence is to modify the FT of the specimen wave, a modification called the 'contrast transfer function' (CTF). As the modification is different for amplitude and phase specimens, we can speak of an **amplitude CTF** (ACTF) or a **phase CTF** (PCTF). The term CTF by itself should mean the particular combination under discussion, though this is often heavily weighted towards the PCTF, since phase contrast is so important.

These effects are elucidated in Fig. 9.8; (a) shows the situation with an 'absorbing' specimen, where the scattered-wave phasor S is parallel to the direct-wave phasor D; (b) shows the situation with a 'refracting' specimen, where S is perpendicular to D. (We draw D horizontal, with its origin at the left, and its right joined to the base of S.) In each case, defocus rotates S by an angle γ that is anti-clockwise with over-focus and clockwise with under-focus. The final wave is the sum of phasors D and S, which has the (fixed) common origin of D at the left and ends with the (moveable) arrow-head of S. The length of this wave is its amplitude. This means a starting amplitude ($\gamma = 0$) of (D + S) in (a); but in (b) it is simply D. Each CTF varies sinusoidally from its starting point, and both CTFs share the same curve, though with different starting-points. So they are combined in (c), though with different positions of the vertical axis.

Thus the dependence of PCTF on γ follows the equation (with the conventional sign):

$$PCTF = -\sin\gamma \qquad (9.3)$$

Having seen how the PCTF varies with γ, we next need to consider how γ varies with defocus (Δf), spatial frequency (X), and electron wavelength (λ). The phase shift γ depends on the path difference (d) between the curved wavefront and the reference plane, so we start by finding how d depends on Δf, X, and λ.

The last two are closely connected. Recall that the sinusoidal specimen gives peaks at $X = \pm 1/p$ and the scattering angle $\alpha = \pm\lambda/p$, where λ is the electron wavelength. So $\alpha = \pm\lambda/p = \lambda X$. Also, the size of the diffraction pattern doubles if λ doubles, or if the period p halves (which doubles the value of X where the peak goes).

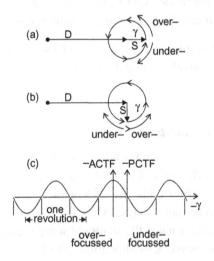

Figure 9.8: Phasor diagrams of the transmitted wave in Fig. 9.7 (image): **(a)** amplitude (ACTF), **(b)** phase (PCTF). In each case, the arrowhead of S rotates, so its distance from the left end of D is a fixed quantity plus a sinusoidal variation. **(c)** Plot of this variation, which measures the visibility of the specimen. The in-focus ACTF/PCTF is given at the corresponding vertical arrow.

The geometry of the γ-lens is outlined in Fig. 9.15 (§9.4) , where the aperture u is (as we have seen) proportional to λX. Moreover, f is either the focal length or at least proportional to it. Consequently the equation $\Delta v \approx -u^2\Delta f/2f^2 = -\alpha^2\Delta f/2$ (see equation 9.7 from the Appendix §9.4) tells us that the path difference Δv is proportional to $(\lambda X)^2\Delta f$. Now the phase difference is (path difference)/(wavelength), so the phase difference γ is $\Delta v/\lambda$ which is proportional to $(\lambda X)^2(\Delta f/\lambda)$.

So we finally get:

$$\gamma = (\text{const.})(\Delta f/\lambda)(\lambda X)^2 \tag{9.4}$$

The constant's sign is (implicitly) settled later when we consider the significance of the sign of Δf.

9.2.6 How Phase Contrast Transfer Function Depends on Defocus, Wavelength and Spatial Frequency

This last equation for γ should now be combined with the plot, in Fig. 9.8 (c), of the PCTF as a function of γ. There is a complication because we now have three independent variables (Δf, λ and X), so the plot of PCTF would require 4-dimensional space. However, the electron wavelength λ is infrequently changed, so the interesting variation only requires 3-dimensional space. When we analyse a single micrograph, the defocus Δf is a constant, so the 2-dimensional plot of PCTF as a function of X (or λX, which is proportional to the scattering angle α) is of some interest. This is shown in Fig. 9.9, where (a) shows the dependence of the PCTF on γ (copied from Fig. 9.8 (c)). The relative stretching (s) of the initial part of the curve, and the relative compression (c) of its later part, are established in (b), which gives

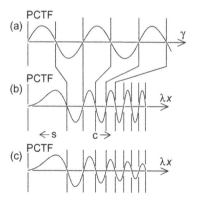

Figure 9.9: The PCTF for pure defocus (no spherical aberration). **(a)** shows it as a function of γ; **(b)** shows it as a function of λX, which is also proportional to the beam angle α. That mapping stretches the beginning of the curve(s) and compresses its end **(c)**. The peaks and troughs are spatial frequency regions that generate visible sine-curves in the image. **(c)** shows the PCTF curve when illumination is partly incoherent. For units of (c), see Fig. 9.10.

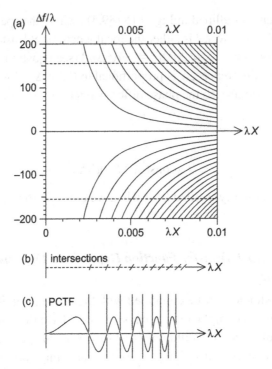

Figure 9.10: (a) Thon's (1966) plot of contrast transfer characteristics, showing the PCTF at different values of defocus (Δf) and spatial frequency (λX). (See below for units and examples.) This summarizes many different plots like Fig. 9.9. Here, the curves are contours of the zeroes of the PCTF when there is *no* spherical aberration (hence their symmetry about the λX-axis). The regions between contours receive phase-contrast, but that has opposite signs in the top and bottom halves. Both defocus (Δf) and spatial frequency (λX) are measured in terms of wavelengths, which makes the curves independent of the accelerating voltage. The lower broken horizontal line shows the position of the curve in Fig. 9.9. **(b)** The region of (a) around the lower broken horizontal line.
(c) The intersections of (b) correlated with the zeroes of the PCTF in Fig. 9.9.
Units: $\Delta f/\lambda$ is measured in thousand wavelengths, and λX measures spatial frequency in numbers of reciprocal wavelengths. The following examples are based on a supposed wavelength $\lambda = 0.01$ Å. If $\Delta f/\lambda = 100$ (on the scale) $= 100{,}000$ (as units are in thousand wavelengths), then $\Delta f = 100{,}000\lambda = 100{,}000(0.01$ Å$) = 1000$ Å $= 100$ nm $= 0.1$ μ. If $\lambda X = 0.005$, the spatial frequency X is $0.005/\lambda = 0.005/0.01 = 0.5$ Å^{-1}, i.e. density bands with a period of $1/(0.5$ Å$^{-1}) = 2$ Å)

the PCTF curve for perfectly coherent illumination. Actual illumination is partially coherent, giving a reduction of the PCTF at higher spatial frequencies, as in (c).

However, this diagram only shows an example of the PCTF for a given defocus Δf. Although including a dependence on Δf transforms the curve into a surface in 3D, we can represent this surface as contours where the PCTF is zero; see Fig. 9.10. Such a plot was first used[8]

---

[8] Thon used the reciprocal of X.

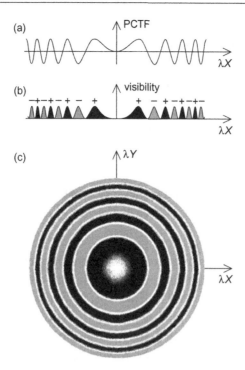

Figure 9.11: (a) The pure defocus PCTF taken from Fig. 9.9, symmetrical about its origin. **(b)** By squaring the PCTF we get its effect on the visibility of a sine-wave of given spatial frequency. Signs are shading-coded here and in (c). **(c)** The PCTF has been rotated about its origin. This shows the effect on a 2D FT of the same defocus but no other aberrations. Both black and grey rings (called 'Thon rings') appear equally bright in optical diffraction patterns.

by Thon (1966, 1968) and labeled by him (1971) *contrast transfer characteristics*. For pure defocus, this plot is symmetrical above and below the λX-axis; the two horizontal dotted lines represent equal amounts of defocus, one being over- and the other under-focus. Both would give identical FTs, if the specimen were a pure phase object (to which cryo-specimens approximate) at resolutions where spherical aberration is negligible.

We return to the case of a single micrograph with a given defocus Δf and an ordinary line curve shown in Fig. 9.9 (b) or, more usually, (c). The imaging conditions would be investigated by finding the micrograph's FT, initially by optical diffraction. The diffraction pattern contains, besides the FT of the specimen, a map of those spatial frequencies where the granular 'noise' in the specimen has been enhanced by the PCTF: an image of the PCTF. This is explained in Fig. 9.11, where (a) shows the PCTF as in Fig. 9.9 (b). Figure 9.11 (b) translates this curve into the visibility of the granular 'noise'. Those parts of (a) above the axis receive enhancement with a positive sign; those parts below it also receive enhancement, but with a negative sign. That means that they are still rendered visible, but their signs should be reversed to give the correct FT of any specimen in that micrograph. (Thus both the black and light-grey parts of the diagram are equally visible or strong in the FT of cryo-specimens.) It shows concentric rings ('Thon rings') of diminishing separations, which are simply contours of the top of a sphere. Curves such as Fig. 9.9 (b) correspond to sections of this diagram.

9.2.7 Phase Contrast Transfer Function Correction

Defocus is not only a virtually universal 'aberration', but also a necessary one, as it is the only practical current method for revealing phase-contrast information about the specimen. However, defocus introduces serious artefacts that need correction, even with only a single image.

To understand the possibilities of correction, consider the character of the PCTF. It is a reciprocal-space filter (§5.1) which changes both amplitude and sign, but has no complex component (equation (9.3)). So the PCTF, at any spatial frequency, is just an ordinary number, though it can be negative as well as positive. Between its positive and negative regions there are therefore zeroes which destroy the signal. This effect is shown in Fig. 9.11 (b), where the black parts are positive and the grey parts negative. Those parts of the image FT affected by the black parts are unchanged, except for amplitude. The grey parts of the PCTF have the same effect, except for a sign change. But the zero (or very small) parts of the PCTF destroy the image's FT.

We can reverse the sign changes, but we cannot replace the missing parts if we have only one image. We could rescale the diminished parts of the specimen's FT, but that would also enhance the noise. Thus, given only one image, perhaps the best we can do is simply to reverse the signs of those parts of the specimen's FT affected by the negative PCTF regions. (This is called 'phase flipping'.) One way to reverse the sign is to *multiply* by the PCTF. (See section §14.4.2 for its application to tilted sheets.) Although the weakened parts of the image FT are thereby further weakened, so too is the noise.

However, the missing parts of the image's FT can only be recovered from other images of the same specimen, taken under different defocus states with different PCTFs, having complementary zeroes. Then we could probably find the 'best' FT using some least-squares method (see section §11.1.5). But a simpler, and generally satisfactory, method is to multiply *each* of the different image FTs by its *own* PCTF and add them. If all the PCTFs were truly complementary, there should be no zero regions in the sum. However, there would probably still be some reciprocal-space regions better corrected than others. This bias could be removed by dividing by a correction factor, the sum of the squares of all the PCTFs. But, in case there should be near-zero regions common to all the PCTFs, we can add a Wiener constant (§5.4.1) to that sum of squares before dividing by it.

9.3 Other Aberrations

9.3.1 Astigmatism

Astigmatism, the most common of aberrations that are instrumental defects (unlike defocus) is the sinusoidal variation of focal strength with the angle φ, as shown in Fig. 7.8. It corresponds to a (small) distortion of lens strength within a plane parallel to the image (§7.1.6). In Fig. 9.10 we would represent astigmatism by two parallel intersections (like one of the dotted lines shown). In the case of the slight astigmatism in a well-corrected microscope, the two lines would be very close together and would describe the Thon ring positions on two perpendicular lines. This is shown in Fig. 9.12, where the diagram (and X, Y frame) is oriented relative to the astigmatism rather than to any fixed direction in the microscope. (Also, to make the diagram clearer, it shows an unacceptably high degree of astigmatism.)

This shows the usual case, where the astigmatism is combined with a bigger amount of defocus. When the astigmatism is bigger than the defocus, different patterns arise that are related to other conics, mostly hyperbolae.

9.3.2 Spherical Aberration

After defocus and its asymmetric equivalent (astigmatism), whose aberration phase-shift γ varies as $(\lambda X)^2$, the next higher (even) power is the fourth, which corresponds to spherical aberration (already discussed in section §7.1.5).

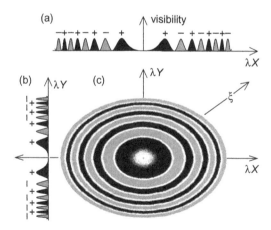

Figure 9.12: (a) The squared PCTF taken from Fig. 9.11. As in that figure, both black and shaded regions are enhanced, but black with positive phases and shaded with negative phases. **(b)** The same plot for a different defocus value. **(c)** The combination of (a) and (b) in a two-dimensional plot with elliptical rings. (It can be viewed either as 'visibility' or PCTF, which are very similar in this kind of diagram.) Note that the amount of defocus varies between two extreme values (shown in (a) and (b)). Each direction, such as the axis ξ, corresponds to an intermediate defocus value; for example, ξ corresponds to the lower line shown in Fig. 9.10.

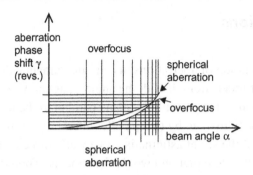

Figure 9.13: Plot of the aberration phase shift against the ray or beam angle α for defocus and spherical astigmatism. The spherical aberration increases slower than defocus at small angles, but at larger angles overtakes it.

Figure 9.13 compares spherical aberration with defocus. In the optical system of Fig. 9.2, we have so far considered 'γ-lenses' that create circularly-curved spherical wave-fronts (converging or diverging). With spherical aberration, the fourth-power dependence leads to a 'hare and tortoise' race in which spherical aberration at first lags behind over-focus but finally overtakes it in a late spurt.

Since the phase-shift of spherical aberration has the same direction as that of over-focus, it opposes under-focus and therefore partly cancels it. Consequently, for minimizing spherical aberration, there is an optimal state of under-focus: **Scherzer** (under-) **focus**, a level where the PCTF is nearly 1 for the biggest range of spatial frequency X (Cowley, 2001). (We came to this conclusion from ray optics in §7.1.5.) This is another example in electron microscopy of a technical deficiency having compensating advantages. For, although the spherical aberration is unavoidably positive, it not only levels out the effects of defocus but does so at a level of under-focus that adds to amplitude contrast.

Since spherical aberration corresponds to over-focus, we know the relative signs of defocus and spherical aberration. Recalling equation (9.4):

$$\gamma = (\text{constant}_1)(\Delta f/\lambda)(\lambda X)^2$$

we add the spherical aberration term to get:

$$\gamma = (\text{constant}_1)(\Delta f/\lambda)(\lambda X)^2 + (\text{constant}_2)(\lambda X)^4$$

What are the two constants? First consider the units. Wavelength λ is a length but X is a reciprocal length, so (λX) is a ratio without units, like $(\Delta f/\lambda)$; thus (constant_1) and (constant_2) must be of the same type. The spherical aberration coefficient C_s is also a length, so we divide it by λ to get the dimensionless ratio (C_s/λ). This gives us $\gamma = (\text{constant}_1)(\Delta f/\lambda)$

$(\lambda X)^2 + (\text{constant}_3)(C_s/\lambda)(\lambda X)^4$. If the two constants are positive numbers, and C_s is positive, then Δf must also be *positive for over-focus*. The total aberration phase shift (in rotations) is:

$$\gamma = \tfrac{1}{2}(\Delta f/\lambda)(\lambda X)^2 + \tfrac{1}{4}(C_s/\lambda)(\lambda X)^4 \qquad (9.5)$$

Finally, we must combine this with the earlier equation (9.3):

$$\text{PCTF} = -\sin\gamma$$

9.3.3 Odd Aberrations: Coma

We have so far discussed aberrations that depend on even powers of the diffraction angle $\alpha = \lambda X$. They are particularly easy to study, as they produce symmetrical phase shifts (the same for $-X$ and X); and we have seen that 'even' shifts convert phase variations into intensity variations. Besides providing phase-contrast specimen images, this leaves an intensity imprint on the noisy fluctuations of specimen density, allowing image characteristics to be studied by FTs (Figs 9.11 and 9.12). But these advantages are lacking from odd powers of α, which produce *anti*-symmetrical phase-shifts and are consequently neither beneficial nor easily detectable (Fig. 9.14).

However, an assessment of overlooked aberrations needs to include the 'odd' ones. We have so far discussed defocus and astigmatism (proportional to $(\lambda X)^2$) and spherical aberration (proportional to $(\lambda X)^4$). The missing aberrations are odd (all higher-order aberrations would tend to have negligible effects at the small aperture sizes allowed by electron lenses). This leaves the aberrations depending on (λX) and $(\lambda X)^3$; could they be causing image damage undetectable by FTs?

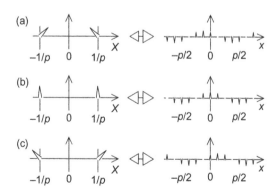

Figure 9.14: The effect of an *anti*-symmetrical aberration phase-shift on phase variations of diffracted beams. Left: the two phasors at $\pm 1/p$ in the reference plane are pure 'imaginary' (pure phase) in **(b)**, but have anti-symmetrical aberration phase-shifts γ added to their FT (right) in **(a)** (phase-angle changed by $-45°$ (left) and $+45°$ (right)) and in **(c)** (phase-angle changed by $+45°$ (left) and $-45°$ (right)) In both cases, the FT (right) remains pure imaginary, so it lacks any real component to give intensity changes when combined with the (real) undiffracted beam.

Figure 9.14 shows that anti-symmetrical phase-shifts have quite different effects on a weak phase image: they only cause *translations* of its Fourier components. In the case of the first power, all the Fourier components are translated by the same amount. So the trivial 'aberration' depending on just α only translates the image, and it would be corrected by moving the specimen stage.

However, the aberration depending on α^3 'shreds' the FT, a non-trivial aberration called **coma**. It is potentially more important than spherical aberration which, depending on a higher power of λX, is likely to be smaller at small aperture sizes, and which is not correctable in ordinary electron microscopes. As an odd aberration, coma has a *direction* (since it changes sign when the aperture angle α is reversed). Thus it is affected by alignment, including a relative displacement of the pole-pieces and, particularly, a tilt of the electron gun. Although coma is invisible in FTs (which reveal only even phase-shifts), it shows up in edge-fringes (Fig. 7.13), when they are unsymmetrical. Unlike those due to astigmatism, the edges that face each other across the hole have different fringe patterns with coma; in extreme cases, one side may be under-focused and the other over-focused. With a well-aligned microscope, coma is likely to become significant at resolutions better than 0.6 nm. Then special alignment procedures (e.g. Zemlin & Zemlin, 2002) are needed to reduce it to an acceptable level.

9.4 Appendix: Aberration Phase-Shift Geometry

In Fig. 9.15, the shaded right-angled triangle, $f^2 = u^2 + (f - v)^2 = u^2 + f^2 - 2fv + v^2$, so $2fv = u^2 + v^2 \approx u^2$, when $v < u$, so $v^2 < <u^2$. Thus:

$$v \approx u^2/2f = a/f, \qquad (9.6)$$

where $a = u^2/2$. If f increases by Δf when v increases by Δv, $v + \Delta v = a/(f + \Delta f) = a(f - \Delta f)/[(f + \Delta f)(f - \Delta f)] = a(f - \Delta f)/[f^2 - (\Delta f)^2] \approx a(f - \Delta f)/f^2 = a/f - a\Delta f/f^2$.

Subtracting $v = a/f$ we get $\Delta v \approx -a\Delta f/f^2 = -(u^2/2)\Delta f/f^2$.

So

$$\Delta v \approx -u^2 \Delta f/2 f^2 \qquad (9.7)$$

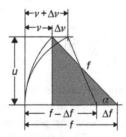

Figure 9.15: The geometry of path-differences in back focal plane. At an aperture u the path difference is v to a converging wavefront radius f. At an extra path Δv the radius is $f - \Delta f$.

General Structural Methods

Symmetry

Chapter Outline

10.1 Principles

Symmetry is important not only as a tool in structure determination, but as a basis for understanding subunit organization in macromolecules. It forms an elegant branch of mathematics, and its intuitive principles are discussed in two classic works: Hilbert & Cohn-Vossen (1952) and Weyl (1952). See also the lively account by Stewart & Golubitsky (1992). Crystallographic symmetry is described by Giacovazzo et al. (1992), and the standard reference is Hahn (2002).

After a general explanation of principles, we describe the symmetry groups with one, two and three translations.

Structural Biology Using Electrons and X-Rays.

© 2011 Michael F. Moody. Published by Elsevier Ltd. All rights reserved.

10.1.1 Introduction to Symmetry

Suppose that identical molecules interact to form an aggregate, and each molecule interacts with its neighbours in a standard (identical) way. Then the final aggregate will have each molecule (monomer or *subunit*) surrounded with identical neighbours arranged identically; its environment will be the same as that of every other molecule. This special relationship (called **equivalence**) is generated by the identical interaction of identical subunits. What are its implications?

Imagine that we have a model of such an aggregate, and support it with a jig that is fastened rigidly to a bench. The jig is shaped to take one subunit in one precise position and orientation, so there is only one way to fasten any of the subunits into the jig. After we have fastened one of them ('a'), we photograph the aggregate from all directions. Then we unfasten 'a' and move the aggregate so that 'a' is replaced by another subunit ('b'). We re-photograph the aggregate from exactly the same directions as before, and compare the photographs with the previous set. What will we find? Since the subunits were *equivalent*, both sets of photographs must look exactly the same, whether with subunit 'a' or subunit 'b' fastened. For, if there were any difference, the aggregate-environment of 'a' would have to differ from that of 'b', which is contrary to hypothesis. If all the subunits are strictly equivalent, the aggregate must look exactly the same when attached to the jig by any of its subunits. If there are n subunits, there will be n different physical positions of the aggregate that are *indistinguishable* in appearance. This indistinguishability is what we mean by **symmetry**.

The argument can also be run the other way, showing that symmetry implies equivalence between the subunits. Thus equivalence and symmetry imply each other, and the difference between them is only one of viewpoint. Equivalence is the 'local' viewpoint of a subunit, symmetry is the 'global' viewpoint of an outside observer. This antithesis between 'local' and 'global' underlies much of biology, where bottom-up processes originating at the 'local' molecular level have implications at the large-scale 'global' level which often affect an organism's fitness and thereby the selection of genes causing 'local' features. Symmetry theory is a useful tool for connecting these viewpoints.

So far, all definitions depend on the meaning of 'indistinguishable'. All corresponding interatomic distances are supposed to be equal, so that physical properties (which depend on these distances) remain unchanged. In mathematics equality is an absolute concept, but in reality it is subject to error-limits. Given the constant thermal movements within protein molecules, no two 'identical' proteins have *exactly* the same structure, even for a moment. But many properties depend on a kind of average structure that is probably close enough.

For simplicity, we start by assuming exact identity, deducing the consequences. We shall explore this mathematical theory of symmetry in the rest of this section, classifying the possible symmetry types as a framework for thinking about symmetrical-looking structures, for classifying them or for deciding when and how symmetry is being broken.

10.1.2 Symmetry Operations

What *kinds* of geometric operation relate the subunits of symmetrical objects? Return to the subunit-binding 'jig', and suppose that the aggregate's subunits get automatically substituted into the 'jig' positions, one by one sequentially. If we stand nearby, what do we see? Since each replacement leaves the aggregate looking exactly the same, we see no change at all despite the impression and sound of movement. So, out of curiosity, we change the machine so that no subunit can be fitted twice into the jig. We start it and wait to see what happens. There are two possibilities: either it stops after a while, having exhausted all the subunits; or else it does not, but continues its replacements for ever. In the first case, the number of subunits is finite; in the second case, it is infinite.

Start with the first case and ask how the aggregate moves between successive replacements. It is a type of movement that, after repeating a finite number of times, stops just before returning the aggregate to *exactly* its original position. Such a movement can only be a *rotation*, so the symmetrical aggregate has been undergoing rotations. Accordingly, its shape must be circular or polygonal in two or three dimensions.

But, in the second case, what shapes can hold an *infinite* number of subunits? Obviously they must be infinitely large in at least one dimension, so there must be at least one type of movement that progressively covers greater and greater distances. That is a movement that keeps moving in the same direction; the simplest form being a **translation**. Strict translational symmetry, which requires an infinitely big aggregate, is impossible, but there are many objects with that symmetry over a limited length. Although their boundary subunits are necessarily special, their interior subunits can be more or less equivalent.

When we come to classifying symmetries, the number of independent translations will be the most important criterion. Clearly it is connected with the dimensionality of the symmetrical object, since (e.g.) a two-dimensional (2D) object cannot have more than two independent translations. Each symmetry operation (like a rotation or translation) applies to the entire structure (within its boundaries), but small sub-regions may have a different, perhaps higher, symmetry than the whole. This is called *local symmetry*, and often occurs when crystals are formed of highly symmetrical units, e.g. virus particles.

Rotations and *translations* are the most important and common examples of symmetry operation, and are also the only ones that can exist in an aggregate containing *identical* subunits that are completely unsymmetrical (like entirely left shoes). The important difference between left- and right-handed structures is called *chirality* (or *handedness* or *enantiomorphism*) and the alternatives are labeled D- and L- (among other notations). Chirality is the aspect of an object's geometry that is altered in a mirror-image, so two molecules of opposite chirality cannot be superposed, even though they have identical distances between corresponding atoms. Biological macromolecules contain monomers, like nucleotides or amino-acid residues, which are chiral. In ordinary nucleic acids or proteins,

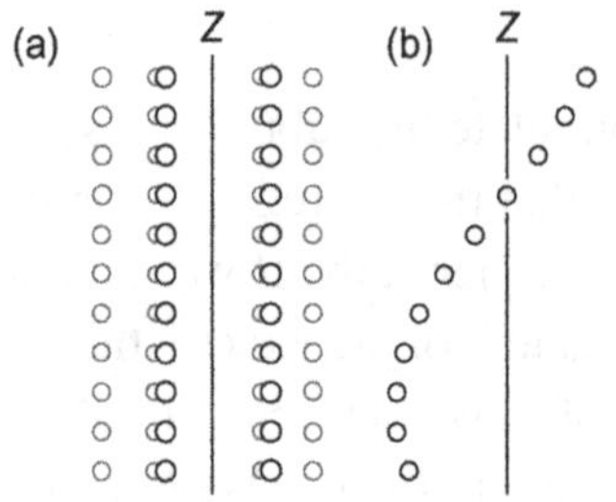

Figure 10.1: (a) Part of a structure with a vertical translation and also a rotation (60°) as symmetry operations. (b) Part of a structure with a vertical screw (see also Figs 10.11 and 10.36) consisting of a rise equal to the translation of (a) and a twist of 17°.

the chirality is the same, being based, respectively, on D-ribose or L-amino-acids. A symmetry element (e.g. a mirror) that reverses chirality would require the existence of equal numbers of D- and L-ribose molecules, so it cannot apply to ordinary nucleic acids. This restriction does not apply to minerals, and indeed the existence of widespread chirality of the same type is a possible sign of life.

Nevertheless, chirality-reversing symmetry elements have limited roles in biology. First, electron microscope images often lack sufficient resolution to show any chirality in the specimen, and the *apparent* symmetries of such images can contain chirality-reversing operations. Second, electron microscope images are projections, and the projections of chiral molecules can be non-chiral. (We shall soon examine some of the implications of projections.) Therefore symmetries that contain opposite chiralities (at least in 1D or 2D) are worth discussing.

Chirality leads to one of the major classification divisions among symmetry operations. **Direct** operations preserve chirality and include rotations or translations: during a day's walk, a left shoe remains a left shoe during all the forward movements (translations) and turning movements (rotations). However, **opposite** operations reverse chirality and include the *mirror*: a left shoe becomes a right shoe in a mirror, which is midway between the opposite images. Chirality has unique aspects, including its effective permanence (for ordinary matter), and also the absence of any intermediate or approximate stage (as a rotation of 179.9° approximates the rotation of 180° in a 2-fold rotation). Since a chiral object exists in only two forms, chiral reversion by opposite operations acts as a 'toggle': one operation reverses chirality so two must restore it.

Some symmetry operations form pairs as 'inseparable hybrids', where neither symmetry operation exists separately. For example, the *screw* combines a rotation ('twist') and a translation ('rise'); but the resulting helix generally has neither rotational nor pure translational symmetry; only 'screw symmetry' (Fig. 10.1 (b), quite unlike (a)). For most twists, the operation is chiral and generates a left- or right-handed helix (though each of these

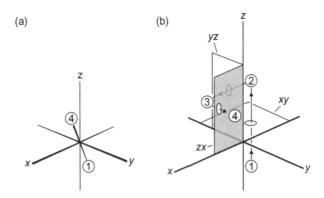

Figure 10.2: (a) Center of symmetry relating two atoms labeled 1 and 4. The straight line joining them is bisected by the origin. **(b)** It can be considered as the effect of three successive reflections, in three mutually perpendicular planes. Marker 1, reflected in the *xy*-plane, gives marker 2 which, reflected in the *yz*-plane, gives marker 3 which, reflected in the *zx*-plane, gives marker 4. Marker 4 is diametrically opposite to marker 1, as in (a).

can incorporate units of any chirality). Although a chiral screw can be applied to a non-chiral object, a chiral helix usually results from the aggregation of chiral units like L-amino-acids or D-nucleotides; for otherwise the left- and right-helices would be equally stable. (Screws are discussed further in Chapter 15.) Another 'inseparable hybrid' is the **glide** (Fig. 10.7 (c)), a combination of a translation with a reflection. (It is a kind of 2D version of a screw.) Yet another is the **inversion center** or **center of symmetry**. This point, usually chosen as the coordinate-origin for a structure, has the property that any atom, joined to it by a straight line that extends an equal distance the other side, ends at a corresponding atom. Figure 10.2 shows that this is generated by three successive reflections, in the *xy*-, *yz*- and *zx*-planes. As each reflection acts as a chirality 'toggle', an *even* number leaves chirality unchanged while an *odd* number reverses it. Hence an inversion center *reverses* chirality in three-dimensional (3D) (with *three* reflections), but preserves it in 2D. The 2D center of symmetry is the projection of a 3D center of symmetry, and is a 2-fold axis.

As mentioned, chirality-preserving operations in 3D can project to give chirality-reversing operations in 2D. Thus a 2-fold axis in 3D gives a mirror for its 2D projection onto a plane parallel to the axis. Also, that intimate hybrid, the 2-fold screw axis (Fig. 10.3 (b)), almost contains a 2-fold axis, and its comparable projection almost contains a mirror; it is actually a glide line (Fig. 10.3 (b)). (For a flat object, the glide and screw axis are indistinguishable.)

10.1.3 Combinations of Symmetry Operations: Groups

When symmetry operations are applied in succession, the aggregate has *both* symmetries. The subject of symmetry really starts to develop with these 'separable combinations', so we examine two simple examples, starting with one without a translation.

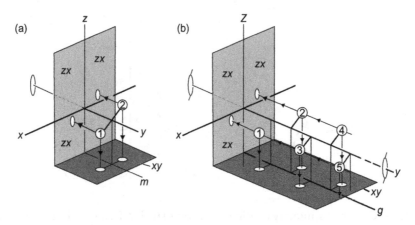

Figure 10.3: (a) Projections of two points (1 and 2) connected by 2-fold rotation. The rotation axis lies along y, and their projections are shown on the xy- and zx-planes. On the zx-plane (perpendicular to the axis) we get a 2-fold rotation axis again. However, on the xy-plane (parallel to the axis), we get two points related by a mirror line (m). **(b)** Points connected by 2-fold screw, giving a glide line (g) projection on the xy plane.

Lacking a translation, we are restricted to rotations and mirrors; and only rotations are allowed if our molecules are chiral. This important branch of symmetry (§10.4.1–3; §16.1.1), is introduced here in 2D. The simplest combination involves the *same* symmetry operation with itself. Figure 10.4 shows the case where that symmetry operation is a rotation by 60°, applied once in (b), twice in (c), and so on up to five times in (f); six applications bring the object back to its starting position. After that, all possibilities have been exhausted, since a combination of any two of the listed operations merely gives a third in the same list. Note that, for completeness, a complete list (as in Fig. 10.4) has to include a 'rotation' of 0° which actually does nothing at all (the 'identity operation').

This list of six symmetry operations is an example of a *group* (here C_6). The *operations* are also called the *elements* of the group, and the *number* of elements (6) is called the group's *order*. The essential feature of a group is its *completeness*: no new elements can be achieved from further combinations[1]. Although rotation groups (like this) are finite, translation groups are infinite, since the translation continues indefinitely. Nevertheless, their operations can easily be listed (they correspond to the sequence: ..., -2, -1, 0, 1, 2, ...,). But we postpone translations until the next section, as there remain important points to be made with simpler groups of smaller order.

---

[1] There is a simple algebraic representation of the group C_6 in Fig. 10.4. We start with the identity operation I (a rotation of 0°), and add the element R (a rotation of 60° clockwise). Then R^2 is a clockwise rotation of 120°, R^3 is a clockwise rotation of 180°, R^4 is a clockwise rotation of 240°, and R^5 is a clockwise rotation of 300°. However, R^6 is a rotation of 360° = 0°, so that $R^6 = I$. Therefore $(R^5)R = I$, so we say that $R^5 = R^{-1}$, the 'inverse operation of R'.

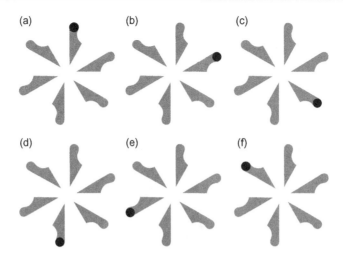

Figure 10.4: The six symmetry operations acting on an object with 6-fold rotational symmetry, illustrated by their effect on (a). (The rotation is revealed by a small black 'marker'.) **(a)** The starting position, i.e. a rotation of 0°. **(b)** The basic rotation, where the object is rotated 60° clockwise. **(c)** Two applications of (b) rotate the object 120° clockwise. **(d)** Three applications rotate 180°. **(e)** Four applications rotate the object 240° or 120° anti-clockwise. **(f)** Five applications rotate 300° clockwise or 60° anti-clockwise.

The 6-fold rotation group C_6, illustrated in Fig. 10.4, is an example of a *cyclic* group C_N. It can be extended if we combine it with a vertical 2-fold axis, in the plane of the paper and intersecting the center, as in Fig. 10.5 (d). That axis flips over the six dark-grey (upper) shapes, so that they become the six light-grey lower shapes. After that, we find there are five new 2-fold axes, making the six shown in the figure. We have found D_6, one example of the simplest set of 3D rotation groups, the *dihedral group* D_N (Fig. 10.5). Other examples in the figure are D_2, D_3, and D_4 (but any positive integer will provide a valid example). Note that, between the 6 2-fold axes in (d) there are another 6 2-fold axes that are slightly different: it is every *alternate* 2-fold that is equivalent. Note also that, if the dark- and light-grey shapes had the same shading, the 2-fold axes would become mirrors of a plane group, the projection of D_N.

Groups can be classified using the concept of a *subgroup*: a subset of the group elements, chosen because it forms a group all by itself. For example, in the 6-fold symmetry group C_6 there are subgroups with 2- and 3-fold symmetry. That is, the group C_6 has subgroups C_2 and C_3, and all have the subgroup C_1 (included for completeness rather than interest). Also, the group C_6 is a subgroup of the dihedral group D_6. Subgroups obviously have a lower order (number of elements), but it is interesting that the order of a subgroup is a *factor* of the order of the group (it divides it exactly; Lagrange's theorem). This applies to C_6, C_2 and C_3, since 2 and 3 each divide 6. So the rotation group of the 17-fold TMV disc can have no subgroup. The opposite of a subgroup is a *supergroup*. But, while the number of subgroups is limited, some groups can have an infinity of supergroups.

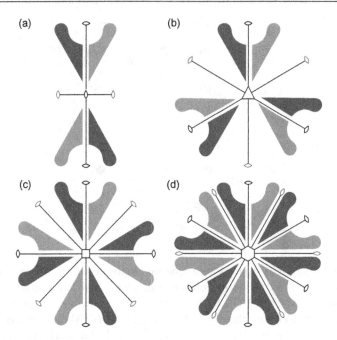

Figure 10.5: View of dihedral groups down the N-fold axis. The different shadings refer to different levels, related by 2-fold axes (ellipses). **(a)** D_2; **(b)** D_3; **(c)** D_4; **(d)** D_6.

10.2 One-Translation Groups

10.2.1 One-Dimensional Groups

Next we look at groups containing translations but, for simplicity, we restrict them to 1D translations at first.

In the simplest case, the structure is itself one-dimensional, i.e. just density variations along a line, like the collagen band patterns that repeat at equal distances along the fiber, the distance between repeats being the period a. In this periodic pattern, one period or *repeat* could also be called a *unit cell*, defined as an exact subdivision of the whole structure. Shifting the fiber by a leaves it looking the same, so that shift is a translational symmetry operation of the group. Two such operations, applied successively, move the fiber by $2a$; and, as it still looks the same (ignoring the ends), $2a$ is also a symmetry operation; and so on. These translations form the simplest symmetry group applicable to a one-dimensional structure that is quite uniform in every perpendicular direction. We illustrate it in Fig. 10.6 (a), (b), and show its abstract representation in (b), which is also called a one-dimensional *lattice* of points. This simplest group is given the symbol p1 (in all sets or kinds of translation group, the very simplest is labeled p1). Here p1 is *polar*, i.e. it has two *structurally* different directions.

The only other symmetry that can be combined with p1, while preserving its strictly 1D character, is a perpendicular mirror (or a 2-fold axis, which is the same in 1D). This gives the

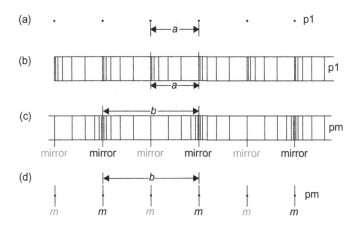

Figure 10.6: Here $b = 2a$ since there are two asymmetric units back-to-back instead of facing the same way. In **(c)**, **(d)**, the mirror could be replaced by a 2-fold axis.

group pm, shown in Fig. 10.6 (c), (d), which applies to fibrils where half the molecules face in each of the two directions. Thus pm is *non-polar*, and both its directions are structurally equivalent. The mirror lines are at a distance (period) b. Any adjacent pair of these mirror lines automatically generates a new mirror line mid-way between them. Thus there is an alternation of the two mirror lines, alternatively shown in black and grey in Fig. 10.6 (c), (d). The smallest part which, with suitable symmetrical repetitions, constructs the whole, is called an *asymmetric unit*. In this case, it is half a period.

10.2.2 Frieze Groups

We can also have a 1D translation group that applies to a 2D or 3D structure. The latter case allows the use of a screw, leading to helical structures that need a separate chapter (Chapter 15). The 2D case is strictly not applicable to molecules, but it is relevant to very thin layers such as fibrous aggregates. It also applies to projections of helices (§15.1.1). These groups are called *frieze groups* (or *border groups*), so-called as they apply to linear decorations.

We shall see that there are just seven frieze groups and they have 'official' labels (see the tabulation by Hahn, 2002), but a more convenient notation was suggested by Coxeter (1989): a line of capital letters having the same symmetry. The three supergroups of p1 (polar) are shown in Fig. 10.7. They must be polar and therefore lack any 2-fold axis perpendicular to their long axis. The only possibility is some axis parallel to the long axis, and the only possible axes are 2-fold and 2-fold screw axes.

The four supergroups of pm (non-polar) are shown in Fig. 10.8. They need a perpendicular 2-fold axis, but it can be oriented either parallel or perpendicular to the paper. If it is *perpendicular*, there could also be a 2-fold axis along the translation. If none, we get

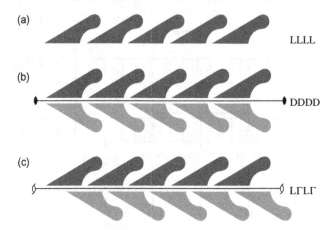

Figure 10.7: The three *polar* groups among the seven frieze- (or border-) groups, appropriate for an object in either 2D or 3D (as here, dark grey being upper and light grey lower; if light grey became dark grey, we would have the 2D projection). **(a)** Translation only: ...LLLL... (*p*1). (This is also present in all other groups, so we only mention any extra symmetry.) **(b)** Reflection (here shown as a half-turn in plane of frieze) along translation: DDDD... (*p*11*m*). **(c)** Glide reflection (here shown as a 2-fold screw axis): LΓLΓ... (*p*11*g*).

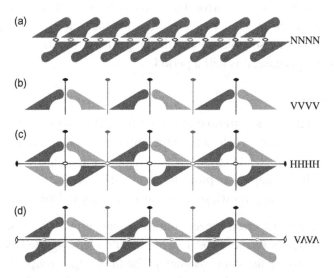

Figure 10.8: The four *non-polar* groups among the seven frieze-groups. Mirrors are here shown as 2-fold rotation axes, and glide reflections as 2-fold screw axes (in plane of the pattern). (This is what their projections produce.) **(a)** Half-turn: ...NNNN... (*p*211). **(b)** Reflection perpendicular to the translation: ...VVVV... (*p*1*m*1). **(c)** Reflection (along the translation) added to (a): ...HHHH... (*p*2*mm*). **(d)** Glide reflection added to (a): ...VΛVΛ... (*p*2*mg*). In (a), (c) and (d), note the half-turns perpendicular to the paper. As in Fig. 10.7, mirrors are here shown as 2-fold rotation axes, and glide reflections as 2-fold screw axes (in plane of the pattern).

...NNNN...; if so, it generates a third 2-fold axis perpendicular to the first two, giving ...HHHH... (Two perpendicular 2-folds automatically generate a third, mutually perpendicular, 2-fold.) However, if the 'non-polar' 2-fold is *parallel* to the paper, it must be perpendicular to the translation. If we also had another 2-fold axis parallel to the translation, we would repeat ...HHHH... So we can either have no other axis, giving ...VVVV...; or else we can also have a 2-fold screw along the translation, giving ...VΛVΛ...

In Fig. 10.7 and Fig. 10.8, *m* is the projection of a 2-fold axis, and *g* is the projection of a 2-fold screw axis. In these figures, the 'near' units are shown dark, and the 'distant' ones light. If this difference is ignored, we have the ordinary 2D frieze groups. These are possible symmetries of the *projections* of 3D objects with one translation, so they apply to the electron microscope images of such structures, especially helices, which we discuss in more detail in Chapter 15. Thus the frieze groups can be interpreted in two ways: using mirrors and glides, when they apply to projections; or using 2-fold axes or screw axes, when they apply to arrangements of molecules. By analogy with the plane groups (§10.3), we could call the projection version '1-sided' and the version applicable to 3D structures '2-sided'.

The symmetry relations between the seven frieze groups are made clear by a 'lattice diagram' that shows which are subgroups of other groups, as shown in Fig. 10.9. The group of highest symmetry is shown at the top, and its subgroups below. Polar groups (with only p1 as a subgroup) are in dark grey; non-polar groups (with pm as a subgroup) are in black. Another 'ancestor' group is the 2-fold screw axis, denoted by its usual symbol 2_1. The arrow from this symbol 'carries' that screw symmetry directly to ...LΓLΓ... and indirectly, via successive arrows, to ...DDDD..., ...VΛVΛ..., and ...HHHH...

The two groups ...LLLL... and ...NNNN..., which use only chirality-preserving operations in projection, are the frieze-group equivalents of the 1D groups p1 and pm. In each case, we can join the frieze head-to-tail to form a circle (though this means losing any symmetry axes relating objects on the outside of the circle to those on the inside). The circles have the

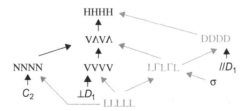

Figure 10.9: A 'lattice diagram' of the seven frieze groups arranged with the groups of higher, above those of lower, symmetry. There are four 'root' symmetry groups: the translation ...LLLL... ; the in-plane 2-folds ($\perp D_1$ & $\!/\!/D_1$) and 2-fold screw; and the perpendicular 2-fold (C_2). Polar groups are shown in grey; non-polar groups (descendants of 2 or C_2) are in black.

Figure 10.10: Packing of pentagons in a double line with the frieze group $p211$ or NNNN. The two kinds of 2-fold axis are marked between the pentagons.

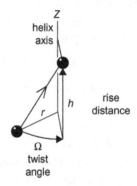

Figure 10.11: A screw operation with the conventional symbols marked.

symmetry C_N or D_N, and can form rings or tubes. For example, the group …NNNN… is a primary way in which pentagons tend to pack together (Fig. 10.10), forming double rows that are units in the tubes of polyoma virus capsids (Kiselev & Klug, 1969; Baker et al., 1983) and in the 'ribbons' in vesicles of tubular crystals of acetylcholine receptor (Brisson & Unwin, 1984). Here the pentagons are good examples of non-crystallographic symmetry.

10.2.3 Helices

We have dealt with symmetries with one translation, first in 1D and then in 2D (the frieze groups). Finally, we must include symmetries with one translation in 3D: the helices. The fundamental operation, the screw axis, is an 'inseparable hybrid' (Fig. 10.11), combining the translation (h) with a twist angle (Ω). (The repeated translation h generates a unique line, the helix axis, defined as the z-axis.) This is the one essential symmetry operation of a helix, and any others must be compatible with the screw. Only rotation axes are permitted, and these must not disturb the helix axis. That is no problem if it coincides with the rotation axis. Otherwise, the rotation axis must be a 2-fold axis perpendicular to the helix axis. Thus the

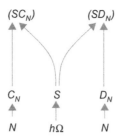

Figure 10.12: A lattice diagram of the group-structure of helices. Although much simpler than the diagrams for regular translation groups, it comprises a great variety of helical structures.

possible point-group is either C_N or D_N, with no restriction on N, but the N-fold axis must lie along the translation. The symmetries of a helix, diagrammed in Fig. 10.12, may appear very limited; but the screw (S) makes the subject remarkably complex and rich (Chapter 15).

10.3 Two-Translation Groups

10.3.1 General Principles

Thin crystalline sheets, with just two translations and almost no symmetry in the third direction, potentially extend indefinitely in two directions, filling a plane. The different possible symmetries (*plane groups*) differ in extra symmetries added to the translations. However, the extra symmetries are subject to numerous restrictions; the first involves *chirality*. Proteins and nucleic acids are chiral macromolecules and, to preserve chirality, the only permissible symmetries are translations, rotations or screws. This restricts the number of thin sheet symmetries to 17.

However, we also need to know the possible symmetries of sheet *projections*, where the restriction to chirality-preserving operations does not apply. Here, 2-fold rotations become mirrors and 2-fold screws become glide-lines. This simple equivalence means that there are also 17 plane groups in projection, each paired with a 3D thin-sheet plane group that has only chirality-preserving operations.

The next restriction is the need to preserve a thin sheet (i.e. to avoid generating a 3D lattice). A screw-axis has a translation which is only allowed *within* the plane. Rotations must also avoid generating any new translations, so the only permitted *in-plane* rotations are 2-fold rotations. These, like the in-plane screws, cause a sheet to have two *identical* sides ('2-sided'), unlike sheets with *distinct* 'upper' and 'lower' surfaces ('1-sided').

Next we consider the rotation axes *perpendicular* to the plane. Although not restricted like in-plane axes, they must be consistent with the translations, i.e. with the lattice. This restricts perpendicular rotations to certain angles. Figure 10.13 (a)–(c) shows three permitted angles, respectively 180° (2-fold), 120° (3-fold) and 90° (4-fold). Each rotates one translation to

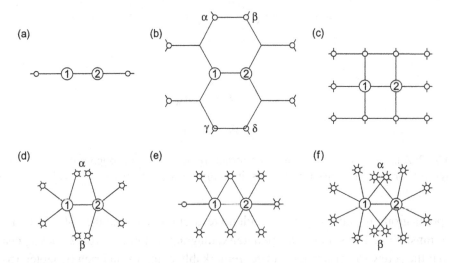

Figure 10.13: The effect of different orders of perpendicular rotation axis on translations within the plane. **(a)**, **(b)**, **(c)** and **(e)** are permissible, but all others generate a smaller translation that is forbidden. (These arguments originated with Barlow, 1901.)

give another permitted translation within the plane. But (d) shows how the next order, 5-fold, creates a problem by generating a much smaller new translation at α and β. If we redefined the lattice to allow this, a similar diagram would generate yet smaller translations, and so on *ad infinitum*; so we have to exclude 5-fold rotations altogether. However, (e) shows that 6-fold rotations (60°) *are* permissible. But (f) shows that 7-fold rotations create a problem similar to that in (d), and higher rotation orders do the same. Thus only 2-, 3-, 4- and 6-fold **crystallographic rotation axes** are permitted *perpendicular* to the plane, though we can also have *in-plane* 2-fold rotation- or screw-axes, creating '2-sided' sheets.

It turns out that there are five '1-sided' plane groups (with point-group symmetries C_N) and 12 '2-sided' plane groups (with point-group symmetries D_N). We now examine the possible combinations of these extra permitted symmetries with the basic translational symmetry, starting with the '1-sided' groups.

10.3.2 'One-Sided' or 'Projection' Plane Groups

To be '1-sided' these must lack any in-plane rotations or screws, so only the 'crystallographic' perpendicular axes are allowed. These symmetries are the same, whether applied to the full (3D) sheet or to its 2D projection; also, these point groups have cyclic symmetry (C_N).

The simplest of the five plane groups is $p1$, with only the common translational symmetry and no other rotational symmetry whatever. Its notation includes the 'p' that stands for 'primitive', i.e. the simple kind of lattice discussed in Chapter 6 as opposed to 'centered'

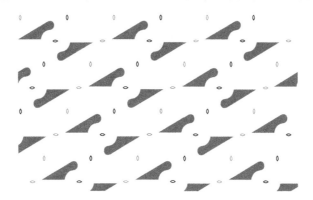

Figure 10.14: One-sided plane group *p*2 with 2-fold rotational symmetry.

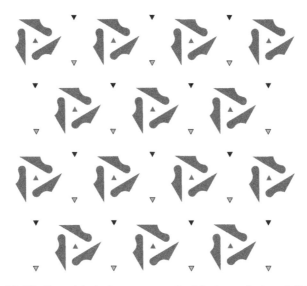

Figure 10.15: One-sided plane group *p*3 with three distinct 3-fold axes.

lattices (see below). The only parameters needed to define *p*1 are the unit cell lengths *a*, *b* and the included angle γ. The remaining four 'one-sided' plane groups have 2-, 3-, 4- or 6-fold axes perpendicular to the sheet. So the second one is *p*2 (Fig. 10.14) with point group C_2, which has four 2-folds with different environments (vertical/horizontal and shaded/unshaded in Fig. 10.14). This number declines with increasing symmetry of the most symmetrical axis: three for 3-folds in *p*3 (Fig. 10.15), two for 4-folds in *p*4 (Fig. 10.16) and just one for 6-folds in *p*6 (Fig. 10.17).

These five are the only plane groups that preserve chirality, so they are the most likely symmetries for a cell membrane or cell wall with a distinct inside and outside, if it should have a regular repeating structure.

Figure 10.16: One-sided plane group *p*4 with two distinct 4-fold axes.

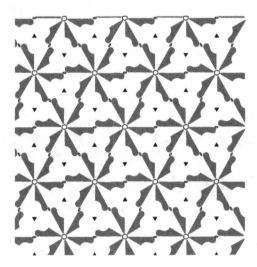

Figure 10.17: One-sided plane group *p*6 with just one type of 6-fold axis.

10.3.3 'Two-Sided' Plane Groups

To be 'two-sided', the remaining 12 groups must allow 2-fold rotation axes and/or 2-fold screw axes in the plane of the sheet. Their projections use 'opposite' (chirality-pairing) symmetry operations: mirrors or glide lines. Consequently, the 'two-sided' groups are distinguishable from their projections. Many centuries ago, those were explored for designing wall patterns, but their symmetries were only classified in the 19th century. By contrast, the 'two-sided' sheets were explored by Holser (1958) shortly before they were needed by electron microscopy. The diagrams show the 'two-sided' groups, for which Holser's notation is used (see also Hahn, 2002). The corresponding projection groups apply if the two shades of

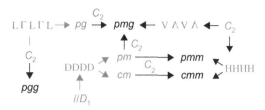

Figure 10.18: Connections between the 'frieze groups' (section §10.2.2) and the plane groups with only 1- or 2-fold perpendicular rotations. The perpendicular 2-fold (C_2) is shown contributing at five places, where the arrows become black. (Besides these, there is just the pair NNNN → $p2$.)

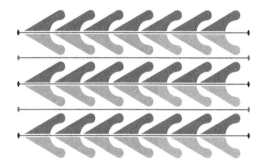

Figure 10.19: 'Two-sided' chiral plane group p12 giving the projection group *pm*.

grey are the same; these are also listed with their conventional symbols. There is a one-to-one correspondence between the 'two-sided' groups, which preserve chirality, and their projection groups, which contain chirality-reversing operations like mirrors. While the former apply to actual specimens, it is more convenient to refer to the latter when deducing the 17 plane groups (below). We shall consider these 12 groups as 'two-sided descendants' of the various 'one-sided' groups, so the perpendicular rotation axis is the basis of our classification. Each 'descendant', with the same rotation axis, is 'two-sided' by virtue of its in-plane 2-fold screw or rotation axes.

Start with the (pure) 'descendants' of p1: *pg*, *pm*, *cm*. Lacking any perpendicular rotation axis, they cannot have more than *one* set of in-plane 2-fold screw or rotation axes. These groups consist of successive lines of 'frieze groups' (section §10.2.2), and Fig. 10.9 shows that only …DDDD… and …LΓLΓL … have in-plane 2-fold screws or 2-folds without a perpendicular 2-fold axis (Fig. 10.18). These frieze-groups have to run parallel to each other, or else they will generate a perpendicular 2-fold axis. There are two ways of arranging the parallel …DDDD… lines: in exact register (*pm*: Fig. 10.19), or displaced half a repeat (*cm*: Fig. 10.20). This centered lattice has lattice points on the corners of a rectangle, while its center has an extra lattice point. (See the same lattice in *cmm*, Fig. 10.25.) The simpler 'primitive' lattice would put the corner and central points in the same oblique line, but then

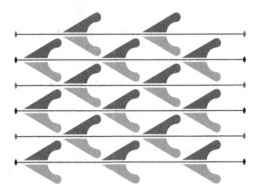

Figure 10.20: 'Two-sided' chiral plane group *c*12 giving the projection group *cm*.

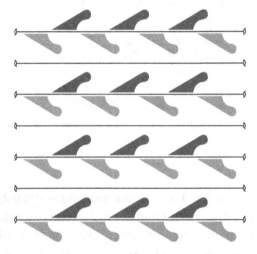

Figure 10.21: 'Two-sided' chiral plane group *p*12₁ giving the projection group *pg*.

the translation would lose its connection with the mirror; this is the reason for 'centered' lattices. As for the parallel….LΓLΓL … lines, there turns out to be only one way of arranging them, *pg*, shown in Fig. 10.21.

In the 'descendants' of *p*2: *pgg*, *pmg*, *pmm*, *cmm*, the perpendicular 2-fold axis (C_2) generates anti-parallel arrangements of the lines …LΓLΓL… (giving *pgg*) and …DDDDDD… (giving *pmg*), or else from the frieze group …HHHH… that already has a perpendicular 2-fold axis. *pgg* (Fig. 10.22) with an added mirror along the glide line comprises the group *pmg* (Fig. 10.23). There are two possible arrangements of the frieze group …HHHH… with a primitive (*pmm*: Fig. 10.24) or centered (*cmm*: Fig. 10.25) lattice. These are also shown in Fig. 10.18.

There are just two 'descendants' of *p*3 and, like their parent, they have three sets of 3-fold axes. The in-plane 2-fold axes must point in three directions (as required by the perpendicular 3-fold axis), and (being 2-folds) they must also bisect the line joining two 3-fold axes of the

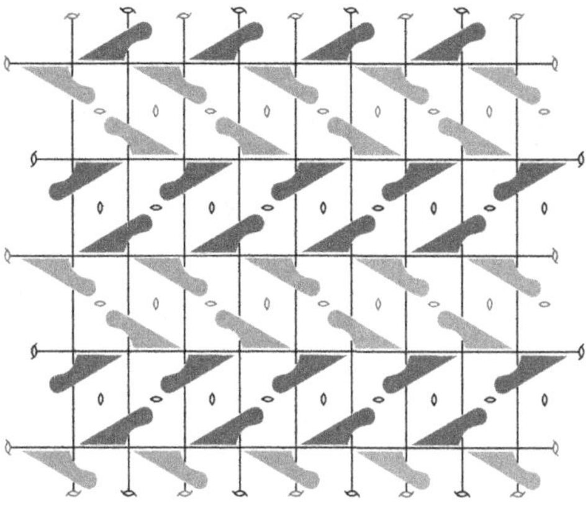

Figure 10.22: 'Two-sided' chiral plane group $p22_12_1$ giving the projection group *pgg*.

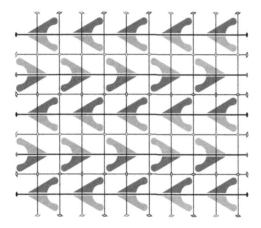

Figure 10.23: 'Two-sided' chiral plane group $p222_1$ giving the projection group *pmg*.

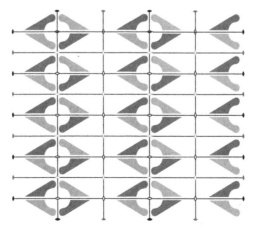

Figure 10.24: 'Two-sided' chiral plane group $p222$ giving the projection group *pmm*.

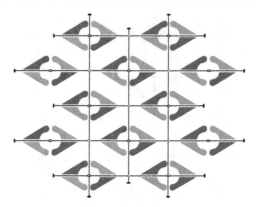

Figure 10.25: 'Two-sided' chiral plane group *c*222 giving the projection group *cmm*.

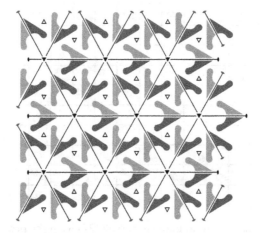

Figure 10.26: 'Two-sided' chiral plane group *p*321, giving the projection group *p*31*m*. Only the *nearest* 3-folds of the first set are joined by 2-folds.

same set. Each set forms a hexagonal lattice, in which the closest pair of axes has the lattice distance *a*, while the next closest pair is $a\sqrt{3}$ apart. If we bisect the pair $a\sqrt{3}$ apart, the 2-folds pass through *only one* of the three sets of 3-fold axes (Fig. 10.26). If we bisect the pair *a* apart, they have to pass through *all three* sets (Fig. 10.27).

There are also two 'descendants' of *p*4, *p*422 (Fig. 10.28) and *p*42$_1$2 (Fig. 10.29). In only one of these (*p*422: Fig. 10.28) do the 2-folds pass through 4-folds.

Finally, there is only one 'descendant' of *p*6: *p*622 (Fig. 10.30). It has the simple structure of 2-folds intersecting the 6-fold axis, forming clusters of subunits with D_6 symmetry. The main subgroup relationships between the projection groups are shown in Fig. 10.31 (p. 213).

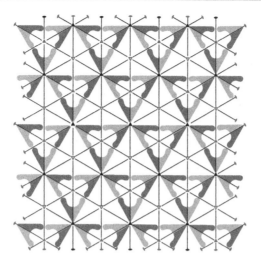

Figure 10.27: 'Two-sided' chiral plane group *p*312, giving the projection group *p*3*m*1. Note that the nearest 3-folds of the first set are *not* joined by 2-folds.

Figure 10.28: 'Two-sided' plane group *p*422, giving the projection group *p*4*m*.

10.4 Three-Translation Groups

Since these groups apply to crystals, they have been thoroughly studied, tabulated and applied by crystallographers. They need three translations to fully exploit 3D space, but translations are by no means their only symmetry operations. Just as we call one-translation groups *line-groups* and two-translation groups *plane-groups*, so we call three-translation groups *space-groups*. They define the geometrical repeats any object must undergo if it is to form part of the crystal structure. As with frieze- and plane-groups, the full group (here, the space-group)

Figure 10.29: 'Two-sided' plane group $p42_12$, giving the projection group p4g.

Figure 10.30: 'Two-sided' plane group $p622$, giving the projection group $p6$m.

contains subgroups for translation, rotation, etc., and their various combinations generate many space-groups. Thus the plane-groups are enlarged by the addition of screw axes and also of two new 3D rotation-groups, deduced below.

10.4.1 Rotations in Three-Dimension

In our survey of 2D plane-groups, we also restricted the rotation groups to those that apply in the plane, or at least in projection. Thus we included not only the cyclic groups C_N, but also the dihedral groups D_N. However, it omitted some other rotation groups that cannot be fitted into a plane. We now deduce these, plus a related 3D rotation group.

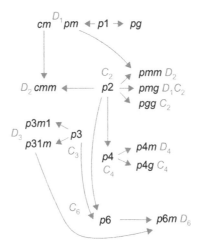

Figure 10.31: Lattice diagram of relationships between the two-dimensional projection groups.

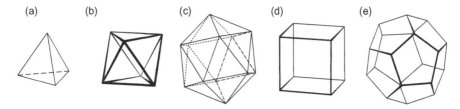

Figure 10.32: The five Platonic solids (regular polyhedra): **(a)** the tetrahedron {3,3}; **(b)** the octahedron {3,4}; **(c)** the icosahedron {3,5}; **(d)** the cube {4,3}; and **(e)** the pentagonal dodecahedron {5,3}.

10.4.2 Platonic Solids

Suppose we have a symmetrical bundle of rotation axes radiating at various angles from a central fixed point, like a kaleidoscope that multiplies any image we place near it. If we place a test-point outside this central point, the various rotation axes will multiply that test-point, but without changing its distance from the central point. So the copies of the test-point will form a spherical ball of points. If each point is joined to its nearest neighbors, the 'ball' will become a regular polyhedron. ('Regular' means that each point, which is now a vertex, is equivalent; and the same for each face.) There is a classical mathematical proof that each of the possible point-groups corresponds to a regular polyhedron. Thus the five regular or Platonic polyhedra (Fig. 10.32) set a limit to the possible point-groups. Indeed, those groups are even more restricted, since it turns out that each point-group corresponds to a regular polyhedron whose faces are *equilateral triangles*. (These are called *deltahedra* by analogy with the triangular capital Greek 'delta' Δ.) This means that the last two polyhedra

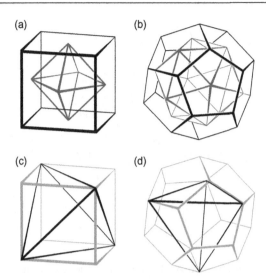

Figure 10.33: (a) Identical groups (*O*) for a cube and octahedron; **(b)** Identical groups (*I*) for a dodecahedron and icosahedron; **(c)** A tetrahedron fits a cube, since *T* is a sub-group of *O*; **(d)** A tetrahedron also fits a dodecahedron, since *T* is a sub-group of *I*.

of Fig. 10.32, whose faces are not triangular, have the same symmetry as some regular deltahedra (Fig. 10.33 (a), (b)).

Thus there are only three extra point-groups, corresponding to the first three polyhedra of Fig. 10.32. Those are (a) the tetrahedron, (b) the octahedron and (c) the icosahedron. (The names refer to the number of faces, in Greek: tetra = 4, octa = 8 and icosa = 20.) So the extra point-groups are tetrahedral (*T* or 23), octahedral (*O* or 432) and icosahedral (*I* or 532), where the letters are Schoenflies symbols, and the numbers are the corresponding Hermann–Mauguin symbols. If each equilateral triangle is decorated with three commas placed symmetrically in the corners, the total number of commas in the polyhedron equals the order of the corresponding point-group. Thus the tetrahedral group's order is $3 \times 4 = 12$, since the tetrahedron has four faces. Similarly, the octahedral group's order is $3 \times 8 = 24$ and the icosahedral group's order is $3 \times 20 = 60$.

The rotation axes of the point-group can be deduced by examination of the corresponding polyhedron. Thus the tetrahedron has four 3-fold axes joining each vertex to the opposite face, and also three 2-fold axes joining the mid-points of opposite edges. Similarly, the octahedron has three 4-fold, four 3-fold and six 2-fold axes; and the icosahedron has six 5-fold, ten 3-fold and fifteen 2-fold axes. All these rotation axes are compatible with three translations, with the exception of 5-fold symmetry. That means that icosahedral symmetry is **non-crystallographic**, and cannot apply to a crystal *as a whole*. (Despite this prohibition, objects like viruses with icosahedral symmetry *can* form crystals, although the 5-fold axis of

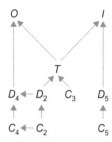

Figure 10.34: Lattice diagram for the 'Platonic' point-groups T, O and I.

each virus cannot apply to the *entire* crystal.) But tetrahedral and octahedral symmetries can apply to entire crystals, and they form the *cubic class* of crystal symmetry.

The tetrahedral group T is a sub-group of both the others (Fig. 10.33, (c), (d)). Diagram (c) shows that half the eight vertices of the cube are occupied by one tetrahedron, leaving space for a second; thus 2 tetrahedra, each with 12 subunits, fit an octahedral structure which accommodates 24 subunits. Similarly, diagram (d) shows that 4 of the 20 vertices of the dodecahedron are occupied with one tetrahedron, leaving space for four other tetrahedra; thus five tetrahedra, each with 12 subunits, fit an icosahedral structure which accommodates 60 subunits. This connection between the new 'Platonic' groups is shown in Fig. 10.34, which also includes their other subgroups.

10.4.3 Crystallographic Groups in Three-Dimension

Here we are concerned with only 65 of the 230 crystallographic space-groups, since we focus on crystals of macromolecules composed of chiral units like L-amino-acids that are present in exclusively one form (the L enantiomer). Therefore they could not fit a crystal packing that required equal numbers of opposite enantiomers (L- and D-), so such symmetry groups can be excluded. This means that we consider only symmetry groups that lack inversion centers or mirror planes, which restricts the possible symmetry operations to translations, rotations and screw-axes. We start by considering the translations.

We have seen that two translations generate plane lattices composed of repeated identical flat unit cells with edges a and b. Different lattice types have cells with different shapes, defined by their side-ratio $a{:}b$ and the value of their included angle γ. Similarly, three translations generate space-filling lattices with solid unit cells. Although these can, strictly, have a wide variety of shapes, we choose for convenience a parallelepiped (all of whose faces are parallelograms) whose shape is defined by the three edges or axes a, b, c and corner-angles α, β, γ (Fig. 10.35).

No special restrictions are imposed on these parameters, in order to form part of a lattice; but restrictions are commonly found in practice. Any two axes might happen to be the same length, or any angle might happen to be a right-angle; but these occurrences would be rare,

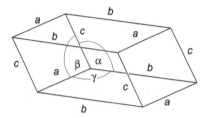

Figure 10.35: Basic 3D cell defined by three sides (a, b, c) and three angles (α, β, γ). Note that the axes are right-handed (like x,y,z), and each plane contains the letters a, b, c, but one of the three is Greek: so we can get either α, b, c or a, β, c or a, b, γ.

if they occurred just by accident. However, here is a systematic way in which they can arise: when a rotation axis (parallel to a cell axis) is perpendicular to a plane defined by the other two axes. Then there are two right-angles. And, if the rotation axis relates the other two axes, they must have equal lengths. So cell-shape is connected with the point-group of the 3D space-group.

We saw, in connection with the plane groups, that the rotational symmetries of lattices must be 1-, 2-, 3-, 4-, 6-fold. These same restrictions apply in 3D, so we get the cyclic and dihedral symmetries found earlier: (no symmetry, C_1), C_2, D_1, D_2, C_3, D_3, C_4, D_4, C_6, and D_6. We found the following lattice-shapes: oblique ($p1$, $p2$), rectangular (pg, pm, pgg, pmg, pmm, cm, cmm), square ($p4$, $p4g$, $p4m$) and hexagonal ($p3$, $p3m1$, $p31m$, $p6$, $p6m$). Not surprisingly, there are rather more shapes in 3D, because a solid cell can have different features in different directions (e.g. square from above but rectangular from the side). However, the additional shapes are relatively few. There are two rectangular shapes, monoclinic and orthorhombic, and also two square shapes, tetragonal and cubic. The last difference is interesting as it concerns the point group. All the rotation groups for plane sheets are cyclic or dihedral, and D_4 guarantees a square lattice but only in one direction; it applies to the tetragonal class. However, there exist additional point groups that are applicable in 3D space (see last section): tetrahedral (T), octahedral (O) and icosahedral (I). The last of these is ruled out because it uses non-crystallographic 5-fold rotations; but the other two are acceptable, and they impose a cubic shape on the lattice (though not necessarily C_4 symmetry).

Thus there are seven unit cell shapes, giving seven *crystal systems*. (i) *Triclinic*: the unsymmetrical extreme: oblique, with three arbitrary angles and axis-lengths; no other symmetry. (ii) *Monoclinic*: with three arbitrary axis-lengths but only one arbitrary angle (the rest being right-angles); C_2 or (2) symmetry. (iii) *Orthorhombic*: with three arbitrary axis-lengths but all angles right-angles; D_2 or (222) symmetry. (iv) *Rhombohedral*: with all lengths equal and all angles equal, though not right-angles; C_3, D_3 or (3, 322) symmetry. (v) *Hexagonal*: with two arbitrary axis-lengths and angles 120° or 60 and 90°; C_6, D_6 or (6, 622) symmetry (subgroups C_3, D_3). (vi) *Tetragonal*: with two arbitrary axis-lengths and all angles

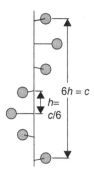

Figure 10.36: A 6-fold screw axis (6_1 in Hermann–Mauguin symbols; the symbol indicates 6 units in 1 turn). The structure repeats exactly after six units, with a rise distance h between each unit.

right-angles; C_4, D_4 or (4, 422) symmetry. (vii) *Cubic*: the symmetrical extreme: all lengths equal and all angles right-angles; T, O or (23, 432) symmetry.

But the most logical unit cell shape sometimes requires more than one lattice-point per cell, i.e. a centered cell. We encountered such cells (*cm*, *cmm*) in plane groups. When the same thing occurs in one crystal plane, it is called 'face-centering' and denoted by C. If all six faces of a cell are centered, it is denoted by F. But another possibility is to have the centering lattice-point in the middle (center) of the unit cell; this is denoted by I. Including centering extends the seven crystal shapes into the 14 three-dimensional Bravais lattices[2]. Centered lattices occur in only certain crystal systems: monoclinic (C); orthorhombic (C, F, I); tetragonal (I); cubic (F, I).

Screw axes, as mentioned in §10.1.2, combine a rotation and translation. They arise both in crystallography and in isolated helical molecules or aggregates, where the viewpoint is different. In crystals, the translation dominates so, if the crystal repeat is c, the translational component of the screw must be a fraction of c. That must be the fraction that the rotation bears to a complete rotation. Thus, if there is a simple 6-fold screw, the rotation must be 1/6th of a revolution, so the translation must be $c/6$ (for a right-handed screw, or $-c/6$ for a left-handed one). (See Fig. 10.36.) However, the Hermann–Mauguin symbols for these are 6_1 and 6_5 respectively; and the related symbols 6_2, 6_3 and 6_4 do not refer to 6-fold screws, but to combinations of a screw with a parallel rotation axis (which is 3-fold in the cases of 6_2 and 6_4, and 2-fold in the case of 6_3, so there are always 6 units per repeat). This way of looking at screw axes proved a liability when crystallographers began to study the screw-symmetry of isolated helical aggregates, where there is no external compulsion for a repeat, so any combination of a rise-distance and twist-angle is permitted in theory (see §10.2.3 and Chapter 15).

---

[2] Conventions add to the complexity. Thus, in a monoclinic lattice, the 2-fold axis is chosen parallel to b; but, in a hexagonal lattice, the 6-fold axis is chosen parallel to c.

Returning to crystallographic screws, where the rise-distance must be a fraction of the translation, the twist-angle must also be a fraction of a revolution, and moreover a fraction compatible with an extended space-lattice. We saw in section §10.2.1 that crystal lattices restrict rotations to 2-, 3-, 4- and 6-fold axes, so these are the only permitted screws that form part of an overall crystal symmetry. (All sorts of different helical symmetries may exist in molecules within crystals, but *their* symmetries don't apply to the *other* molecules.)

With 3-, 4- and 6-fold axes, screws must be left- or right-handed. Any space-group containing these is therefore itself chiral, and consequently also exists in the opposite hand. These enantiomorphous space-groups can, of course, pack enantiomorphous molecules. Then the L-group containing the L-molecule is the mirror image of the R-group containing the R-molecule, but is quite different from the L-group containing the R-molecule, which is the mirror image of the R-group containing the L-molecule.

Counting only space-groups suitable for completely asymmetric molecules of one type, there are just 54 different space-group arrangements, but 11 of these are enantiomorphous, i.e. there are another 11 space-groups that are mirror images. So there are 65 altogether. (If we made no restrictions about suitability for asymmetric molecules, we would get 230 space-groups.)

10.5 Fourier Transforms of Crystallographic Symmetry Operations

The FTs of symmetry groups have special conditions imposed by the group, and these conditions help to assign the plane- or space-group from the diffraction pattern; they are tabulated in Hahn (2002). Here we only discuss the effects of isolated symmetry operations.

10.5.1 Translations

The FT of a 1D fiber, a 2D crystalline sheet or a 3D crystal is dominated by the pure *translational* component of its symmetry. We can get an image of this component if we imagine marking each (translational) asymmetric unit by a point. Then the 1D structure has a line of points, the 2D sheet has a flat sheet of points arranged in a regular lattice and the 3D crystal has an imaginary 3D lattice of points. This lattice can be viewed as the *repetition function* of the original structure; it determines how the asymmetric unit will be repeated (translationally).

As usual, the deduction runs:

(original structure) = (one asymmetric unit)★(repetition function).

Taking FTs, we get:

FT(original structure) = FT(one asymmetric unit) × FT(repetition function)

Now FT(repetition function) is a function that is zero over most of space, and it *samples* the FT of the asymmetric unit at the non-zero points; so we call it the *sampling function*. The 2D sampling function of a 2D lattice is the same lattice, rotated by 90°, in reciprocal space: the *reciprocal lattice*. However, because the lattice is confined to a plane, the reciprocal lattice points run perpendicular to the lattice plane, extending into long lines of density, an array of needles or rods (Fig. 14.1). The 3D sampling function of a 3D lattice is a 3D reciprocal lattice (in which vector of the reciprocal lattice is perpendicular to a plane of the real lattice).

The sampling function samples the asymmetric unit's FT. That typically has a strong point at its origin, and varies continuously, forming 'blobs' of roughly the reciprocal of the asymmetric unit's width. All this FT density extends only out to a boundary that defines its resolution. This gets sampled by the sampling function.

10.5.2 Rotations

The *rotations* of the symmetry group form a point-group and, by the rotation theorem (§6.1.2), a rotation's FT is the same rotation, so the sheet's point-group applies equally to its FT. For 1D and 2D translation groups, by far the most common rotation axis is the 2-fold axis, either by itself, or as part of a dihedral point-group or simply as a sub-group of 4-fold or 6-fold rotation. Now a 2-fold axis causes part of the FT of a real object to be real. The way this arises is shown in Fig. 10.37. Thus, in the FT of a real object, the central plane perpendicular to a 2-fold axis has entirely real numbers (b). This leads to the point group D_2 giving rise to an FT with three intersecting perpendicular planes that are entirely real (c). That is the effect of the 2-fold axis in a perpendicular plane; on a parallel plane (as discussed in §10.1.2 and Fig. 10.3 (a)) it generates a mirror.

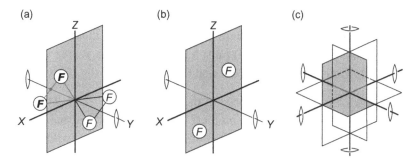

Figure 10.37: Effect of a 2-fold axis (along *Y*) and *XZ*-mirror on the FT of a real object or image. **(a)** An arbitrary point of the FT, with value F, has an identical point generated by the 2-fold axis but also its conjugate F* (shown as *F*) generated by Friedel symmetry; and that too is copied by the 2-fold axis. **(b)** A point in the *XZ*-plane, perpendicular to the 2-fold axis, is also copied by the 2-fold axis; but, in addition, the two points are related by Friedel symmetry. So they are real[3]. **(c)** FT of a real object with the point-group symmetry D_2 has three mutually perpendicular planes, perpendicular to the 2-fold axes, on which the FT is real.

---

[3] If the number is $a + ib$, its conjugate is $a - ib$ and, when they are equal, $b = 0$.

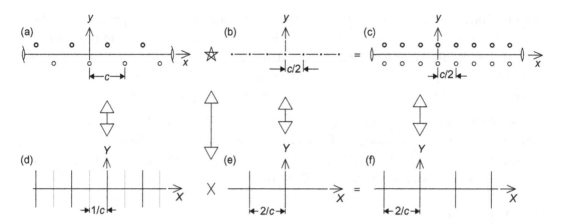

Figure 10.38: The pattern made by a 2-fold screw axis **(a)**, when convoluted with a line of points with *half* the repeat **(b)**, becomes a combination **(c)** of a translation (with half the repeat) and a 2-fold rotation axis. The FT (below) contains equidistant planes perpendicular to X.

10.5.3 Screw Axes

Lastly we consider the screw axes. A 2-fold screw axis is the commonest in 3D, and it is the only type that can exist in flat sheets. Figure 10.38 shows one (a), with a repeat c which causes its FT (d) to exist only on planes $1/c$ apart; (c) shows the same screw-generated pattern (a), convoluted with an infinite line of points (b) having the half repeat, $c/2$. This convolution converts the screw pattern into one, (c), generated by the combination of a 2-fold axis with a translation of $c/2$. Multiplication of (a)'s FT, (d), by (e) is equivalent to filtering (a)'s FT by passing only the planes at multiples of $2/c$. We thereby get the FT of a 2-fold axis, so each plane of (f) must have a 2-fold axis. But these planes are precisely those in (d) at multiples of $2/c$. Thus each alternate plane in the FT (d) must have a 2-fold axis. As the plane through the origin also has a 2-fold axis, its FT values are real (if the source of the FT was real).

We find out what happens on the other planes of the FT (d) in Fig. 10.39. This shows the effect of 180° rotation (upside-down), (b), on the 2-fold screw axis pattern (a). This rotation is also produced by shifting (translating) the pattern by $c/2$, through convolution, (c), with a point at $x = c/2$. This allows us to deduce the FT (g) of the rotated axis's pattern (d). (The FT of the point at $x = c/2$ is, as in section §3.4.2, the phasor wave (f) which is $+1$ at $X = 0$, or $2/c$, and consequently at all even multiples of $1/c$; but it is -1 at all odd multiples of $1/c$.) Thus 180° rotation of the FT leaves unchanged the FT planes at 0, $\pm 2/c$, $\pm 4/c$,...; but it changes all signs on the intervening FT planes at $\pm 1/c$, $\pm 3/c$,... On those planes, the origin point is not moved by the 180° rotation, yet its sign is changed, which means that its value must be zero. Thus, on the line $X = 0$, the FT is non-zero only at 0, $\pm 2/c$, $\pm 4/c$,... This line gives the FT of the projection onto the x-axis of (a), or any pattern with a 2-fold screw. And, indeed, the projection onto a parallel plane of a 2-fold screw has a halved repeat.

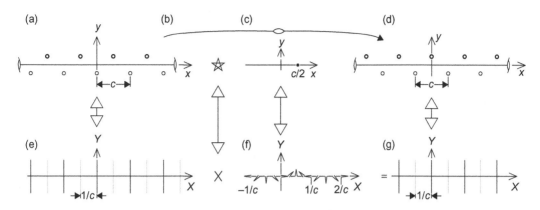

Figure 10.39: The pattern made by a 2-fold screw axis **(a)** can be turned upside-down, either by rotation through 180° **(b)**, or by shifting to the right by $c/2$ **(c)**. Thus the effect **(g)** on the FT **(e)** is the same, if we rotate it by 180° or change the signs on alternate sheets (through **(f)**).

The salient features of FTs of higher-order screw axes can be deduced along the lines of helical diffraction theory (§15.2). Thus the 6_1 screw axis (Fig. 10.36) is unchanged by convolution with a vertical line of points spaced c apart, so the FT is unchanged by multiplication with a perpendicular set of parallel planes spaced $1/c$ apart, i.e. the FT only exists on these planes. Moreover, the projection onto the vertical axis repeats every $c/6$, so the parallel line-section of the FT has points every $6/c$. This is another example of 'systematic absences'.

Statistics and Matrices

Chapter Outline

These rather technical matters are essential topics, since macromolecular structures only emerge from noisy data with the help of powerful algebraic methods. Moreover, structural biology, in adopting the methods of physical chemistry, adopts their statistical treatment of experimental results and also the statistical-thermodynamics basis of macromolecule structure.

11.1 Statistics

11.1.1 Basic Principles

Symmetry, where everything has a precise position, is closely related to statistics, the study of randomness. Symmetry prescribes which objects are equivalent to others, so they should have exactly the same experiences. And, as long as strict symmetry is preserved, there should remain no difference between their experiences. However, circumstances can prevent symmetry-preservation. The exactly symmetric arrangement can be unstable: a coin can only maintain the 'heads':'tails' equivalence by landing on its rim. However, in such 'symmetry-breaking' situations, a vestige of the original symmetrical equivalence remains. The coin has to land on one of its faces, but these faces are *equally likely*. This is (Pierre) Curie's principle (Stewart & Golubitsky, 1992) that, when symmetry is broken, it is *preserved statistically*.

Thus we expect, after many coin-tosses, about equal numbers of 'heads' and 'tails'. But how does this come about in detail? How can the coin 'know' when there have been too many 'tails' so that a 'head' is needed? Can it have some form of memory? But, if there's no memory, so 'heads' and 'tails' are just equally likely, isn't the equal sequence ...HTHTHTHT... no more likely than the unequal sequence ...HHHHHHHH...?

This last observation is quite correct, and it is indeed the reason for roughly equal numbers of 'heads' and 'tails'. Suppose we toss the coin four times, and get some sequence, e.g. HTTH. This has the ratio 2H:2T, and the sequence can be regarded as a 'route' to this ratio. But of course HHTT, HTHT, TTHH and THTH are also 'routes' to the same ratio. Each of these 'routes' is equally likely, and as likely as other sequences that are 'routes' to other ratios. Indeed, *all* such 'routes' of the same length are equally likely. The crucial question is this: are there more 'routes' to nearly equal ratios than to unequal ratios? And, since every 'route' is equally likely, whatever ratio which has the most 'routes' will occur most frequently. (Every different 'route' can be viewed as a kind of 'popularity vote'.)

The effect of this is shown in Fig. 11.1 which plots all the routes in a coin-tossing. The topmost '1' has a pair of arrows below it that represent 'heads' (left) and 'tails' (right). The consequences are shown in two squares, each with 1 (as they are equally likely). Each has another pair of arrows below it that mean the same as before. The third line of squares, 1, 2, 1, represents the consequences of HH, HT + TH, TT. Every route leading to this line has two tosses, as we can see from either of the two oblique line of squares numbered 1, 2, 3,....

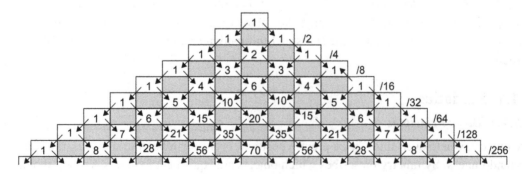

Figure 11.1: Each square contains the number of routes that reach it. The probability of a ball reaching the square is that number, divided by the total number of paths at that level (shown on the right, where the number under the oblique stroke is the sum of the numbers in all the squares).

Indeed, *every* square can get votes from squares above it, provided they lie within a V-pair of 45° lines radiating upwards from it. This 'V' gathers more squares if it radiates from a central square.

The numbers in the squares easily give us the H:T ratios. From any square, follow any route that leads to the topmost square and count the number of left turns (T) and right turns (H). We can use these ratios to test if a coin is 'unbiased'. Thus if, out of seven tosses (on the horizontal line starting 1, 7, 21,...), we had all heads (first square), the probability is 1/128 (/128 at the far right), which is significantly unlikely. This is the way in which hypotheses are tested against noisy experimental data: we find the probability of the agreement arising from chance alone. In this case, we could use the H:T ratios to test if a coin has an unsuspected bias (see examples in Keynes, 1921).

If we run along a horizontal line, the squares contain a series of numbers (of 'routes'), like 1,4,6,4,1, that rise to a maximum and then fall. A plot of the corresponding ratios in Fig. 11.2 resembles the Gaussian curve, which it approximates when the number of throws becomes very large. This curve often recurs in any quantitative work. Many measurement processes involve random processes that add or subtract from the ideal or theoretical value. Each of these random processes acts rather like the random choice of alternate paths in Fig. 11.1; so they all add together rather like the alternate routes, giving a distribution of errors on each side of the mean value, as in Fig. 11.2.

This curve is also connected with the 'random walk', a path based entirely on chance, and most easily discussed in 1D. Suppose many people are lined up, entirely side-by-side and

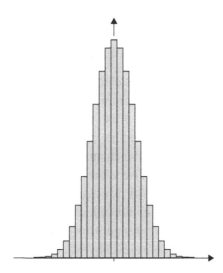

Figure 11.2: The distribution of route-numbers (or probabilities), at the level with 51 squares, closely resembles the Gaussian curve (in Fig. 5.4 (d)).

facing the same way, on a squared pavement. Before each step, each throws a coin and walks forwards after a 'head' or backwards after a 'tail'. Each person would follow a 'route' in Fig. 11.1 and, since each route would be equally likely, the distribution of people would follow a Gaussian curve. This is how molecules diffuse away from a sharp line of dye. Each molecule moves randomly (like tossing coins) under the impact of random collisions with water molecules. So the colored line becomes more blurred and broad with time, approximating a Gaussian curve. But just how rapidly does the distribution broaden with time? Is its breadth proportional to time? It turns out that it is only proportional to the *square root* of the time (see Appendix, §11.4.1).

The random walk is the concept underlying **Wilson statistics** in crystallography (Wilson, 1942). Consider the diffraction pattern of a molecule that is so complicated that its atomic distribution might be considered random. A typical diffraction spot derives from a phasor which is the sum of contributions from every atom, the contribution depending on the atom's coordinates and the diffraction spot's coordinates. Thus the contribution of each atom to the phasor might also be viewed as random, so the contributions will add together to form a long chain, following a random walk. Thus the length of the walk is expected to be proportional to the square root of the number of contributions, i.e. of the number of atoms in the molecule.

A special form of randomness affects measurements that involve the *counting* of quanta or particles of high-energy radiation. Suppose that a certain diffracted X-ray beam has an average of 100 quanta per second. The *expected* count after t seconds is $100t$, but of course the *actual* number will differ from this. Then the difference between $100t$ and the total counts, which is the error, is equally likely to be positive or negative; so it moves as a random walk and increases as $\sqrt{t}$. Thus, although the signal is proportional to the total counts, the error is proportional to their square root. The relative error (error/counts) therefore decreases as the square root of the counts. This is generally true: the signal:noise ratio increases as the square root of the number of measurements averaged together.

The random arrival of particles like quanta or electrons ('shot noise') is only one example of that universal phenomenon, the random contamination of data, called 'noise'. Two important characteristics of noise are worth reviewing. First, its essential feature is unpredictability, so *any* kind of regularity (beyond what would be expected by pure chance) is avoided. Second, any data-stream is a 'communication channel' of given capacity (determined by data-frequency, etc.). When present, the noise takes its share of the channel, restricting the capacity for data. It also adds the problem of distinguishing signal from noise; a completely predictable sequence added to the signal could, at least, be removed by calculation, although the remaining signal would have been weakened. However, while it's easy to describe the channel capacity of digital channels, this is less obvious with analog channels like voltage or wave intensity. Then we may need a method of digital interpolation, such as discussed in §5.3.1.

11.1.2 Fourier Transforms of Noise

Noise occurs in all EM images and is therefore unavoidably included in image-processing activities, many of which start by calculating a Fourier Transform (FT). What, then, is noise's FT? It is easier to deduce this if the noise is *complex*, i.e. when at each pixel there is a phasor, with random amplitude and random phase. (In practice, there is a maximum value to the amplitude set by energy considerations, but the phase can cover the entire angular range.) The advantage of complex noise is that its FT isn't changed by a shift of origin. By contrast, the FT of real noise has Friedel symmetry about the origin. So, if both noise and FT are complex, there is no general property to reveal if one of them has been shifted. In that case, neither is significantly changed by shifting origin.

That makes the FT of noise sound completely uniform, which cannot be true, for the FT of a uniform distribution is a delta-function; it is concentrated at the origin, and not at all uniform. Here, by contrast, both noise and its FT are quasi-uniform. In any case, a uniform distribution is entirely predictable and, as we have seen, noise is best defined by its unpredictability (though that is limited by restricting amplitudes). This implies that its FT must also be quite unpredictable, which implies that it must be noise. Thus we conclude that the FT of noise must itself be noise.

A crucial defining parameter of noise is its *pixel size*, the biggest region which is *by definition* uniform. Real noise has the most probable distribution (like the distribution of heads/tails mentioned above) of items that correspond to pixels. An obvious example of noise is TV 'snow', where the smallest pixel is the image pixel, whose linear spacing corresponds to the scanning raster. No spot on the TV screen can be smaller than this, so what we see when there is no real signal is a random distribution of these pixels, presumably switching at the same rate as TV frames. But another example, with a very different 'pixel', is a glass of water. Here the 'pixels' are water molecules, whose distribution approximates to randomness, allowing for some short-range structure imposed by hydrogen-bonds. The reason for the apparent disappearance of this fine random structure, is fundamentally the same as the predictable frequencies of coin-sequences with roughly equal numbers of heads and tails: there are vastly more micro-states in arrangements that appear uniform on our scale. By contrast, a highly unequal distribution, such as ice filling half a glass and boiling water the other half, has far fewer distributions and is thus *far* less likely to occur by chance. Thus water is apparently quite uniform on our size-scale, whereas X-rays, which operate on the atomic scale of their wavelength, 'see' its microscopic structural fluctuations as noise. The resulting band of diffraction around 3 Å marks a critical region in the resolution of an X-ray diffraction pattern.

However, we must conclude that, while the FT of noise is of course noise, its amplitude declines steadily beyond the reciprocal of the pixel size.

11.1.3 Average Value of Measurements

We saw how the 'symmetry model' of a perfect coin could be tested from the statistics of its agreement with experiment. But suppose we should find a slight bias in favor of heads. Then we might refine the model by supposing that the coin has (say) a 55% chance of heads. So we would have to refine the mathematical model (giving the **binomial distribution**) to calculate the expected heads:tails frequencies for this ratio. Indeed, we might test the data against a range of different inherent heads:tails probabilities to find the best fit and, more important, the probability of it differing from this (i.e. the confidence levels). This is a very simple example of a very common procedure: adjusting a model to fit experimental data.

The commonest example of this is taking the mean of a series of measurements. Consider the simple example of calculating an average count-rate from successive 10-second counts of a random particle-stream. We instinctively calculate (total counts)/(number of count-frames). If pressed, we might justify this by requiring that the 'average' count-rate should differ as little as possible from the individual 10-second counts. So we calculate the individual errors (count minus average) and then are tempted to estimate the *total* error. As individual errors will be negative as often as positive, they cancel each other most efficiently when we use the 'true' average; and that is one way to define it. Another way uses the idea of *fluctuations*, so we must make all the errors positive (e.g. by ignoring their signs or by squaring them) before adding them. The 'squaring' alternative is chosen as it has led to a rich mathematical theory. So we can *define* the 'mean count' as the number that minimizes the total of these squares, or (which is effectively the same) minimizes the *mean square error*. But is that the same as our error-cancelling 'average'?

Of course this is easily proved by algebra, but a physical 'proof' is more illuminating (Fig. 11.3). (a) Shows points with different y-coordinates representing different counts. To minimize the sum of squares of y-differences, represent the possible mean by a horizontal rod (b) from which we attach a different spring (from a box of identical springs) to each point (represented by a nail), then adjust the rod's height to minimize the elastic energy of the springs (which we suppose have a zero resting length). Indeed, the springs make this

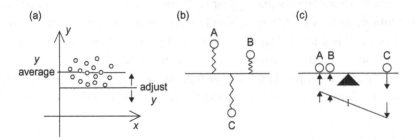

Figure 11.3: (a) The probable mean (horizontal line) of a series of points... **(b)** ...corresponds to minimizing the elastic energy of springs... **(c)** ...or to positioning weights on a balance.

adjustment for us, minimizing their total energy to reach mechanical equilibrium. Then all the springs' forces balance, as springs attached to lower nails pull the rod downwards as much as those attached to higher ones pull it upwards.

This balancing equilibrium minimizes the total energy of the springs. Now the energy of a stretched spring is (approximately) proportional to the square of its extension (see p. 237, Fig. 11.10 (f)), so minimizing the total energy of the springs means minimizing the total of squared spring-lengths, which is the sum of squared deviations of points from the mean.

This same equilibrium is also reached by a balance (c) but, as it works by gravity, the *forces* are *constant* with displacement but *energy* increases *linearly* like elastic force. Thus, in Fig. 11.3 (c) the gravitational *energies* balance; and, if the beam is tilted, the energy released by the weights that fall pays exactly for raising the rest. If this were not true, the beam would be pulled down on one side and we would move the fulcrum towards that side, until the beam stayed horizontal, showing that those energies were equal.

11.1.4 Accuracy of Measurements

As well as the mean, we also need a measure of how well all the measurements differ or agree. We have already estimated the total error by squaring the individual errors before adding them and finally dividing by the number of measurements, getting the mean square. But the (mean squared) errors depend on the assumed mean, which we adjust to minimize them. Now, that minimized mean squared error is a good measure of accuracy, called the **variance**. So we first need to know the mean, and then (by subtracting it from each value) we force the measurements to have a zero mean. After that, we proceed as before, summing the squares of the error values and then dividing by the number of measurements to get the variance. Another (mathematically equivalent) way to get the variance uses the formula: 'mean square minus squared mean'. First we find two means: the mean of the values themselves (the average value), and also the mean of their squares. To get the variance, we take the second mean and subtract from it the *square* of the first mean. (To remember the formula, suppose the mean is zero; the variance will remain positive.)

However, the variance is unsuitable for measuring the error in a physical parameter like length. It has the wrong units: a mean in centimeters would have a *variance* in *square* centimeters, an *area* instead of a *length*. To correct for that, we take the square root of the variance, getting the **standard deviation** or 'root mean square deviation' (r.m.s. deviation).

11.1.5 Straight-Line Fits

So the value of one parameter is summarized by its mean, and its accuracy by its variance or standard deviation. Consider now the case where *two* parameters (like length and weight; or, for a two-pixel picture, the density of each pixel), that are possibly connected, need summarizing *together*. If we calculate the means of each, we lose the information about

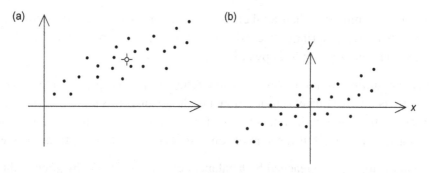

Figure 11.4: Weight plotted horizontally and height vertically. **(a)** The means (open circle) have been calculated for a data-cloud. **(b)** Subtraction of the joint mean from each point gives an origin-centered data-cloud.

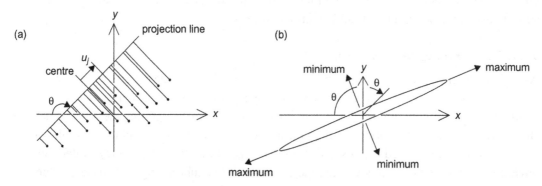

Figure 11.5: (a) Calculating the variance of the points in an arbitrary direction. Project all the points onto a line (angle θ) and calculate the variance of the projected points. Repeat for lines with different θ **(b)** The calculated variance, plotted as a radius from the origin at the angle θ. (The short line is parallel to the projection line in (a).) The ellipse's major axis is marked 'maximum', the minor axis 'minimum'.

their connection or **correlation** apparent in a plot of the raw data, like Fig. 11.4 (a). We plot a cloud of points whose shape indicates the kind of correlation. A spherical cloud, or one distorted along the horizontal or vertical axis (i.e. symmetrical with respect to both axes), indicates no correlation between the parameters. Correlation shows up in an oblique plot, as in Fig. 11.4 (a), and we want its direction.

We first subtract both means to get an origin-centered cloud (b). To find the orientation of this cloud, we try projecting all the points onto a straight line and calculating the variance of the projected points (Fig. 11.5 (a)). We repeat this with many different orientations of the line. If we make a polar plot, with each variance fixing the radial distance at different angles, we get an ellipse Fig. 11.5 (b). The ellipse's major axis gives the orientation of the straight line. So now we know both the gradient, and the constant term, of a straight-line approximation

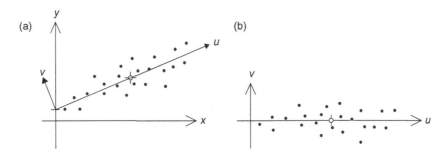

Figure 11.6: (a) The original data-cloud with the least-squares line. **(b)** A re-plotting of the data, using the 'natural' coordinate system found in (a).

to the data. This is a **least squares** line (or *regression line*), which often summarizes the data satisfactorily. It is shown in Fig. 11.6 (a).

Thus the simplest connection between two parameters is a straight-line, which we fitted by minimizing the variance, i.e. using a least-squares criterion. This straight-line fit is just the simplest example of fitting a theoretical curve (or plane, etc.) so that the sum of squares of deviations is minimized. Note that, for this to be possible, there must be a surfeit of experimental data, of mediocre precision, so we sense the need for some kind of average or 'best fit'. What we then need is almost always a least-squares fit.

11.1.6 Principal Component Analysis

The procedure just described could have a different purpose. After processing the data-cloud, as described, to get the ellipse (b) of Fig. 11.5, its two axes give 'natural' coordinates which can be used to re-plot the data in (b) of Fig. 11.6. Whatever quality we are plotting along the u-axis is the most important distinguishing character for describing a member of this population. We might investigate this character by choosing points that differ as much as possible in their u-coordinate (while having a minimal a v-coordinate). (Similarly, the v-character could be investigated by examining specimens with identical u-indices.)

This statistical method, called **principal component analysis**, was first introduced into psychology by Hotelling (1933). It subsequently spread (also in modified versions) to many fields because it can be applied to many dimensions, i.e. to populations characterized by many different measurements that are more or less correlated. It is an important tool of **multivariate statistical analysis (MSA)** (i.e., the statistical analysis of data that have many variable parameters). This is a general statistical tool, though it is also specifically associated (like principal component analysis) with single-particle image-analysis (section §17.3).

Note the leading features of the method (as seen in this simple example). It is multi-dimensional, here 2-parameter, so correlations between the different parameters are very important. Through the method, the statistics of a complicated aggregate (Fig. 11.4 (b)) are condensed into an ellipse (Fig. 11.5 (b)) which is itself abbreviated into just two parameters, its axes. We have sorted the data by finding first the character responsible for most of the observed variation, and then the next most important character. If other variations had no significance, all important information in the original data would be condensed to one coordinate, and the procedure would have achieved *data-compression*. But the other axis, though smaller, may not be negligible; points with an extreme value of the second parameter may be recognizable 'types', particularly if there is a clustering of the population along the axis.

Obviously, this can be extended to situations with 3, 4, etc. parameters, leading to the same number of principal characteristics. In this way, the population could receive a natural classification which would also involve data compression, since the number of 'types' will be much less than the number of items in the population. Of course, the greatest data-compression would be achieved if we could ignore several of the later (minor) axes.

11.1.7 Interpolation of Fourier Transforms

In Chapter 5 (section §5.3.1) we looked at the question of interpolating isolated values to get a smooth curve. We found that, for the important problem of interpolating the FT of an image of maximum dimension D, there can (fortunately) be an exact solution, provided by the sampling theorem (§5.3.2), which requires the FT samples to be available at intervals of $1/D$.

However, suppose we only have a 1D FT sampled at *arbitrary* points. If we need to calculate the FT anywhere, we want a kind of universal interpolation licence; and that is only provided by the sampling theorem, which requires a knowledge of the FT's values at *regular, equidistant* intervals of $1/D$. But how can we get from the arbitrary samples to the regular ones? It is as if we must solve the original problem (and thus already *have* the universal interpolation licence), in order to find the values at those regular intervals needed to get the sought-for licence. But this paradoxical situation, of apparently needing the answer before we can start the calculation, is typical of many algebra problems. One labels the required answer and writes down the required properties of the answer, thereby getting an equation that is soluble by simple logical processes. Crowther et al. (1970) found a satisfactory solution by inverting the usual interpolation procedure. We shall call the arbitrary sampling points 'available values', and the required equidistant points 'defining values'. Then, if we knew the defining values, we could calculate the available ones; so we need 'only' to reverse the calculation, and use *'reverse interpolation'*.

Consider the simple example of a real 1D FT illustrated in Fig. 11.7 (a). The two 'defining values' $F(0)$ and $F(1) = F(-1)$ (equal and thus effectively just one variable[1]) allow us to find

---

[1] As $F(1) = F(-1)$, we can simplify the full equation: $F(X) = F(1)[\text{sinc}(X + 1) + \text{sinc}(X - 1)] + F(0)\,\text{sinc}(X)$ and write it as: $F(X) = F(0)u(X) + F(1)v(X)$. Besides some values of $F(X)$, we also know the interpolation functions $u(X) = [\text{sinc}(X + 1) + \text{sinc}(X - 1)]$ and $v(X) = \text{sinc}(X)$.

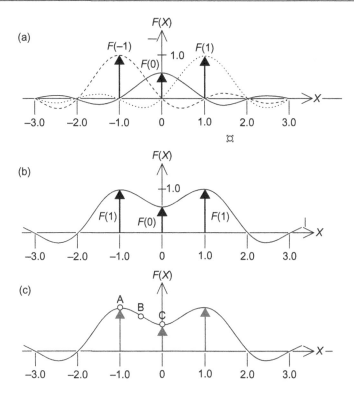

Figure 11.7: Using the sampling theorem, a symmetric FT is constructed from three defining values, two of which are equal: $F(-1) = F(1)$ and $F(0)$. Later, we need to get these values from sampled values of the FT. **(a)** shows the defining values and their contributions to the FT. (As the sampling theorem shows, these are sinc-functions.) Note that the contribution of a defining value is maximal at its position (indicated by a vertical peak), but becomes attenuated at more distant places. **(b)** shows the resulting curve of the FT, together with its two defining values. **(c)** Suppose we have sampled the FT at three different points, A, B and C. What do these sampled values tell us about the two defining values $F(1)$ and $F(0)$? See text and Fig. 11.8.

$F(X)$ anywhere in between by convolution with the sampling theorem interpolation-functions ($u(X)$ or $v(X)$): $F(X) = F(0)u(X) + F(1)v(X)$ (Fig. 11.7 (a), (b).) 'Reverse interpolation' would give us $F(0)$ and $F(1)$ from different 'available values' of $F(X)$. Thus, in Fig. 11.7 (c), we can see that point A gives $F(1)$, point C gives $F(0)$ and point B gives partial knowledge of both. This situation is shown in Fig. 11.8, where the four graphs plot our knowledge of the two 'defining values'. Points A and C give knowledge about only one of them (a), but point B gives us a connection between their values in the form of an oblique line (b). By itself, that line is insufficient, but it gives us both values if we know either A or B. (Similar lines would apply if we knew the FT $F(X)$ at other points.)

It seems that, since any two non-parallel lines should intersect to give the solution, *any* two 'available values' of the FT should suffice. But here we encounter the second aspect of the

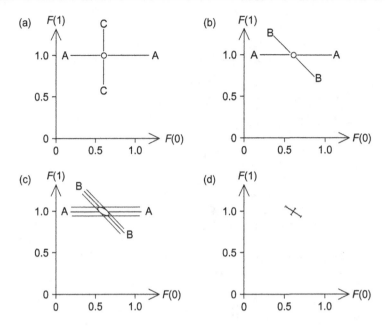

Figure 11.8: (a) The amplitude at point A of Fig. 11.7 (c) gives us the horizontal line AA in a diagram of $F(0)$ and $F(1)$. Similarly, point C gives us the vertical line CC in the same diagram. **(b)** The point B, between A and C, corresponds to an intermediate gradient which intersects AA at the correct values of $F(0)$ and $F(1)$. **(c)** The 'error-bars' in measuring the amplitudes of points A and B are represented by parallel lines. Where they meet, a parallelogram indicates the error ellipse of our estimate of $F(1)$ and $F(0)$. **(d)** The axes of this error ellipse give the error bars of measuring $F(0)$ and $F(1)$.

problem: reliability. We cannot expect to get a reliable estimate of a 'defining value', if it lies too far from any 'available value' of the FT. Plot (c) shows the result of intersecting the lines from (b) when the measurements have errors. These widen the lines with error-bars, and their intersection in (c) shows the accuracy of our estimates of $F(0)$ and $F(1)$. Estimates of equal accuracy lie along a small tilted ellipse which makes $F(1)$ the better-determined quantity. (This is to be expected, since the data were from point B and point A on $F(-1) = F(1)$.)

In order to simplify this example, we restricted the number of data values (two: A and B) so as to equal the number of unknowns (two: $F(0)$ and $F(1)$). This leads to a unique solution, because a solution is only just possible. In practice, we need a good excess of experimental data to help reduce the inevitable errors; and then we need a kind of 'best-fit', i.e. a least-squares solution. Apart from providing a more accurate result, the least-squares solution also has the merit of giving an estimate of the variance.

11.2 Matrices

The last two examples were cases combining statistics with multi-parameter problems. Fitting a straight-line requires knowledge of both the angle (gradient) and shifts (intercepts) of the line; and interpolating FTs needed, in even the simplest case (two defining values). It's very common to have several parameters describing a problem, the conditions of which affect all parameters simultaneously. So the parameters need to be disentangled mathematically to get their individual values.

When the conditions of multi-parameter problems take the simplest (i.e., linear) form, we need the mathematics of vectors and matrices. These are second only to Fourier transforms in importance for structural biology, and they are also (to begin with) just simple extensions of ordinary algebra. Some more detail is included in Part V, Chapter 20. That simple introductory chapter should help to clarify and amplify the intuitive descriptions given here.

11.2.1 Uniform Distortion and Matrix Eigenvalue Methods

A good way to approach matrices intuitively is through distortion, a topic important in image analysis (see Chapter 18). The very simplest distortion (*uniform distortion*) is the same everywhere, so it is *linear* (i.e. it transforms straight lines into other straight lines). Figure 11.9 shows examples of the same uniform distortion on different figures. In (a) a square with an inscribed circle which has been shear-distorted into an ellipse inside a rhombus; the square has become linearly distorted (uniformly stretched along one diagonal and uniformly compressed along the other). In (b) the four arrows at the corners show the directions of stretching or compression, and diagram (c) shows how the points on a circle move so as to lie on an ellipse.

Diagram (b) shows how the shear distortion moves *every* point to a new position by a shift-vector, and all these vectors constitute a **vector field**: at each point there is a vector, rather like the wind direction and speed. To visualize it, we fill space with a square lattice of points, each of which is then joined to its distorted position. (But note that the central point remains fixed.)

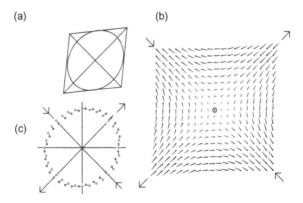

Figure 11.9: The effect of the same symmetric matrix on **(a)** a square with inscribed circle; **(b)** a square lattice of points and **(c)** points symmetrically placed around a circle. (In (b), the original positions are shown as large dots from which lines of distortion point away.)

The movements of the points in (b) follow a simple rule: the new (x,y) positions are related to the original ones by a pair of equations similar to those for $F(X)$ in the previous section. The rule is expressed very simply using a square array of four numbers[2] called a **matrix**. To describe distortion, we need only three *different* numbers (so two of the four are the same), and we have a **symmetric matrix**, a very important type. These three parameters control a uniform 2D distortion everywhere, so a knowledge of them (which could be obtained from the initial and final positions of just two points) allows us to predict the movement of any point. The effect of multiplication by a particular symmetric matrix can be seen from Fig. 11.9 (b). The entire pattern resembles the flow of a viscous fluid, entering at the top-left and bottom-right, and leaving at the top-right and bottom-left.

These diagonal directions have a special significance. The line joining each point to the center is a vector, and multiplication by this symmetric matrix usually changes its direction, except for points lying along the diagonals, where vectors change only their lengths. These special directions, defined by the symmetric matrix, are called its **eigenvectors**. The distortion can be described as a compression along one of these eigenvectors, and a stretching along the other. Together with the eigenvector directions, the compression or stretching ratios completely define a uniform distortion. Each ratio is called the **eigenvalue** of its eigenvector. Finally, the matrix has a reciprocal (the **inverse matrix**) which converts the final positions back to their initial positions. This is portrayed in the same diagram, if one reverses the direction of all vectors.

It is generally true that a 2D symmetric matrix (as portrayed here) has two perpendicular eigenvectors (also called **principal axes**). The entire diagram (b) is a kind of portrait of solving equations like those in the previous section. If we knew the movements of two points lying along the eigenvectors, we could easily find the reverse movements that define the inverse matrix. But multiplication with the inverse matrix may not always reverse the transformation exactly. If the original distortion shrank vectors, it could be hard to measure the shrunken vectors precisely, so their correction might only be approximate. Thus it would be possible to correct precisely along one diagonal (eigenvector), but only inaccurately along the other, if the former eigenvalue were big and the latter small. Pairs of linked equations can give solutions of varying accuracy, depending on the **eigenvalue spectrum** of their matrix. This concept is particularly important when there are not just two, but many linked equations, with the same number of eigenvalues.

When solving such equations, the eigenvalue spectrum is an important 'quality control' check (see e.g. Acton, 1970; Press et al., 2007). However, one well-established method for calculating an eigenvalue is already intuitively clear from Fig. 11.9 (b). Notice the flow directions (dot to line-end) which move any particle out towards the diagonals at top right

---

[2]   That is the very simplest square matrix, 2×2 or 2D, of which an introductory account is given in Chapter 20. Square matrices can have any number of dimensions, and there are also rectangular matrices.

or bottom left. This means that any point, repeatedly multiplied by the matrix, will move towards the outflow directions along the eigenvector with the biggest eigenvalue. Thus alternate multiplication by the matrix, and rescaling the vector, moves a point onto the biggest eigenvector. This is useful for getting the eigenvectors with the biggest eigenvalues, which are often the most important.

11.2.2 Multidimensional Energy Surfaces

Theoretical studies of macromolecules often extend simple one-dimensional physical principles up to molecules with thousands of structural parameters, and one-dimensional physics becomes thousand-dimensional. It's obviously impossible to imagine so vast a 'space', and it might seem that, by comparison, 2D space is too close to 1D space to be worth consideration. However, that 'modest' extension reveals many of the new concepts in higher dimensional space[3].

For our 2D example, we continue with distortion and the associated topic of elasticity. One-dimensional distortion is the simple stretching or compression of a spring (Fig. 11.10). Unlike a weight (a), where each small rise of the weight costs the same small energy (b), an elastic force on a weight (d) increases with the distance x it is moved. Thus, while gravitational work is found by adding the constant energy-strips to give a rectangle (b), elastic work involves finding the area of (half) a triangle (e). So gravitational energy increases *linearly* with distance (c), but elastic energy increases *quadratically* (f). This differs most

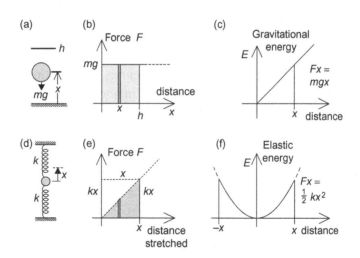

Figure 11.10: 1D force ((b), (e)) and energy ((c), (f)) curves for constant gravitation ((**a**)–(**c**)) and variable elasticity ((**d**)–(**f**)).

---

[3] Or is it that our mathematical imaginations already start to become exhausted after 2D? What important concepts might become significant only after 100D?

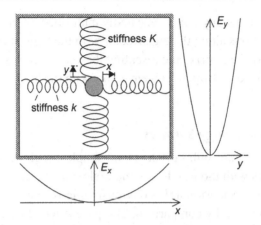

Figure 11.11: Energy curves for two perpendicular springs along x (weak) and y (strong).

fundamentally in being always positive (*negative* energy would mean a free ride!) whether the distortion is positive or negative. The consequence is that the gradient must change sign when the distortion does; and, to do so, the gradient must pass through zero. Therefore the energy changes very slowly when the gradient changes sign. So it's harder getting precise values for the minimum of energy, than of force.

The easiest way to extend this to 2D is to put two such 1D systems at right angles, as in Fig. 11.11, where the x and y directions carry springs of different stiffness, leading to different energy parabolae along those directions. However, a better picture of the energy surface is shown in Fig. 11.12 (b) and (c). It is a 'paraboloid', that can be visualized as a parabola rotated around its vertical (energy) axis, yielding a circularly-symmetrical surface that is then distorted along perpendicular directions. The resulting curve, when sectioned perpendicular to the energy axis, yields ellipses (b) with their major and minor axes along the directions of the eigenvectors of the (symmetric) distortion matrix.

11.2.3 Energy Minimization in Simple Systems

Now consider what the energy surface can tell us about the *dynamical* behavior of the system. This depends on the system's position on the surface. If it is not already at the bottom, energy can be released (as motion) when the system drops down. And, once energy has become motion, part at least gets converted into heat, an irreversible process; so the system drops further and further, ending at the very bottom, where it has no further potential for motion. This bottom state is easy to recognize in a 1D energy parabola or a 2D paraboloid (Fig. 11.12 (b),(c)). But suppose we are building a macromolecular model and wish to calculate its energy as we change its conformation. Its complicated multidimensional energy-surface would have many different minima of different depths. How should we find the lowest, corresponding to the conformation of lowest energy?

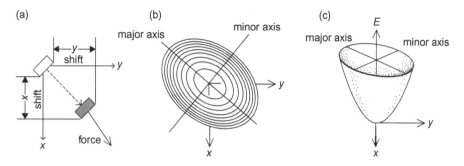

Figure 11.12: (a) Looking down on the top of a stick of rectangular cross-section fastened below. To displace the stick to the position shown, the shown force must be applied (not exactly along the displacement line). **(b)** The potential energy diagram of the energy required to push the stick to a given position. **(c)** A surface where the same potential energy is needed to push a ball to the given position.

The minimization process divides into two phases: trying to locate roughly the deeper minima, and finding the exact position of the chosen, deepest minimum. The second phase is easier and has been most thoroughly studied. It is particularly important in adjusting a macromolecular polymer chain, of known sequence, to fit the details of an experimental image. Then energy-minimization may be a way to improve the image's resolution. The simplest way (the **method of steepest descent**) is to imagine a small ball at the system's current position on the energy surface, and to follow its natural process of energy loss. We test every direction to find 'downhill', and make a small movement in that direction. On a 2D 'energy surface', our path would resemble the downhill flow of water.

Although conceptually simple, this method is not the most efficient. How far should we move downhill? If we go too far, we may miss the minimum entirely and have to turn back, oscillating to and fro before we find it. However, if we take too short steps, our journey will be safe but slow. Moreover, the steepest downhill direction may not point at the local minimum. So, before moving, it is better to get more information about the energy surface around us. In the simple parabola of Fig. 11.10 (f), three nearby points can define the entire parabola and allow us to calculate the minimum directly without any further searching. More sampling points are needed to map a 2D surface like Fig. 11.12 (c) properly, and increasingly more for 'surfaces' of even higher dimensionality. So, for problems (like macromolecule conformations) in very many dimensions, a compromise is made between: (i) undertaking a thorough exploration of the local surface before making each step, and (ii) making more progress by venturing steps based on less complete knowledge. And the higher the dimensionality, the more work it takes to study the locality. The **method of conjugate gradients** is a particularly successful compromise (see Press et al., 2007 for details and McCammon & Harvey, 1987, section §4.4 for a simple explanation).

11.2.4 Optimization of Alignment Parameters

This sort of 'energy minimization' is also useful for finding inter-particle alignments (see next chapter for the relevant parameters). We first define an error-function of the required parameters; this is constructed to have a single minimum when the parameters have their correct values but is bigger at all other values. Some parameters may be restricted to only integers or alternatives (in the case of chirality or polarity); then we must try each value in turn. However, the parameters are usually continuously adjustable (e.g. translations and rotations required to bring two images into coincidence).

Finding the minimum might be trivial or hard, depending on how many minima there are, and how well the error-function helps us to find them. Dimensionality (number of parameters) is the most important factor determining the ease of our search. Merely increasing the number of parameters from two to three means expanding our search from an area to a volume, and every pixel must, in principle, be searched.

There are two ways to facilitate such a multi-parameter search. We can simplify the problem by *intuitively* selecting the approximate minimum, so that only its exact position needs calculation, and a short slide downhill need not be too slow. But, if we start in complete ignorance of the solution, we should try if possible to separate the multi-parameter search into two (or more) successive searches that each involve fewer parameters. The convenience is similar to when we first look for the motorway, then for the exit, and last for the town, instead of searching every small square of the entire map for the town.

Finally, the value and gradient of the error-function at the minimum indicate how much trust we can place in our results. A deep, steep minimum inspires more confidence than a high, shallow one.

11.3 Structure Optimization and Simulation

11.3.1 Energy Minimization in Complex Systems

The refinement methods of the previous section are suitable for finding the single bottom of a smooth surface, but not for distinguishing several minima, nor for finding the bottom of a rough, irregular surface, nor for the more subtle problem of finding the energy minimum of a protein at room temperature. For its total energy U only reaches a minimum at absolute zero; at room temperature, a macromolecule has random motion that is maintained through the bombardment of solvent molecules. (This motion is of course connected with the entropy S and temperature T, and thus the equilibrium state minimizes, neither the internal energy U nor even the heat capacity H – which includes corrections for work against atmospheric pressure – but the Gibbs free energy $G = H - TS$.)

To take account of entropic processes, we need to simulate molecular movements in solution. The successive atomic positions are calculated iteratively using a simple approximation to the rules of classical mechanics (**molecular dynamics**). The attractive and repulsive forces are calculated quickly from compact formulae; they produce accelerations, leading to movements, of the atoms; and these movements produce new forces. Of course, each cycle must cover a period too short ($\sim 10^{-15}$ second) to include any second-order effects from movements. All these feedback cycles result in complex (and hence *chaotic*) overall changes. Their unpredictability forces us to look at ensemble averages, so molecular dynamics calculations are much used to estimate thermodynamic quantities. The rules of mechanics maintain the conservation of energy (corrections are made to ensure this), and the average energy defines the 'temperature' of the simulation (see Allen & Tildesley, 1987).

Through this chaotic motion, the molecule can briefly acquire, from the environment, extra energy to help it out of one energy minimum and into a deeper one. So molecular dynamics is a way to find the stable states of a macromolecule. To do this, the search-point must acquire energy. At first we want the program to do mostly exploration and relatively little minimization, so we give it plenty of energy – a 'high temperature'. To change its activity to minimization, we lower the 'temperature': we 'anneal' the system, like gradually cooling molten glass to remove sources of strain (regions of elastic energy). Minimization involving this regime of changing the simulation-temperature is called **simulated annealing** (SA). The structural adjustments can be achieved through molecular dynamics or by the Monte-Carlo method.

In molecular dynamics the minimization of energy is achieved, not directly, but through the complicated actions of the forces to which it gives rise. Cartesian coordinates are used for the equations, and bond lengths are kept constant by an iterative method ('SHAKE'). But this is less convenient than using torsion angles, with only one parameter per bond. Also, because of the necessary short cycle-times, long calculations are required to simulate the chaotic dynamics of macromolecules.

If the only purpose is energy minimization by simulated annealing, a possible alternative is the Monte Carlo method. This calculation also runs in cycles covering brief time intervals. In each cycle, small random changes are made to the configuration, and the new potential energy is calculated and compared with the previous value. If the new value is smaller (or equal), the new configuration is accepted. But, if the new energy is bigger, the new configuration is not immediately rejected. It is given a 'second chance', in which the probability of acceptance is a declining exponential function of the increased energy. (That function has the effect that every 'standard increase' of energy halves the acceptance probability, the 'standard increase' being an adjustable constant related to the simulation temperature.) So this method includes an adjustable simulation temperature, and can thus be adapted to a regime of simulated annealing.

11.3.2 Structure Refinement: NMR Compared with Images

As we have seen, establishing the conformation of a polypeptide chain falls into broadly two phases: the *starting phase*, where the general layout of the chain is established; and the *refinement phase*, where its detailed positioning, and the arrangement of side-chains, are all optimized (§1.2.1).

Although crystallography and microscopy can provide images for the *starting phase*, NMR cannot, as it is not an image-based method. Thus an organized starting phase must be replaced by a random ensemble of possible starting points. To avoid bias, each of these is often put far from the final solution from which it is separated by a barrier of local energy minima. To cross these, simulated annealing is needed at low resolution.

However, the *refinement phase* is broadly similar in crystallography or (higher-resolution) microscopy, and NMR. The *experimental data* (e.g. lists of neighboring H-atoms) are incorporated into an 'error function' that must be minimized. However, it is desirable to include purely structural features, which would be optimized by energy minimization (*molecular modeling*). Therefore the experimental 'error function' is given an energy value and added to the calculated potential energy of the unrefined molecular structure. Of course, there is no fixed 'energy exchange rate' for the error function, so it is possible to adjust it during the refinement. Thus we can start by making refinement more dependent on experimental data (usually stronger at lower resolution), and end by making it more dependent on purely structural features. Each has essential interests that must be respected.

The wealth or paucity of experimental data shift the relative contributions of new experiments and old model-building data. At one extreme, ultra high-resolution crystallographic data (e.g. Jelsch et al., 2000) are sufficient to define an accurate structure with no extraneous information from small molecules. At the other extreme, NMR or microscope studies sometimes give only few data that sketch the rough structural organization, so much additional information needs to be incorporated and fitted to give a convincing model. This changes the character of the last stages of refinement.

Historically, structural biology started with simple molecules giving rich data at good resolution, and has subsequently mostly progressed towards large, flexible or peculiar molecules (like membrane proteins) for which the best available methods give data that are few compared with the molecule's complexity. This is particularly severe with NMR, which provides the maximum separations of specific hydrogen atoms (derived from NOEs), since these atoms must be identified from their spectral frequencies ('chemical shifts'). However, poor resolution of the spectra leaves many (even most) NOEs unassigned. The necessary extra assignment information is available from a molecular model (which decides which H-atoms are close and which are distant). Thus the structure must be found through alternating cycles of model-building (yielding bundles of conformers) and NOE assignment (Herrmann et al., 2002). The final 'structure' consists only of the latest, most refined, bundle of conformers, from which the mean gives the 'structure' and the variations define the 'resolution' (§12.3.2).

The same structural optimization technique has been applied to other structural constraints (e.g. shape data from ultracentrifugation and electron microscopy) in order to find the nuclear pore structure (Alber et al., 2007).

11.3.3 Dynamics and Normal Vibration Modes

As we have just seen, the energy minimum of a macromolecule is not a state of zero energy (which could only exist at absolute zero, if then), but one of constant, chaotic motion needing long calculations to simulate it. A more ordered view of this state *apparently* avoids chaos (but see below), by using recurrent cycles as in planetary dynamics. Instead of gravity, the main force driving these dynamics is the chemical bond, which has mechanical effects similar to those of elasticity. So we return to the 1D parabolic energy-curve (Fig. 11.10 (f)). Consider the state of a particle on a spring, represented by a stationary point at the top left of the parabola, where the system has the maximum potential energy. The gradient of the energy-curve implies a force on the particle and, when it is released, that force moves it, converting potential into kinetic energy. So the point falls down the energy curve, reaching the bottom where all the potential energy has become particle-motion. If this is not converted into heat, it carries the system up the other side of the curve, reversing everything exactly (except for direction) until the point is back at the top again, now on the right side. Apart from that change, everything is just the same as at the start, and therefore repeats again, oscillating from side to side.

Thus in 1D we have a periodic, cyclic behavior that repeats exactly (apart from friction). How does this differ in 2D? Figure 11.12 (b) shows that the energy surface has an elliptical cross-section, a shape with two unique axes (the distortion matrix's eigenvectors) bisecting the ellipse. So the entire 3D surface has two mirror-planes, one of which is shown in Fig. 11.13 (a), (b). If the ball is released at one of these planes, its symmetrical position should remain undisturbed. So, staying in this plane, it runs down the parabolic curve of the surface and rises to (nearly) the same height on the other side. Thus it will oscillate to and fro along this path. The same result applies if the ball is released on the other, perpendicular, symmetry-plane, though the vibration frequency must differ because the ellipse's axes have different lengths. In the plane of the longer major axis, the gradient is gentler, so the acceleration is less and the vibration frequency lower.

Both of these are the **normal modes** of vibration, whether of the small ball falling down the paraboloidal energy-surface in Fig. 11.13, or of the vibrating stick in Fig. 11.12. Such modes repeat cyclically; each has a characteristic frequency[4], and one mode never excites another. However, we see a more complicated result if we release the ball away from both symmetry

---

[4] Sometimes two eigenvalues coincide, as for example if the cross-section is circular rather than elliptical.

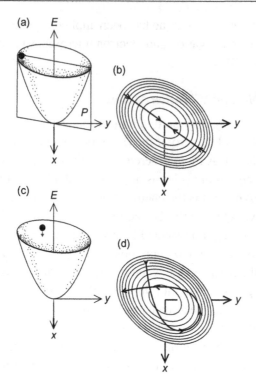

Figure 11.13: **(a)** A small ball is dropped down the surface. If it is released on a symmetry plane, its motion **(b)** is essentially 1D, like Fig. 11.10 (f). **(c)** If it falls away from the symmetry plane, the force is not parallel to the displacement (see Fig. 11.13 (a)). The consequence of the oblique force is a long curved path **(d)** that combines the two characteristic motions for this potential energy diagram.

planes (Fig. 11.13 (d)). Then its fall involves a combination of the two simple paths, but in perpendicular directions and at different frequencies. So we get a more complicated path, especially if the frequencies are not in a simple numerical ratio. We shall see the same effects if we release the stick in Fig. 11.12.

The elliptical shape of the elastic energy surface can be represented (as in distortion) by a matrix whose eigenvectors are the axes, and whose lengths (the eigenvalues) define the vibration-frequencies. These modes, the eigenvectors, act like orthogonal functions (such as sines and cosines). Recall that combinations of these can imitate almost any function. Thus, when a sharp collision with another molecule excites a very localized vibration, it is equivalent to the composite vibration of very many modes which happen to cancel out elsewhere at that moment. But, because of the different frequencies of every mode, the cancellation soon disappears, giving place to a very complex series of changes. In this way the apparently simple, cyclic motion of very many cycles can be equivalent to chaotic motion.

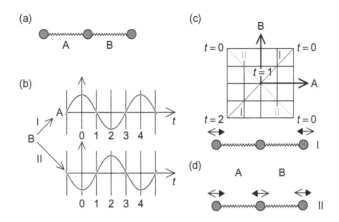

Figure 11.14: The simple case of a symmetrical tri-atomic molecule **(a)** in 1D, with two modes (I, II) with the vibrations **(b)**, where A follows the upper curve and B follows the mode arrow; **(c)** and **(d)** show the vibration coordination in modes I, II.

In a macromolecule, the eigenvectors (i.e. the patterns of vibration we call modes) are very complicated. The underlying principles are more clearly shown in a simple linear structure. Figure 11.14 shows the simplest example of this, the normal modes in a linear triatomic molecule (a), where the terminal atoms are identical and so too are the bonds A and B; (b) shows how these can lengthen/shorten in phase (I) or anti-phase (II), made explicit in (c); (d) shows the two resulting modes in 1D. (This applies to a triatomic molecule like CO_2 in 1D and mode (II), together with the two bending modes in 2D/3D, are responsible for its 'greenhouse gas' properties.) Some vibration patterns leave portions of the structure quite stationary. In (I) the central mass is stationary, so it plays no role in this particular vibration, where there are effectively only the outer masses joined by a spring of double length. Stationary parts of the structure are called **nodes**. Vibration (II) is more complicated and has a higher frequency.

A very long straight system of many identical masses and springs, of which Fig. 11.14 (a) is only a short approximation, would have a vibration mode like (I) where the center is a stationary node and the movement is bigger towards the ends (Fig. 11.15 (a)–(c)). This (lowest) mode approximates a simple cosine wave, and the next higher mode (d)–(f) has a cosine wave of shorter wavelength and two nodes. Thus the eigenvectors of a vibrating system can approximate the trigonometric functions that form the basis of the FT. Similar eigenvectors (or eigenfunctions, as they might be called) arise in the statistical analysis of single particles (section §17.3).

As with the characteristic resonances of a room (section §7.4.4), the vibration modes of a molecule may be excited by an impulse (hitting the molecule), by a steady vibration or by noise (Brownian movement). (However, the last process is also involved in damping a molecule's vibrations.) Only a few of the modes of a macromolecule are of interest: those

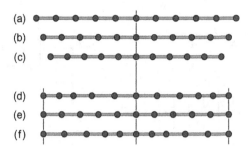

Figure 11.15: Two lowest normal modes of a long line of equal weights joined by equal springs. (Nodes indicated by dark lines.) (b), (e) neutral positions during the vibrations. **(a)**,…,**(c)** Lowest mode. **(d)**,…,**(f)** Next lowest mode.

with the lowest frequencies, in which the distortions are consequently the smallest and the vibration involves the greatest numbers of atoms moving in unison. These involve distortions big enough to be prominent at the lowest resolution. If the structure is symmetrical, like many biological macromolecules, then the vibration pattern can have the symmetry elements of the structure. The use of this fact leads to more efficient methods for calculating the normal modes of highly symmetrical structures, and these methods have been applied to viruses (van Vlijmen & Karplus, 2005).

11.4 Appendix

11.4.1 Random Walk[5]

If, after n tosses, the net number of heads ($=$ total heads $-$ total tails) is ΔH_n, then it is ± 1 more than the previous number ΔH_{n-1}, so $\Delta H_n = \Delta H_{n-1} + c$, where c is the difference ± 1. Squaring, we get $\Delta H_n^2 = (\Delta H_{n-1} + c)^2 = (\Delta H_{n-1})^2 + 2c\,\Delta H_{n-1} + c^2$. Now take averages $<\ >$: $<\Delta H_n^2> = <(\Delta H_{n-1} + c)^2> = <(\Delta H_{n-1})^2> + <2c\,\Delta H_{n-1}> + <c^2>$. The average $<2c\,\Delta H_{n-1}>$ is zero as c (± 1) is as often positive as negative. So $<\Delta H_n^2> = <(\Delta H_{n-1})^2> + <c^2>$. Consequently $<\Delta H_{n-1}^2> = <(\Delta H_{n-2})^2> + <c^2>$, so (substituting for $<\Delta H_{n-1}^2>$) we get $<\Delta H_n^2> = <(\Delta H_{n-2})^2> + 2<c^2>$. Proceeding this way, we find $<\Delta H_n^2> = n<c^2> = n$. Thus the deviation between heads and tails, which is proportional to $\sqrt{<\Delta H_n^2>}$, is proportional to $\sqrt{n}$. If the 'tosses' occur at a regular rate, then n is proportional to t, so the deviation between heads and tails is proportional to $\sqrt{t}$.

---

[5] The square-root increase of a random walk is a well-known theorem of statistical physics (e.g. Feynman, 1963).

CHAPTER 12

The Third Dimension

Chapter Outline

12.1 Depth Through Tilting

12.1.1 The 'Viewing-Sphere' and its Fourier Transforms

We ordinarily perceive 3D structure through stereo-vision or, with very close objects, from the narrow depth of focus (a method developed by the microscope's 'optical sections'). But both methods use a small tilt angle, and therefore depend upon extremely sharp, clear (i.e. 'very high resolution') images. No depth-information is obtainable without viewing the object from different directions[1]; but, to perceive depth from nearly parallel directions, the very small parallax effects from two views must be both distinguishable and interpretable. Distinguishability requires that the different levels of the image must give separate peaks in

---

[1] In astronomy, where only one viewing direction is available beyond neighboring stars, there are a few interesting exceptions like plotting the spiral arms in our own galaxy. But these depend on a prior knowledge of the type of structure present.

Structural Biology Using Electrons and X-Rays.

the cross-correlation function (CCF) of the two views. Since the shifts are small, the matched detail must therefore be extremely small. Also, to be interpretable, the CCF peaks must not yield ambiguous matches, as are often given by random distributions of points.

Both requirements imply *high* resolution combined with *low* information content. Unfortunately, macromolecular structures combine *limited* resolution with *high* information content (so this combination must be artificially reversed to create interpretable stereo-images). To perceive the details of complicated, blurred 3D structures, we need the full range of viewing directions. (We shall soon examine more precisely what this means.) Having got these, there is the further challenge of organizing the data. A 'single-particle' microscopist (Chapter 17), faced with innumerable different projection-images of the same particle-type seen in every possible orientation, needs to fit the images together correctly. We might call the ideal assembly a projection-covered 'viewing-sphere' (Fig. 12.1). This (purely conceptual) device would be a large polyhedron, the center of which would contain a model of the particle. Each of the device's polygonal windows would contain a photograph of the particle, taken from that window. So each window would show a different photograph, and rotating the viewing sphere would reveal the particle's full 3D structure, just as if we were looking at it instead of its images.

This 'thought-apparatus' could contain any structure and, if that were an ordinary object (like a china ornament), all the photographs would indeed be different. But molecules, as seen with X-rays or electrons, are almost completely *transparent*. When viewed from opposite sides, a glass ornament looks different, but a molecular image (a pure projection) looks almost exactly the same (more precisely, it shows the exact mirror-image, like the shadows of ordinary objects). Consequently, half the viewing-sphere would be (essentially) identical to the other half, so the 'well-covered viewing-sphere' for (transparent) molecular particles has a center of symmetry and each half conveys the full information of the 'viewing-sphere'.

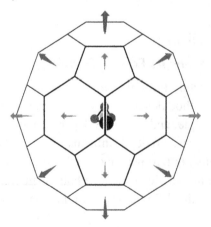

Figure 12.1: The projection-covered 'viewing-sphere': a polyhedron with viewing 'ports' surrounding the central object (shaded); each 'port' gets a different view of its projection, a photograph of which is hung from the 'port'.

Thus we need only half the viewing-sphere to find the particle's structure. Can this fraction be reduced further? We have already seen how the number of 'views' could be reduced substantially if the resolution were extremely high, but we are concerned here with what is practical with macromolecules. However, we can gain a new perspective on this problem by considering the viewing-sphere's Fourier transforms (FT). By the projection theorem (§6.1.2), each photograph in a polygonal window has a 2D FT that is a parallel section of the object's 3D FT. Since the central object is supposed real, its FT has Friedel symmetry (§6.1.2): on opposite sides of its center, the phasors have opposite phase-angles but the same amplitude (or intensity). Thus if, as in X-ray diffraction, we can record only intensities, the diffraction pattern has the same central symmetry as the molecule's viewing-sphere.

If the initial goal of a 'single-particle' microscopist would be a 'well-covered viewing-sphere', that of an X-ray crystallographer could be called a 'well-filled FT'. Using electrons, or X-rays with an extremely small wavelength, the Ewald sphere (§8.4.3) is flat and each record consists of a thin flat slice passing through the center. Such slices must be packed so as to have each reciprocal lattice point on at least one sheet. Of the many ways to arrange the slices, the simplest has them all intersecting at a common axis, from which they fan out, as in Fig. 12.2 (b). But, within the resolution limit of the FT, the slices (or sheets) must not be separated by any gap big enough to hide a reciprocal-lattice point.

This shows us the 'well-covered viewing-sphere' from a different perspective. That sphere was covered all over with polyhedral 'viewing ports', but the viewing directions corresponding to Fig. 12.2 (b) constitute a fan of rays lying in a plane (a). So the 'well-covered viewing-sphere' need only have 'viewing ports' around an equator, provided there are sufficient that are evenly distributed. It is not *essential* to have views from many directions, as in Fig. 12.1, though there are important advantages, especially if the object is wide and flattened as in Fig. 12.2 (a). An isometric object is probably best studied symmetrically from many different directions.

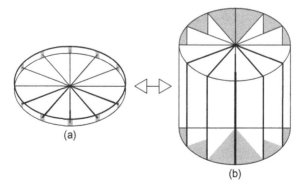

(a)

(b)

Figure 12.2: (a) A series of views of a central object, similar to Fig. 12.1, but arranged only around an equator. **(b)** Their 'well-filled FT' consists of thin flat sheets with a common vertical axis.

12.1.2 Tilts and Three-Dimensional Resolution

The case where the FT is sampled with a single-axis tilt allows us to find a simple formula for the resolution of an FT. The essential condition, that the sheets must not be separated by any gap big enough to hide a reciprocal-lattice point, sets a minimum number to the intersecting record-sheets required to reach a given resolution d. The separation of reciprocal-lattice points is (from the sampling theorem, §5.3.2) roughly the reciprocal of the molecular width, D. If we had been looking at a single molecule's FT, the fan of n sheets in Fig. 12.2 (b) would correspond to Fig. 12.3. Beyond a radius of $1/d$, the gap between the sheets becomes too great to determine the FT. Now the maximum separation of the transform planes (corresponding to projections) is approximately the radius $(=1/d)$ times the angle between the planes $(=\pi/n$ radians). That separation must equal the reciprocal of the object size (D), so we have $\pi/(nd) = 1/D$, showing that the resolution of the reconstruction $d = \pi D/n$. Alternatively (Crowther et al., 1970), the number (n) of projections needed to reach some desired resolution (d) in the reconstruction is:

$$n = \pi D/d \qquad\qquad (12.1)$$

This general rule of course needs modification if the particle (and hence its FT) are symmetrical. Thus an object with C_N symmetry, with the axis coinciding with the tilt axis, would only need $1/N$th of the tilt images to reach the same resolution.

This formula gives only a minimum value, and it relates to a single-axis tilt method that is used only in crystalline sheets and (effectively) in helices. Other geometries need other criteria, but they are always important. They are based on the conditions for filling the FT, and therefore apply as much to collecting crystallographic data (Chapter 13) as to Fourier 3D reconstruction (section §12.4).

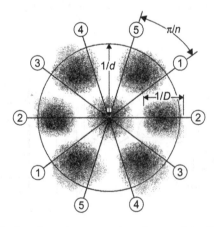

Figure 12.3: Diagram for deducing the rule connecting the particle diameter (D) and the number of equally-spaced tilt images (n), with the resolution of the three-dimensional reconstruction (d). (See text.)

An alternative condition can be deduced for filling real space. Figure 12.4 shows how we can divide the image into a square array of pixels and put the pixel density at the centers of each (Crowther et al., 1970). Then each projection gives a series of projection-lines ending in line-pixels. Each of these is the sum of all pixel densities along the projection line.

If the particle width is w and the reconstruction resolution is r, then the number of pixels needed is (particle area)/(pixel area) = $(¼)\pi w^2/r^2 = (\pi/4)(w/r)^2$. This is the number of unknowns. The average projection, at resolution r, contains about $w/2r$ line-pixels; so n projections give us $nw/2r$ line-pixel densities. This is the number of knowns, i.e. the number of equations. Now a set of equations is soluble only if there are at least as many equations as unknowns. So $nw/2r \geq (½)\pi(w/r)^2$, or $n \geq (½)\pi(w/r)$, or $r \geq (½)\pi(w/n)$. This is the same equation as (12.1), but with $\pi/2$ instead of π. (This is not unexpected, as the arguments about resolution are uncertain within a small arbitrary constant.)

12.1.3 Tilts from Symmetry

The information needed to determine a particle's 3D structure can be obtained from different views of an asymmetric particle, or from the internal rotations of a highly symmetrical one. Help from symmetry was very important at first, and is still most welcome. We survey in Part IV all the symmetrical structures, ordered by diminishing numbers (3, 2, 1, 0) of independent translations. But these are important for raising the signal:noise ratio, not for producing tilts. The largest number of independent rotations (potentially useable for tilts) is (24, 12, ∞, 60), respectively, showing a measure of complementarity between signal:noise and tilts. Crystals (three translations and ≤24 rotational orientations) are not available for electron microscopy, but thin crystalline sheets (two translations and ≤12 rotational orientations) are suitable.

Rotations are useful only when the rotation axis is not parallel to the beam (it is best in the grid-plane), which restricts the symmetry value of thin crystalline sheets. However, helices

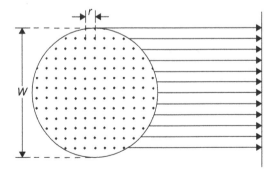

Figure 12.4: A circular object, diameter W, is divided into pixels r apart (r = resolution). Each pixel's density is concentrated in a point, and the sum of all densities along a line of points is registered on the vertical line at the right. This shows how much projection-data we need to calculate the pixel densities.

have a screw axis that could generate very many views, and icosahedral particles can show 60 different views. In this way, helical (Chapter 15) or icosahedral (Chapter 16) particles can generate almost sufficient tilts for a 3D reconstruction from one particle (though good resolution would require very many copies). This can be illustrated by stereo-images of highly symmetrical structures' projections, which reveal the internal symmetries through depth effects. The following images show how two copies of a single image, positioned so that the copies differ by a reflection or shift, can generate a stereo image showing structural details. The first two (Figs 12.5 and 12.6) exploit helical symmetry, the former image showing

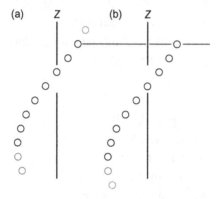

Figure 12.5: Stereo picture of a discrete right-handed helix, both images being identical copies of the same helix. The vertical line shows the helix axis. **(a)** image for right eye; **(b)** image for left eye. Place a finger between your nose and the picture, focus on the finger and try to fuse the two z's at the top of the picture. (The following stereo pictures should be viewed in the same way, except for Figs 12.6 and 12.8, which allow either method of viewing.)

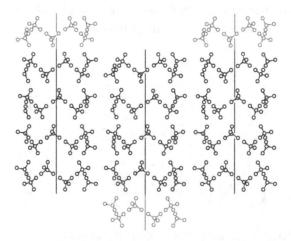

Figure 12.6: Stereo picture consisting of two identical helix images, shifted along the axis (vertical line) as in Fig. 12.5. The tetrahedral geometry should be visible when the images are fused 'cross-eyed' as in Fig. 12.5.

only the geometry of a helix while the latter shows structural details within it. In the second two (Figs 12.7 and 12.8) the single image is a part-projection of a structure with icosahedral symmetry.

12.1.4 Collecting Tilt Data

The theory of using tilted particles is clear and simple, and its application to negatively-stained specimens at low resolution is not too difficult. But detailed study is made hard by the extreme radiation-sensitivity of particles bearing high-resolution data. While under this tight constraint, we need, first, to record projection-images from particles in a sufficient range of orientations and, second, to determine those orientations afterwards. There are three general, distinct ways to collect tilt data so as to find orientations: from 'internal' tilts generated by the particle's symmetry; from particles already tilted, where the tilt angle is deduced from the image; and from one particle's *pair* of tilt images, related by an angle known from a goniometer stage, the second image (bearing the resolution loss) being used solely for alignment purposes. (We can call these general methods 'internal tilts', 'deduced tilts' and 'double tilts'.)

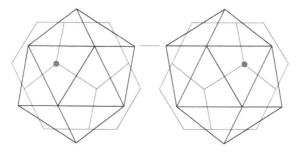

Figure 12.7: Stereo picture formed from two identical images related by a vertical mirror between them. Black: top half of an icosahedron; grey: top half of a pentagonal dodecahedron whose vertices are nearly at the centers of the icosahedron's faces.

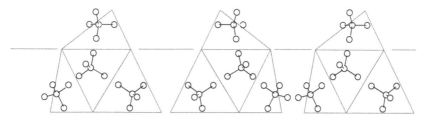

Figure 12.8: Stereo picture of detailed molecular structures fitting the arrangement of Fig. 12.7. Each of these identical (mirror) images is a projection of a structure with icosahedral symmetry, which can be approximately seen.

All these methods can accept particles in different orientations relative to the grid, except that a long translation is effectively restricted to lying along the grid. Thus grid-tilting is essential for crystalline sheets, but not for helices which can use the single tilt-axis geometry of §12.1.2. However, particles lacking translations benefit greatly from tilt data, which thus contribute to all except perhaps helices, where the 'internal tilts' are so generous.

Determining orientations is a major problem, especially with smaller particles. It is mitigated with larger particles, and becomes straightforward when they also have a complex 'symmetry framework' whose orientation can be found. That is easiest when a translation(s) improve the signal:noise ratio. 'Single particles' with low or no symmetry pose the biggest challenge. Although 'deduced tilts' ('angular reconstruction', §17.4.2) are used for these, they are also candidates for 'double tilts'. Here the most practical scheme is the **random-conical** data-collection method (Radermacher et al., 1987), which uses two successive exposures (e.g., $5\,e/\text{Å}^2$ for the first exposure and perhaps $15\text{--}20\,e/\text{Å}^2$ for the second). The *second* (low-resolution, high-exposure) image is taken with the grid in the standard position perpendicular to the beam,

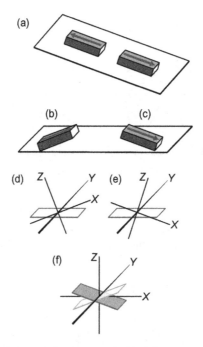

Figure 12.9: Random-conical data collection method. **(a)** Two specimens on the same grid, so tilted the same way but pointing in opposite directions. **(b),(c)** The left specimen (b) has been turned to point right, so now the specimens point the same way but tilt opposite ways. The intensity 'spikes' of their corresponding FTs correspond with the specimens' edges (**(d)** to **(b)**, **(e)** to **(c)**), and the rectangular projection planes are parallel to the image plane. If these FTs are straightened (so that their axes coincide), and then added together, we get **(f)** showing the two section planes, the shaded one corresponding to that in (d).

after the *first* (high-resolution, low-exposure) has already been taken with the grid tilted. This method is particularly useful when specimen particles all attach by the same surface to the grid. Thus a photograph of an untilted grid would contain many 'copies' of the same particle image, but with random in-plane orientations. Such a field is suitable for particle averaging, but not for structure-determination, since each image is essentially the same, after turning the paper. However, that is changed by tilting the grid. Now each different particle-rotation becomes converted into a different particle orientation. The geometry of the process is shown in Fig. 12.9. Two different particle orientations on the tilted grid are shown in Fig. 12.9 (a); they are anti-parallel, so the grid tilt has the effect of rotating them opposite ways, as shown in (b), (c). The FT of each particle is oriented parallel to the particle's internal coordinate system (from the 'rotation theorem', §6.1.2), as in (d), (e). In each case, the section plane (shown horizontal) samples different parts of the FT. Which parts are sampled becomes clear if all FTs are first made parallel, and then combined into one diagram (e) with the two section planes keeping their relation with the coordinate axes. We see that the part of the FT that is recorded is an equatorial wedge around the X-axis. (Of course, with a range of different particle rotations on the tilted grid (a), that wedge gets rotated about the Z-axis to form a 'toroidal wedge'.)

In order to construct either the 'viewing-sphere' or the FT (or to average the data from symmetrical particles), we need to know the spatial relations of different particles. So we must understand the numbers and types of the orientation parameters required.

12.2 Aligning Particle Images
12.2.1 Overview of Alignment Methods

In electron microscopy, we need to align particle images in order to average them; in crystallography, we need to use a structure solved in one crystal to elucidate a related structure in another crystal. In the former case, our data are in real space; in the latter, in reciprocal space. In each case, we need to know what movements will bring two structures into exact coincidence. Two methods are widely used to bring this about, using respectively FTs and correlation. The first method, better adapted to symmetrical structures which have an inherent coordinate frame, first calculates the FT of each structure relative to that common frame. Then the two FTs can be simply added together. The second method regards the two structures as different functions that need to be matched together, as discussed in Chapter 2, §2.3. This is done using the CCF (or, better, the related correlation-coefficient, in which the CCF is scaled according to the size of the two functions). Both methods work best with a single rotation, which has no 'termination errors' that accompany the FT or CCF of short functions (e.g. the triangular shape of the correlation histogram in Fig. 2.1). With translations, the CCF is apparently better; it is clearly better when searching a long function with a short function to find a match.

The task of alignment is made more challenging by the presence of noise, which is different in the two cases. In electron microscopy, this is mostly a high frequency 'shot noise' from low electron exposure (to minimize specimen damage). We have seen in section §2.3 that, without noise, the CCF of two identical 1D functions would give the function's autocorrelation function (ACF); and the position of its high origin peak would indicate the correct movement. But a high noise background nearly submerges this peak, leading to the need for a low-noise 'reference' image, and problems in finding a sequence of images to get it. (See Chapter 17.) Such translational and rotational alignments of images are particularly important when the particles lack symmetry.

In crystallography (See Chapter 13), such alignments have new problems. First, the pairs of structures we need to align are not both simple images. One (at least) will be known only through its diffraction pattern. That limitation is equivalent to knowing only its ACF (called the 'Patterson function' in crystallography, §3.4.6), which is a map of the image's vectors, rendered very 'noisy' by the manner of their superposition. So we need to find the optimal overlap between two Patterson vector-maps (one of which may be calculated from a known structure). Although these ACFs are not as good as the actual images, they can give satisfactory overlap peaks. They can be used to find both rotations and translations. But finding translations is complicated when the crystal's symmetry framework contains rotations. Each trial translation moves the molecule into a new place relative to the crystal's rotation axes, which then make many new copies of it. When searching for overlaps, we need to take account of the vectors between all these copies and the unknown molecule within the crystal.

As we shall see, we need a total of six parameters for an alignment in 3D, so the problem is in principle a CCF search in 6D, involving many small increments of each parameter. That would be a very heavy calculation, which is much simpler if it can be separated into separate stages (as we saw in the last chapter, §11.2.4). Fortunately, rotations can be separated from translations. Moreover, CCF calculations can be greatly accelerated by using the FFT (§4.3.2). This applies to both translations and rotations in 1D, and the FFT works well with translations in higher dimensions. However, that is not so for rotations (see the discussion in §16.1.3), although there is a (rather complicated) 'fast rotation function' (Crowther, 1972).

12.2.2 Three Translations and Three Rotations (3T + 3R) Define a Particle's Orientation and Position

Suppose there are two copies of the same coordinate frame, (x,y,z) and (x',y',z'). What movements (position and orientation parameters) are required to bring them into exact coincidence? They are shown in Fig. 12.10. (i) First, we need to make the origins coincide (a). To bring the (x',y',z') origin into coincidence with the (x,y,z) origin, we shift it first by Δx so that it lies in yz-plane (shaded), then by Δy so that it reaches the z-axis, and finally by Δz so that it reaches the origin. Now the (x',y',z')- and (x,y,z)-axes have a common origin, but they still have quite different orientations, as in (b). (ii) So we rotate the (x',y',z') axes about the x-axis until z' lies in the xz-plane (shaded). This rotation is $[R_x]$. Now the situation is shown in (c), with the y' and z' axes still in arbitrary positions but the z' axis is in the xz-plane.

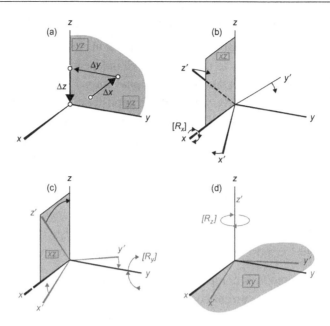

Figure 12.10: Alignment of two coordinate systems. The three translational parameters are needed to superpose origins (a) and the three rotational parameters $[R_x]$, $[R_y]$, $[R_z]$ are needed to make parallel coordinate axes **(b)–(d)**. (See text.) (This resembles Euler's angles (§12.2.3)).

(iii) So, in (c), we can use a rotation $[R_y]$ about the y-axis to bring it, within that plane, into coincidence with the z-axis. Now the situation is shown in (d), with the z- and z'- axes in complete coincidence, including their origins. The x' and y' axes are perpendicular to it, so they must both lie in the xy-plane. (iv) Therefore only one final rotation $[R_z]$ about the z-axis is needed to bring the x' and y' axes into coincidence with the x and y axes.

Thus, to record the exact position and orientation of any object, we need six position and orientation parameters (**alignment parameters**): three translations (Δx, Δy and Δz) and three rotations ($[R_x]$, $[R_y]$ and $[R_z]$). The fact that we need precisely as many rotations as translations reminds us that there is a continuum between rotations and translations; a rotation approximates closer to a translation as the distance from the rotation axis increases[2]. We shall refer to these three translations and three rotations by the shorthand (3T + 3R) = (T_x + T_y + T_z + R_x + R_y + R_z).

---

[2] Good examples of this are the rotation of the earth about its axis or (even better) about the sun. The rotational components are so small that the associated Coriolis forces are very hard to detect – so hard that these rotations were (notoriously) long denied. The diurnal forces can now be easily demonstrated by Foucault's pendulum; but the annual forces?

12.2.3 Representing Rotations: Eulerian and Spherical Polar Angles

These three translations and three rotations could be defined in alternative ways. Thus, although we usually define the total translation as the sum of three independent Cartesian ones (as in Fig. 12.10 (a)), we could make the total translation in *one movement* defined by its length and direction, as with vectors. Similarly, we could accomplish the overall 3D rotation in one operation about a defined axis. Figure 12.11 (a) shows how, given two arbitrary points A and B on the surface of a sphere, we could join them by a line (geodesic) of shortest length (obtainable by pulling a cord tightly over a metal sphere), shown thickened in Fig. 12.11 (a). This would have to be part of a great circle (shown shaded), whose central perpendicular would be the rotation axis, and a rotation of ω would move A to B. The axis direction can be defined by two angles in spherical polar coordinates, as in Fig. 12.11 (b). The three angular parameters (θ, ϕ, ω) in (b) depend on the *rotation formula* (Jeffreys & Jeffreys, 1962; Goldstein, 1980). These **spherical polar angles** are perhaps the most intuitively 'obvious' method for representing rotations.

However, the simplest method for computation (**Euler's angles**) uses three successive rotations, each about one of the current coordinate axes of the object. Generally the starting axis is labeled z, followed by either x or y, and ending with a z-axis rotation again (Fig. 12.12). These are called the x-convention and y-convention, respectively. The x-convention is usually chosen in molecular dynamics and crystallography, but the y-convention is the usual choice in single-particle analysis (i.e., rotations are in the order z, y, z). Note that all methods use *three* angles.

12.2.4 Classifying the (3T + 3R) Alignment Parameters

We are here considering the orientation problem in both electron microscopy and X-ray crystallography. We start with the latter problem, which is much simpler as translations have no effect on diffraction pattern intensities (§3.4.7). So (3T + 3R) → 3R for X-ray diffraction patterns. Besides needing only the rotations, one rotation (within the plane of the film or detector) makes

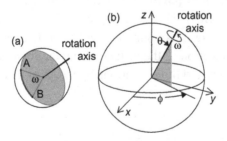

Figure 12.11: (a) Any point A can be rotated to any other point B (on the surface of the same sphere) in *one rotation*. The shaded plane bisects the sphere, and thus contains its center and meets the sphere in a great circle. **(b)** The three spherical polar angles (θ, ϕ, ω) for rotating a rigid body in any required way.

no change to the diffraction pattern (excluding any 'anomalous scattering', §13.3.4). Thus only the two 'out of plane' rotations can affect the character of the X-ray diffraction pattern.

Microscope images, which preserve phases, allow no such simplifications. Before we can average two images, we need to find the (3T + 3R) alignment parameters of their particles, but we get them from different sources. The (3T + 3R) can be divided into three different groups: $(3T + 3R) = (2T + R) + (T_z) + (2R)$.

(i)   The $(2T + R) = (T_x + T_y + R_z)$ are in-plane adjustments. The translations $(T_x + T_y)$ are in the (x, y)-plane of the detector, and the rotation axis of R_z is along the optic axis z, perpendicular to the detector plane. (This is the 'in-plane' rotation for diffraction patterns.) These 'in-plane' adjustments are for averaging different images of the same particle type, presenting the same direction to the observer.

(ii) The remaining translation (T_z), perpendicular to the image-plane, occurs along the z-axis, the optic axis of the microscope. It therefore causes defocus, which changes the phase-contrast transfer-function (PCTF, section §9.2.6).

(iii) The remaining $(2R) = (R_x + R_y)$ are the two 'out of plane' rotations that we encountered with crystals. They generate genuinely new images, and finding these parameters for each image is a major challenge with all methods of image analysis in electron microscopy. Because of the importance of these adjustments, the representation of these rotations is discussed in the next section.

It is useful to distinguish between *mutual* alignment parameters and *absolute* alignment parameters. Absolute alignment parameters fit all the images to a single coordinate frame, defined by one object. Mutual alignment parameters merely specify how to fit two projections to each other. We can always calculate the mutual alignment parameters of two particles

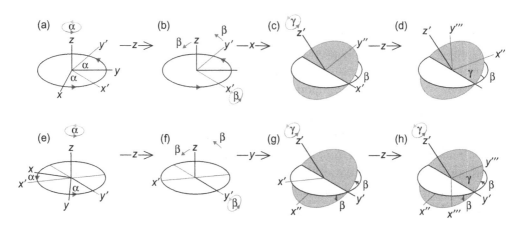

Figure 12.12: Euler's angles method of rotating a rigid body in any required way with three angles, α, β, γ. (a) x-convention; (b) y-convention.

whose absolute alignment parameters are known. And, if we had all the mutual alignment parameters that join all particles in a chain then (ignoring cumulative errors) we could calculate their absolute alignment parameters.

12.3 Information Content of Particle Images

12.3.1 Original Images

Obviously, all structural information in electron microscopy comes from images of 'particles'. (The term is ambiguous, meaning anything from a monomer to an asymmetrical or symmetrical aggregate.) But one particle's image never suffices for a high-resolution 3D structure. Noise must be reduced by averaging many identical views, and depth information obtained by combining many different views.

Averaging or combining views depends on knowing the particle's alignment parameters, whose accuracy determines the final image resolution. Of the $(3T + 3R)$ parameters, just one (T_z) can be found from background information since it affects the CTF. The remaining $(2T + 3R)$ depend exclusively on the information in particle images. If that is inadequate, different images will be badly aligned, so their average will be blurred or, worse, systematically wrong. This will make a poor basis for aligning other particle images, blocking any progress towards high resolution, or locking the initial approximate structure into a completely false one. We might have a great wealth of particle images that, well averaged, could yield molecular details of the structure; but, lacking the crucial alignment information, none of this is available to us; our project resembles an overladen plane that could fly very well, once in the air. To achieve 'take-off', each particle must contribute a *threshold* quantity of electron signal to the average. Its contribution depends on the image's content and quality of information.

Particle Size

We start with information content, which is mostly limited by the number of transmitted electrons, although it isn't simply proportional to it. For the error in the electron count n is proportional to the *square root* of the count: error $= k\sqrt{n}$. Thus the signal, which is count/error, is $n/k\sqrt{n} = (\sqrt{n})/k$, so it is proportional to the square root of the number of electrons.

Radiation damage limits the exposure of any particle (typically to 5–30 electrons Å^{-2} or 500–3000 per nm^2). If we assume that all images are recorded at this limiting exposure, the number of transmitted electrons from a particle is proportional to the particle area. Thus the signal (proportional to its square root) is proportional to the *linear dimensions* of the particle (e.g. to the diameter of a circular particle). (However, the signal wouldn't be proportional to the *length* of a helix, but to the *square root* of the length.)

These parameters apply to typical electron microscope particles that are thin enough to transmit electrons easily. However, when radiation is more penetrating (as in X-ray diffraction), the total scattering is also determined by the specimen thickness. Therefore the important quantity in crystallography is the volume of crystal within the X-ray beam, relative to the unit cell volume: in other words, the number of unit cells in the beam.

Contrast

Suppose we included a large background area with the particle. That would not only add irrelevant electron numbers; it would also make less accurate our count of those electrons coming strictly from within the particle itself. The reason is that the extra electrons' count would not be a simple constant; it would fluctuate because of 'shot noise'. We can subtract the extra electrons, but not their count-fluctuation. Exactly the same would apply if those extra electrons were spread uniformly over the signal electrons. That is the situation when there is an electron 'background', e.g. of inelastically-scattered electrons. This effect is usually described as a loss of **contrast**, so contrast is obviously another factor contributing to the information in a particle image. (Another way of expressing this is that the noise level must be exceeded by the amplitude of the structure curve.)

Contrast is measured by the relative difference in the numbers of electrons transmitted by the brightest and darkest parts of an image. This kind of effect is seen in *neutron scattering experiments*, when the particles' contrast is varied by changing the hydrogen:deuterium ratio in the solvent. In that case, the intensity of scattered radiation is proportional to the *square* of the scattering factor difference (particle minus solvent). Figure 12.13 shows why this is so. Changing the contrast reverses the band-pattern, which one would think implied a linear change. But the images on each side of (d) look almost the same, despite having reversed contrast. Thus they give the same diffuse scatter, which must therefore be a *quadratic* function of the contrast. (For the same reason, contrast always enters as the square.) This contrast term shows the importance of using defocus to add contrast through the PCTF,

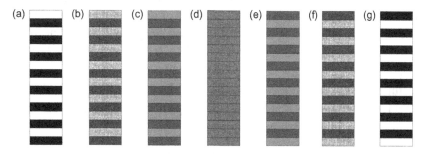

Figure 12.13: Grades of darkness represent different scattering power (or refractive index), white meaning low and dark meaning high. Two gradients alternate down each column. The top small rectangle in **(a)** is white (lowest); in **(b)** it is slightly darker (higher) and so on to **(g)** where it is black (highest). The rectangle below the top one in (a) has the reverse series. Consequently, the scattering power gradients are maximal in (a) and (g), and zero in **(d)**.

when particles are preserved in ice (and so have very low amplitude contrast). (But that contrast can be increased, after an initial reduction, by adding some negative stain to the buffer.)

Rosenthal & Henderson (2003) analyzed how the scattering amplitude of particles varies with resolution. At low resolutions (worse than about 1 nm), the density fluctuations within the protein are (relatively) 'invisible' and the main factor is contrast between the average protein scattering and the surrounding solvent. We shall call the resultant low-resolution high-contrast structure the 'solvent-contrast structure'. It is important because it forms the basis for particle alignment, whose accuracy determines how much of the weak, noisy information at higher resolutions can get averaged well enough to extract its precious contribution. For, unfortunately, the high-resolution information lies in a region where the aggregate's shape no longer contributes significantly to scattering, which is now mostly caused by internal density fluctuations (i.e. by protein structure). Here, the (much lower) scattering level is set by Wilson statistics (§11.1.1), and it would not be expected to decline very much from that level at higher resolutions, provided the images were perfect. But Rosenthal & Henderson found a *further* steady attenuation, thought to be caused by image blurring due to specimen movement induced by radiation damage. Thus it seems at present that this problem of specimen movement may represent the main barrier to higher microscope resolution.

Conclusions
Combining these two factors we find (Saxton & Frank, 1977) that the overall signal is proportional to (linear dimensions of the particle) $\times$ (contrast)^2. (This is a semi-quantitative relationship, showing the effect of doubling a parameter.) What are its implications?

We have seen that the quality of the final structure is limited mainly by the accuracy with which we can align (in translation and rotation) the many images we need to average to reduce the noise level. Now alignment depends on finding a clear, sharp maximum peak in the structures' CCF. As discussed in Chapter 2 (section §2.3.2), when we compare two signals by their CCF, we need strong signals with lots of detail. And we have just seen that a particle's signal depends on the *square* of the contrast, magnifying the contribution of the low-resolution 'solvent-contrast structure' of a particle. Indeed, that feature becomes the key to access to the interesting (but low-contrast) high-resolution structure. So this key is a low-resolution feature where we nevertheless need lots of detail, and a particle cannot contain lots of detail at low resolution unless it is *big*. (It also helps if its 'solvent-contrast structure' is not too smooth, but rough with projections and cavities.)

So high-resolution electron microscopy depends on getting big particles or aggregates, of which there are broadly two types: those with, and those without, translations. An aggregate with translations (a flat sheet or a helix) is potentially enormous, giving a very good alignment signal. That is why the early successes of image-processing came from helices and sheets. Particles without translations are compact, and it is here that the problem of particle size can become critical. The biggest compact particles tend to be those with high symmetry or pseudo-symmetry, i.e. the icosahedral viruses and similar polyhedral enzyme complexes. However, asymmetrical big particles (like the ribosome) are also satisfactory; symmetry is less important than size and surface contrast. (It is fortunate that the ribosome has such an irregular shape.)

To summarize, high resolution depends on averaging which in turn depends on accurate alignment, that must be based on the low-resolution 'solvent-contrast structure'. This leaves us with two useful specimen-types for high-resolution electron microscopy. First, there are the compact 'single particles' of two types. (i) The icosahedral particles (Chapter 16), especially viruses, which have the additional advantages of symmetry-averaging. Here a resolution of 3.3 Å has recently been achieved with a 90 MDa reovirus (Zhang et al., 2010). (ii) The low- or no-symmetry particles studied by 'single-particle' methods (Chapter 17 and Frank, 2006). Here, reaching molecular resolution is rarely achieved but is apparently possible, and therefore a challenge (Henderson, 2004). With both types, many thousands of images need to be aligned and averaged.

The second useful specimen-type is the 'enormous' aggregate with translational symmetry (especially crystalline sheets, Chapter 14, and helices, Chapter 15). Here the sheer size and convenient translational symmetry allow the use of FTs for structure-determination and internal averaging. They possess a 'symmetry framework', easy to find and thus invaluable for getting accurate alignment parameters. These structures were the subject of the 'electron crystallography' techniques (Glaeser et al., 2007). However, these 'enormous' helices and sheets have their own problem: distortion. As we shall see in Chapter 18, the solution to this requires that we select small segments from very big particles, and to align these segments like smaller 'particles'. Consequently, part of the apparent advantage of 'enormous' aggregates is illusory. Nevertheless, there have been several high-resolution images yielding molecular structures, such as bacteriorhodopsin (Henderson et al., 1990; Mitsuoka et al., 1999), the photosynthetic light-harvesting complex from chloroplasts (Kühlbrandt et al., 1994), tubulin (Nogales et al., 1998) and aquaporin (Gonen et al., 2004) from sheets. From helices, there have been bacterial flagellum (Yonekura et al., 2003), acetyl choline receptor (Miyazawa et al., 2003) and TMV (Sachse et al., 2007).

Finally, we should really conclude this list with the *macro*scopic (though small) crystals studied by X-ray diffraction (Chapter 13). Some molecules (e.g. glycoproteins) give only small crystals (tens of microns rather than hundreds), and others (e.g. viruses) contain very large 'molecules'. In both cases, relatively few unit cells are available, so radiation damage becomes a dominant factor. As the intensity of X-ray beams from storage rings increases (using wigglers, free-electron lasers, etc.), crystals with even fewer unit cells are giving recordable data, which need extensive averaging from very many crystals, like electron microscope specimens. This is developing into a further step in the convergence of microscopy with diffraction.

12.3.2 *Information, Resolution and Fourier Transforms*

The concept of information, discussed in Chapter 5, is closely connected with that of **resolution** – a kind of quality badge, both for the diffraction pattern and for the structure deduced from it. As applied to images, the simplest definition of resolution is: the minimum separation distance d of the points, in a line of equidistant points that can just be distinguished from a continuous line. The line of points will give rise to diffracted rays at angles corresponding to multiples of $1/d$ (in reciprocal space), so all trace of the separate points disappears within the region where $X < 1/d$. Therefore, if a diffraction pattern extends up to $1/d \, \mathrm{nm}^{-1}$ (i.e. $1/10d \, \text{Å}^{-1}$), it is said to have a resolution of $d \, \mathrm{nm}$ (i.e. $10d \, \text{Å}$, remembering that $1 \, \mathrm{nm} = 10 \, \text{Å}$). Thus resolution is a quality measure judged, paradoxically, by its *reciprocal*: a resolution of $1 \, \text{Å}$ is ten times *better* than one of $10 \, \text{Å}$. Moreover, it is a nonlinear function of experimental effort: to get a resolution of $1 \, \text{Å}$ we need, not ten times the amount of data for one of $10 \, \text{Å}$, but a *thousand* times as much.

The connection between resolution and information is clear if we imagine a structure's image divided into pixels (or *voxels*, for a 3D 'image'). What size voxel is small enough to transmit the image without serious loss of detail? When the voxel is too big, the image loses essential detail (as when a face is 'pixellated' to ensure anonymity) while, below a certain voxel size, further image improvements are invisible. Between these extremes lies the region where the voxel size is approximately the same as the resolution.

The information content of a (3D) image is a list of the densities of each resolution-sized voxel. The length of this list is (volume of object)/(volume of voxel) = (volume of object)/ (resolution)³. So the number of voxel-densities required increases as the *cube* of the (inverse) resolution: at $1 \, \text{Å}$ resolution we need 1000 times the number of densities that suffice for $10 \, \text{Å}$ resolution. If 10 particles yield a reconstruction at $20 \, \text{Å} = 2 \, \mathrm{nm}$, then $10 \, \text{Å} = 1 \, \mathrm{nm}$ resolution will require 80 particles at the very least. This cube-law dependence of information on resolution should also apply to the FT, which is easily confirmed. $2 \, \mathrm{nm}$ resolution implies an FT extending to $0.5 \, \mathrm{nm}^{-1}$ (i.e. a sphere of $0.5 \, \mathrm{nm}^{-1}$ diameter), while $1 \, \mathrm{nm}$ resolution implies an FT extending to $1 \, \mathrm{nm}^{-1}$, filling a sphere with a volume eight times bigger, and thus containing eight times as many diffraction spots.

However, the boundaries of a given resolution are not as sharp and simple as this analogy suggests. If the spots of a diffraction pattern disappear at spacings $>1/d$, they will show a pronounced weakening at $1/2d$. (Thus, referring to the filter curves in Fig. 5.4, the curve of diffraction-versus-resolution typically resembles the Gaussian curve, not the rectangle.)

---

[3] The precise definition of the Debye–Waller temperature factor B matters since it measures the resolution of a diffraction pattern. It is usually defined (Woolfson, 1997) so that the structure factor F attenuates approximately according to $F(\theta) = F(0)\exp(-B\sin^2\theta/\lambda^2)$, so the intensity must attenuate according to $I(\theta) = I(0)\exp(-2B\sin^2\theta/\lambda^2)$. Since $2d\sin\theta = \lambda$, $\sin^2\theta/\lambda^2 = 1/d^2$ and thus, at a resolution d, we have $I(d) = I(0)\exp(-Bd^2/2)$. (Here $\exp(x)$ is the exponential function e^x.; see Chapter 19.)

This makes the precise number we assign to resolution somewhat arbitrary. In crystallography, where such matters have been studied longer, the best single number (the *B*-factor) describes the coefficient of a Gaussian curve[3]. Moreover, the resolution can be anisotropic, so the voxels are not perfect cubes. If they resembled the shape and packing of ordinary bricks in a wall, the resolution would be finest vertically and coarsest along the length of a wall. Anisotropic resolution is particularly pronounced in the electron microscopy of large structures which are very thin in the beam direction and thus of asymmetric shape.

Information is of course easily lost and consequently resolution, which measures information content, is easily lowered. Its optimum is set by the original data, most obviously (in the case of structures with translational symmetry) by their diffraction pattern, whose resolution is reduced by anything that lowers the precision of the translational repeat. This is true of thermal motion, which continually shifts into different positions the atoms of different unit cells. Thus the *B*-factor that measures the pattern's resolution was originally called its 'temperature factor'.

Though universal, this contribution to disorder is a minor factor with macromolecule crystals. The outer parts of macromolecules are almost in a state of aqueous solution, and therefore more or less mobile. Since these are the parts that maintain the macromolecules in the crystal lattice, precise repeats can hardly be expected. Radiation damage, by randomly selecting different molecules for different chemical changes, reduces the repeatability still further. And these factors, which reduce the signal, are supported by other factors (such as photon shot-noise and solvent diffraction) that increase the noise.

All this is true not just for X-ray crystallography but also for electron microscopy, though this also suffers from further factors. The tiny specimen volume causes distortion and leads us to impose severe electron shot-noise through low exposures to minimize increased radiation damage. Moreover, there are instrumental factors like inadequate phase-contrast at higher resolution and specimen-charging. All these either increase the noise, or require extra image-correction. However, data-processing has two powerful tools for restoring some of the lost information: the averaging of data to reduce noise, and the incorporation of additional information from other sources (such as macromolecule sequences and the conformations of peptides or oligonucleotides). But even these 'curative' stages need controlling to check their progress. In this matter of statistical controls, crystallography has always led the way (see Chapter 13).

Most important is the problem of estimating the reliability and accuracy of the final structure. This is usually viewed as the problem of finding that structure's resolution. Thus, with almost any well-found structure, there is *some* resolution below which all the features are reliable, and another (better) resolution beyond which all details are conjectural. The *data* resolution is easily assessed from the diffraction pattern, both in crystallography, and electron microscopy when the specimen has sufficient translational symmetry to give an electron diffraction pattern (or the image to give an optical diffraction pattern). But the lack of repeats demands more ingenuity in devising resolution tests. One commonly divides the structure's FT into concentric spherical shells of uniform thickness, calculates the reliability of each shell's contents, and finally judges at which shell that reliability disappears. Measuring that reliability is based on some intrinsic symmetry: we estimate the deviations between measurements required to be the same. In crystallography, rotational symmetry is the basis of various R-factors. This is also applicable to most of the plane groups of thin crystalline sheets. Rotational symmetry is also helpful in helical and icosahedral specimens.

The greatest challenge is faced when assessing the resolution of particles with no symmetry at all. This is similar to the problem of estimating the reliability of NMR structures, which usually lack any symmetry. In both cases, lacking intrinsic symmetry, we can generate its essential feature (i.e., different measurements that should be the same) by repeating the same measurement on identical specimens. In NMR, we start with a random coil that we proceed to fold by calculation, under the action of artificial 'forces' imposed by the distance-data. It is then usual to make 10–20 different starting extended chains, giving the same number of different folded chains. This ensemble can be analyzed statistically. In electron microscopy, however, we divide the data into only two (theoretically identical) sets of measurements, and then measure their disagreement over some important quantity. Perhaps richer statistics could be got from more numerous (arbitrary) divisions of the data.

As already noted, the data need to be divided into spherical FT shells. But the data 'votes' are not strictly democratic: data that are more 'substantial' by virtue of their structure-factors F are made to count for more. That has been found necessary even when the data are inherently weight-free: phase residuals, used in the *Differential Phase Residual* (DPR), devised by Crowther (1971) to assess icosahedral reconstructions. Alternatively, we use the product $F(1)F^*(2)$, which is already weighted towards high-F data. The numbers 1 and 2 refer to the two measurement sets so, if $F(1) = F(2)$, this product becomes $F(1)F^*(1) = F(1)^2$ which is always real and positive. When the sets differ, this product is complex and smaller. So we always take its real part, which is diminished by differences. In this way we get the Fourier Ring (or Shell) Correlation (FRC) method (Saxton & Baumeister, 1982; van Heel et al., 1982). (The FRC applies to *rings* of data in 2D, the FSC to *spherical shells* of data in 3D.) A better measure, $C_{ref} = \{2FSC/(1 + FSC)^{\frac{1}{2}}$, which estimates the correlation between the calculated and a perfect map, and is equivalent to the crystallographic figure-of-merit, was proposed by Rosenthal, Crowther & Henderson (see appendix to Rosenthal & Henderson, 2003).

These are the basic resolution measures used to assess data and data-processing in electron microscopy. In general, however, scientific knowledge gets validated in two different ways: from internal tests (within the field), whose rigor and reliability are always improving through experience; and from the measure of agreement between new results and complementary evidence from other branches of science. It has been one of the great strengths of crystallography to have always had a great body of comparative data from chemistry, and this support continued during the development of macromolecule crystallography. But the basis of this comparison, atomic and molecular structure, was lacking in the various forms of microscopy because of their poor resolution. All the earlier developments of image processing lacked this important validation, which was only achieved when cryo-specimens and improved data-processing yielded molecular resolution (starting with bacteriorhodopsin, Henderson et al., 1990).

This is perhaps the main value of extending the resolution of electron microscope structures. However, even 'sub-molecular' resolution can be compared with chemistry via molecular structures from crystallography, if those structures are sufficiently complex and the electron microscope images sufficiently detailed. Here the outstanding examples have been, first, the ribosome, where the breakthrough of atomic crystallographic structures has been combined with improved electron microscope methods for extracting detailed structures from single-particle images (see review by Schmeing & Ramakrishnan, 2009); and, second, the aqua reovirus (Zhang et al,, 2010), where the electron microscope images are actually better than those that might be expected from X-ray crystallography.

12.4 Three-Dimensional Reconstruction: Principles

It had long been appreciated that electron micrographs of thin specimens are projections, and amateurish attempts were made quite early to combine them by 'model building' that led to some insights, though compromised by uncertainties and controversies. These attempts were eventually replaced by an objective procedure, **3D reconstruction**, in which the structure was *calculated* from its projections. This is the reverse of the model-building method, where a hypothetical structure was checked against the data by calculating its projections[4]. But, while projections are easily calculated from a 3D density distribution, it wasn't so obvious how to calculate the density distribution directly from its projections.

12.4.1 *Fourier Three-Dimensional Reconstruction and Back-Projection*

Nevertheless, to anyone familiar with Fourier transform theory, even at the simple intuitive level of Chapter 3, there *is* an obvious way to do this[5]. Since structures are easily calculated from their FTs, finding the structure's FT is effectively the same as finding the structure itself. We are given various projections of the structure. The projection rule (§6.1.3) tells us that a projection is the FT of a central section of the 3D FT. So a collection of projections is effectively a collection of FT sections. These planes (or lines in 2D) all intersect in the

---

[4]　However, in those 'bad old days' there was usually no evidence as to the *directions* of any projections.

[5]　The essential discoveries were made in the 1920s by crystallographers, especially the two Braggs (father and son), who related crystal structure to Fourier series, the 2D series giving projections and the 3D synthesis giving the full structure (Hunter, 2004).

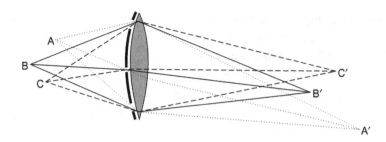

Figure 12.14: A wide aperture lens has a narrow depth of field for each point of the object, and hence preserves depth information about the object's points (A, B, C) in the image (A′, B′, C′). The lens achieves this by superposing a series of different pinhole images; three such images are isolated by the mask covering the left of the lens.

FT center from which they fan out. Sufficient projections give us enough lines (or planes) to define the FT, at least close to the center where they are closer. This gives us the **Fourier 3D-reconstruction** method (De Rosier & Klug, 1968). It can be classified, in the terms of Chapter 5, as a 'reconstruction' method (§5.1.2) for reversing a natural filter (i.e. projection) which destroys information (i.e. the distribution of density along each projection line).

The Fourier method is intuitively clear, and much used for exposition and discussion. However, we do not need to enter reciprocal space to reconstruct 3D structures. We can directly solve the projection equations (Crowther et al., 1970; also §12.1.2). But the simplest real-space solution to the reconstruction problem emerges if we consider the origin of the difficulty: the electron microscope's large depth of field. How do (light) microscopes with a *narrow* field-depth create their 'optical sections'? Field-depth is determined by the lens aperture, so we consider the image formed by such a lens. If we were to cover the lens by a card with a pinhole, that tiny aperture would much increase the field-depth, and objects at any field-depth would now give an equally sharp (pinhole) image. This would be true irrespective of where the pinhole were placed, but different positions of the pinhole would give different pinhole images (Fig. 12.14). When the entire lens aperture is used, the image is the sum of all the different pinhole images (ignoring diffraction effects). Fig. 12.14 shows how, with three simultaneous pinholes, the three different pinhole images from any object combine exactly at only one distance. Objects (A, B or C) at different distances from the lens give pinhole images that superpose exactly at different field depths (A′, B′, C′). The narrow depth of field of a wide-aperture lens is thus a consequence of combining many different pinhole images. Each shows (approximately) a projection of the scene from the viewpoint of the corresponding pinhole.

So the wide-aperture lens is an analog device for recombining many different projections to yield a 3D image. It should be quite feasible to simulate this process using different electron micrograph projections. But we can simplify this process, as one feature of these pinhole images is unnecessary: the different directions of the rays from A, B or C when they reach a

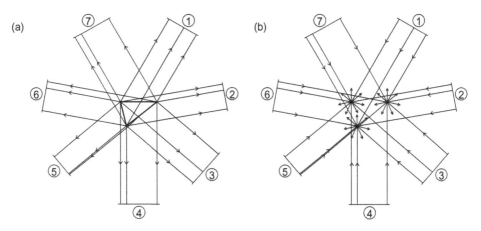

Figure 12.15: Real-space depth reconstruction by 'back projection'. **(a)** A central triangle generating seven projections in different directions. (The seven destinations correspond to the three pinholes in Fig. 12.14.) **(b)** Back-projecting from these seven projections, in their correct orientations and positions, reconstructs a representation of the original triangle, of which the three corners are shown.

pinhole. If the object ABC were very small, each pinhole would receive nearly parallel rays from all parts of the (small) object. So the reconstruction could just as well be accomplished with groups of parallel rays, which would give us the **back-projection** method of 3D reconstruction. To represent this on paper, as in Fig. 12.15, we reduce dimensionality, starting with a 2D structure of which we have only its 1D projections.

12.4.2 An Example in Both Real and Fourier Space

The (unsymmetrical) triangle in Fig. 12.15 can be unambiguously reconstructed from seven different views, but what is the minimum necessary number? That turns out to depend on the resolution we expect in the reconstruction. We get more insight into these issues by taking a very simple case: the symmetrical equilateral triangle. Figure 12.16 shows that, because of this symmetry, three different views are identical. The triangle's symmetry is giving us three views for the price of one – indeed, six different views, as each group of arrowed lines can also give a projection in the opposite direction. Let us follow how these could be used in both the Fourier and back-projection methods.

First we use the *Fourier method*. The projection consists of three equidistant peaks of equal height. Their FT, shown in Fig. 12.17 (c) (bottom right) is a partially-submerged cosine curve; this is a central section of the two-dimensional FT of the triangle. Because of the symmetry, we can start to assemble that FT from its central sections. We simply take three copies of this section oriented at 120° (or, because of the symmetry, at 60°). This is shown in Fig. 12.18, but how should we interpret it?

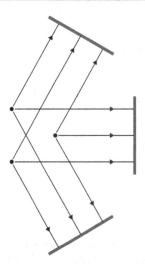

Figure 12.16: Three (identical) projections of an equilateral point-triangle.

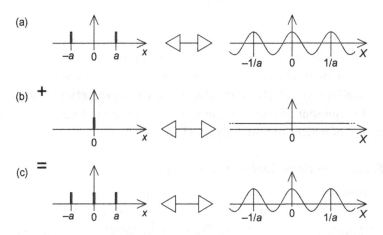

Figure 12.17: The FT of three identical equidistant peaks is in **(c)**. It is derived in three stages. **(a)** shows the FT of the two outer peaks; **(b)** the FT of the central peak, and the sum of (a) and (b) is shown in **(c)**. That is the projection onto the vertical y-axis of the middle triangle in Fig. 12.16, and the distance a here corresponds to $A/2$ in Fig. 6.18(b). The FT consists of large positive peaks alternating with smaller negative valleys.

The very center of Fig. 12.18 is unambiguous: it is the positive central maximum of the FT sectioned by three planes. The black sections near the ends of the six arms could be interpreted as the middles of six similar maxima, so we might get the impression that the FT consists of a central maximum surrounded by six similar maxima, all floating on a shallow negative sea. Indeed, it might be guessed that the FT consists simply of hexagonally-packed peaks protruding from that sea, rather like Fig. 6.17 (b). The accuracy of our guesses is

Figure 12.18: Three copies of the FT derived in Fig. 12.17. Black: positive; grey: negative.

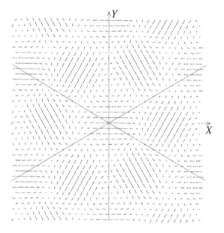

Figure 12.19: The FT of an equilateral triangle point-polyhedron (see Fig. 6.18 (b)). The three lines correspond to the FT sections revealed in Fig. 12.16 and Fig. 12.17.

revealed by Fig. 12.19, and we see that we were partly right. The hexagonally-packed positive maxima there are shown as circles[6] with inscribed '0' in Fig 6.18 (a), but there are also two other similar maxima with phase-angles of $\pm 120°$ (shown as grey circles in Fig. 6.18 (a)). We completely missed these because they were not present in our FT sections. And we had too few sections because we had too few *different* views of the triangle.

Next, we use the *back-projection method* to reconstruct the equilateral triangle. Figure 12.20 shows how the one view we started with yields six pencils of rays converging onto the central

---

[6] The triangle in Fig. 6.18 (b) (oriented just like that in Fig. 12.16 (b)) gives, when projected parallel to the *x*-axis, a projection that corresponds to the section of the FT (right of Fig. 12.17 (c)), i.e. along the *Y*-axis in Fig. 6.18 (c). Along this axis, the only maxima are those containing a central '0'. (Remember that $A/2 = a$.)

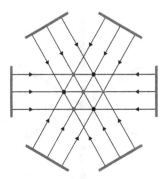

Figure 12.20: The periphery consists of six copies of the projection obtained in Fig. 12.16, yielding a reconstruction in the middle. Of the seven (equally significant) intersection points, the three correct ones are shown in black.

back-projected synthesis. Out of seven intersection points there, only three are correct (black) but that could not have been deduced from the data. The seven points form a hexagon with an added central point. If we omitted that point, we would have the point-hexagon whose FT was obtained in Fig. 6.17 (b). And this FT has already been noted as corresponding to our guess-interpretation of the Fourier reconstruction. (Adding the central point in real space merely adds a modest uniform density to the FT.)

This example shows the underlying connection – one might almost say identity – of the two reconstruction methods. Both essentially just add together the fragmentary information about the overall structure, at least at the initial stage. The Fourier method adds central section planes of the FT, and back-projection adds the projecting rays themselves. Then these first products need correction for their bias towards the center. But we see that this common underlying method has two problems. We can get wrong structures from insufficient projections. And, even when sufficient projections give us essentially correct structures, we get poor quality images. This second problem is less serious, so we consider it first.

12.4.3 Improving Naïve Reconstructions

Even when we have enough projections, there remains a defect of reconstruction that is seen more clearly in the real-space back-projection method. Figure 12.15 (b) shows the reconstruction of the triangle vertices. They emerge, not as points, but as 'stars' of reconstruction lines. Consider what sort of image of a point object we get, if we use back-projection from innumerable rays distributed evenly over all angles. The incoming rays might arrive from all directions (as in Fig. 12.15 (b)), or they might arrive from one semi-circle and depart from the other (as with a tilt protocol). In either case, the point-image is surrounded by a multitude of radiating lines, like the lines of force around a point charge (in 2D). The average density of these lines varies inversely as the distance from the point-image ($1/r$ dependence). Every point of an object would be imaged the same way, so the reconstructed image would (apart from magnification) be the object convoluted with a spherically symmetric $1/r$ density distribution (the 'point spread function'). However, the point-spread function will not obscure the image of a point very much. Furthermore, a line of density will, after convolution, remain a line (though with a more diffuse cross-section). The images of distant points will be so blurred that they cause only slight interference. This explains the success of the light microscope in surveying the 3D structure of thick Golgi sections containing a few, dark, widely separated cell processes. But more diffuse objects (such as most specimens for high-resolution electron microscopy) will be poorly reconstructed.

Thus even an optically perfect electron microscope could not give the best possible images, and some form of image processing is inescapable for determining the 3D structure of the object. But it is quite easy to correct for the imperfections of back-projection. In the case of a tilt-series, the sections in Fourier space fan out, so their separation increases proportionately to the radius R from their common intersection. Thus, for a back-projected image, the density in Fourier space drops off as $1/R$; so multiplying by R can restore the correct density. This **R-weighted back-projection** method (Gilbert, 1972) is the most common method in computer-assisted tomography (Herman, 1980; Natterer, 1986). It is also the method of choice when reconstructing the 3D structure of particles from tilt-series.

12.4.4 Refinement and Chirality

In any refinement, there is a crucial threshold (called the 'blob threshold') that is crossed when the image starts to become interpretable in molecular terms. (This is around 2 nm resolution for nucleic acids and around 1 nm for α-helices.) Refinement is needed before and after crossing this 'blob threshold', but its character and methods differ. Before crossing it, 'pre-molecular' refinement aims at getting interpretable images, but progress is made in the dark. One guesses what problems are hindering further clarification and tries to remedy them. Better CTF correction, more distinct particle views, improved accuracy of alignment parameters, further correction of particle distortion – all should be tried, but there is little to guide or encourage this search by indicating the critical problem. A sharper *uninterpretable* image might be merely another kind of noise, and it's particularly hard to convince readers of the contrary. It is this phase that benefits most by measures of resolution and reliability.

After crossing the threshold, however, there's a built-in test for improvement during 'post-molecular refinement'. An improved image must become not only more interpretable, but more in agreement with other sources of information, as the 'decoding' starts to make sense. Moreover, if the raw data are rich enough, it may be possible to make progress through testing structural guesses, by comparison with atomic models. (This is usually easier in X-ray crystallography, where the data have higher resolution, than in electron microscopy.)

In the 'pre-molecular' phase, it may be necessary to make a conscious guess as to chirality (e.g. in the hand of a helix), but it has no immediate consequences. If not guessed, chirality has to be found experimentally, but it's unlikely to matter. All this starts to change during the 'post-molecular' phase, when chirality may be testable by the handedness of helices. However, it is always better to have experimental evidence for chirality. In helices, handedness has gross, large-scale implications that may be easily tested (e.g. by shadowed specimens). But this is exceptional, and many single particles (symmetrical or not) give low-resolution images which would not indicate chirality even after tilting. After achieving sufficient resolution to test chirality, it is worth doing so by tilting experiments. Rosenthal & Henderson (2003) developed a method in which two images were recorded of 50 particles, each before and after a tilt of 10°. Then models with each of the two alternative chiralities were projected, in the computer, through a ± 15° range of angles (at 1° intervals), and the agreement with images calculated (using phase residuals). The correct chirality gave an optimum with the known direction and known magnitude of tilt, and simultaneously validated the accuracy of the tilt angle and the tilt axis determination.

Symmetry-Based Methods

After the broad issues discussed in Part III, we now look at how those general principles are applied in specific cases. These cover a wide range of symmetry types, from three to zero translations, as well as different levels of rotational symmetry. We present these symmetries in declining sequence of the number of translations, so the sequence ends with completely asymmetrical aggregates and with the correction of distortions that vitiate high-resolution FT methods.

Obviously, different symmetry types require different detailed methods of analysis. But, deeper than this, there is a more fundamental change in outlook from the older, FT-based 'crystallographic' methods adapted to translational symmetry, and the newer 'single particle' methods, anchored in real rather than reciprocal space and particularly influenced by modern computer methods for optimization. So the reader needs to adapt to a cultural difference, almost as when reading a book on art-history which passes from the Renaissance through to the Impressionists and Modern Art. The earliest chapters of this part are very 'FT-centered', and consider it axiomatic that the whole problem of structural analysis is to find a structure's FT; because its inversion to give the actual structure is so routine as to appear trivial. By contrast, at the end the FT is hardly mentioned, as its only functions are defocus-correction, and a hidden role deep within the computer calculation of real-space correlations.

However, a scientific pragmatist will learn all potentially useful methods and be prepared to change weapons as the problem demands.

CHAPTER 13

X-Ray Crystallography

Chapter Outline

13.1 Introduction

13.1.1 Thick Specimens

Every image is 2D but every specimen is 3D so, the thicker it is, the more information must be extracted from it. Thus thick specimens are associated with either low resolution, or increased complexity of data-collection and processing.

Here we consider thick *symmetrical* specimens. Every translation leads to an extension of at least tens of unit cells along the translation. Three independent translations thus imply a specimen of at least tens of unit cells in every dimension. This usually leads to a specimen too thick to transmit electrons of the voltages used in most electron microscopes (100–300 kV). There are two solutions to this problem: reducing the expected thickness by sectioning, and increasing the voltage much further. The latter possibility is not only very expensive, but also faces interference and other problems with thick specimens (Chapter 15 of Glaeser et al., 2007).

The process of cutting thin sections reduces the resolution to a pre-molecular level. That is nevertheless still most informative in the case of muscle, and there has been considerable study of well-ordered muscles by thin-sectioning. The most obvious defect of sectioning is to remove parts of the lattice outside the section, and the need to restore this has led to ingenious computational methods[1].

Very high voltage microscopes exist in special locations, but they appear not to have been much used for forming images of 3D lattices. Instead, the study of these has been the virtual monopoly of X-ray crystallography. The following brief and fairly superficial survey of this large and complex subject is intended to indicate how X-ray diffraction fits into a general pattern of diffraction-based structural techniques, and also to serve as an elementary introduction to newcomers. Those interested in going further, though without getting too deep, can consult Blow's general introduction for biologists (Blow, 2002), or (especially for those needing to *use* crystallographic structures) Rhodes, 2006. There are also many more thorough (though mathematical) textbooks, including Giacovazzo (1992) (especially for general crystallography) and (for protein crystallography) the new book by Rupp (2010).

13.1.2 X-Ray Study of Thick Crystals

When a thick specimen is studied by radiation, it must first penetrate, and then interact with the specimen without destroying it. Finally, it must be possible to extract the specimen structure from the changes in transmitted radiation.

In view of the proverbial penetration of X-rays into ordinary matter, it might seem surprising that considerable work was needed to explain how X-rays are diffracted by ordinary crystals without attenuation, that would be expected from the scattering power of most atoms. This is only possible because the lattices aren't perfect, and the imperfections lead to small regions diffracting at closely similar, but not quite identical, angles. Consequently, they don't attenuate exactly the *same* beam, so the recorded diffraction spots have an inherent 'size' (**mosaic spread**), besides that of the beam, etc. This is just one of the remarkable pieces of good fortune that, together, eventually made possible the early crystallographic study of protein structure, using millimeter-sized crystals of relatively small proteins. However, more intense X-ray sources allow the study of much smaller crystals (less than 50 μm thick), comparable in size to the ordered regions themselves.

Interaction with the specimen, without destroying it, is really one problem: how much information is obtainable before specimen-destruction is becoming unacceptable? This is the problem of radiation-damage considered below (§13.2.1).

However, for many years the overwhelming problem was that of extracting the specimen structure. Bernal & Crowfoot (1934) showed that protein crystals can convey a great wealth

---

[1]   Unfortunately, there is neither time nor space to describe these here.

of information, but it was essentially undecipherable before 1953. Then Perutz's discovery[2] of the applicability of the **isomorphous replacement** method (Bokhoven et al., 1951) to proteins (Green et al., 1954) led to the full development of this first *ab initio* method for finding protein structures, which revealed their astonishing variety and complexity. However, structurally related proteins were soon found (especially the globins). So a second approach was already proposed in the first decade (Rossmann & Blow, 1962, 1963): to use an existing structure as the basis for solving another one. Obviously at first most structures were completely novel, so the *ab initio* methods were essential. But quite new structures are rare nowadays, so the 'copying' method (**molecular replacement**) is very important. Its relative importance will presumably continue to grow, as a more comprehensive knowledge of protein structures (or at least of domains) is gradually acquired. Thus there are two fundamentally different approaches to solving protein structures: *ab initio* methods based on adding a very few strongly-scattering centers, and *molecular replacement* methods based on information from similar known structures.

However, most of the experimental techniques are common to both methods, and they are described first.

13.2 Specimen and Data Collection

13.2.1 *Protein Crystals*

Nature of Proteins

Globular proteins are quite different from the ordinary molecules of organic chemistry. The long polypeptide chain folds so that, on balance, hydrophobic side-chains are predominantly on the inside, with hydrophilic ones mostly outside, so the entire structure is held together by the pressure of the hydrogen-bonded water molecules outside it. However, the detailed packing of specific groups within the protein is organized and maintained through numerous medium-weak electrostatic bonds (hydrogen bonds and salt links). The peripheral hydrophilic groups are virtually dissolved in water, whose pH, ionic composition etc. affects the delicate balance maintaining the protein's native state. In particular, a protein's tendency to crystallize is easily altered by those solutes and by substances (like polyethylene glycol, PEG) that change, even slightly, the properties of water. Crystals can sometimes be grown if that balance swings gently towards adhesion, and if sufficient time is allowed for these giant molecules to ease themselves into a lattice. Once in that loose federation, they are attached by only a few weak intermolecular bonds between adjacent groups. Most of their surface remains dissolved

---

[2] This was his realization that one heavy atom could change the diffraction pattern of a much heavier protein molecule. This (another piece of unexpected luck) is because the extended molecule's diffraction amplitude develops as a 'random walk' from the contributions of its pseudo-randomly distributed light atoms. This results in their contribution increasing only as the *square root* of their number (Wilson, 1942 and §11.1.1), in contrast to the concentrated scattering by the heavy atom.

in the solvent that constitutes at least a third of the volume of the protein crystal, which is a delicate object, easily broken by attempts to grasp it and shrinking under the mildest dehydration.

However, many corresponding specimens for electron microscopy are no more robust, but this technique imposed, for many decades, far rougher conditions on them. While protein crystallography started with the discovery of appropriate methods for specimen presentation, electron 'crystallography' only found them after decades of development.

Specimen Presentation

Both protein crystals and electron microscope specimens are self-supporting inside the solvent, and they only require additional support when it is thinned to reduce its background scattering.

The X-ray diffraction of small-molecule crystals rarely encounters radiation damage, giving the false impression that X-rays are little more damaging than light, in sharp contrast to high-energy electrons. However, X-ray photons are actually about 1000 times more damaging than electrons, per elastic scattering event; for, although they carry much less energy, they need to inflict more 'collateral damage' to get the same information. Moreover, early protein crystallography techniques used wasteful data-collection methods.

The apparent immunity to radiation damage during early X-ray protein crystallography depended on the choice of big crystals of small, robust protein molecules. But the most interesting proteins are relatively big and sensitive, and usually yield small crystals. So their study required advances in instrumentation (see below) but also in specimen presentation.

Since, as in microscopy, the limiting factor is the radiation-sensitivity of biological specimens, similar measures have been adopted for extending specimen lifetimes. Freezing crystals reduces the decay-rate of diffraction spots by a thousand-fold, so the lifetime of a $100\,\mu m$ crystal at room temperature is that of a $10\,\mu m$ crystal at $-190°C$. The frozen crystals can, like electron microscope specimens, be transported under liquid nitrogen in cryo-containers, and they are kept cold during the X-ray exposure by a stream of very cold nitrogen gas, condensation being prevented by an outer current of dry air.

13.2.2 Instrumentation

The main technical improvements concerned the detector and the incident beam. Before turning to these, however, there are two more similarities relating to the specimen. First, there is a need to reduce background noise from air scattering (in microscopy, an uncovenanted consequence of needing a vacuum to transmit electron beams). In X-ray diffraction this need is less urgent, allowing the use of nitrogen gas-cooling (see above) and of helium-filled (as well as evacuated) beam-paths.

In both techniques there is also a need to control the specimen's orientation. In the X-ray diffraction of crystals of only moderate radiation-sensitivity, this involves the rotation method (see below) for ensuring that all diffracted rays are recorded without overlap. (When crystals are extremely radiation-sensitive, only a single exposure is possible, as in 'low-dose' electron microscopy.)

In both techniques, the only detector was originally film (double-sided with X-rays because of their greater penetration). But this has largely changed, though the replacement of film by electronic detectors started earlier with X-ray detectors than with electron microscopes.

However, the biggest change in X-ray instrumentation has concerned the X-ray source. Although 'white' (polychromatic) X-radiation does give diffraction patterns, they are very complicated, as most of the diffraction pattern is present in a single exposure. (Difficulties of disentangling its different components mean that this involves unnecessarily high radiation-exposure.) Since the early years (not quite the earliest), monochromatic radiation has been used in crystallography. Like monochromatic light, it is efficiently made by exciting the appropriate type of atom. Because of the frequency, and hence energy, of X-ray photons, they were emitted by transitions involving the innermost electrons, falling to fill a gap created when the atom had an electron ejected. The traditional atom source is copper for the water-cooled anode which is bombarded with electrons, and its radiation is made effectively monochromatic by passage through a nickel filter. Later laboratory sources used crystal diffraction to monochromatize the beam more thoroughly, and also allowed the copper to endure a stronger excitatory electron beam by moving it rapidly so it did not melt.

The first signs of any new technology were in the 1960s, when it was realized that the 'waste' X-rays produced by electron synchrotrons are both intense and 'white' (and therefore tuneable). Traditional synchrotron radiation has a broad peak whose wavelength decreases with the synchrotron's energy. This needs monochromatization and focusing (e.g. by mirrors) to avoid divergence over long beam-paths. Synchrotrons' X-ray emission has since undergone many technical developments: wigglers increase its output and undulators and free-electron lasers reduce its wavelength spread, with concomitant increases in intensity. These developments let us collect data from big molecules and small crystals; just as important is the opportunity to exploit powerful phasing methods using anomalous scattering (§13.3.5).

13.2.3 Collecting a Data-Set

The last chapter (§12.1.1) explained why all-round views (either images or Fourier Transform (FT) sections) are needed to get the data required for a full 3D structure. Thus the crystal(s) must, in total, present a complete view of the molecule, though the number of degrees of rotation required for data collection is reduced if the crystal lattice has rotational symmetry.

The problem of collecting a *complete* dataset is seen most clearly when collecting the data of a crystal's diffraction pattern. Each diffracted ray corresponds to one of the

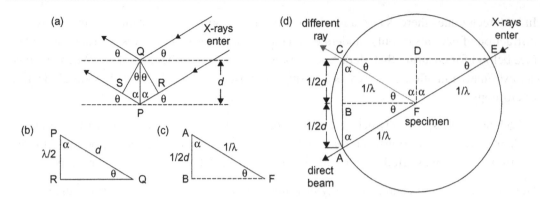

Figure 13.1: (a) Bragg's theory of diffraction from layers of atoms in a crystal, shown as horizontal broken lines, d apart ($=$PQ). The angle of incidence θ gives a diffracted ray when the path difference (RPS) between rays from successive planes is a multiple of wavelengths λ. The angles $\alpha + \theta = 90°$. **(b)** The triangle PQR, showing the condition for diffraction. **(c)** Triangle ABF is (b) with all lengths multiplied by $1/\lambda d$. **(d)** Four copies of triangle ABF packed into one double-sized copy of it (ACE). The angle at F is $2\alpha + 2\theta = 180°$, so AFE is a straight line, while the angle at C is $\alpha + \theta = 90°$. Thus C is the locus of a right-angle subtended by a straight line, i.e. a circle (Euclid).

lattice-points of an imaginary *reciprocal lattice* which acts like a filing-system, providing a set of precisely-defined numbered pigeonholes. Thus they have a definite starting-point and a set of clearly prescribed positions, but a vague end or boundary at high-resolution (i.e., the outer parts of the pattern). Thus, although the total may be unclear, there's no doubt if an inner component is missing, and a clear distinction between whether it's merely overlooked or actually non-existent. The essential link between this abstract organizer and experiment is provided by another abstraction, the **Ewald sphere**. This greatly simplifies the problem of predicting a crystal's diffraction pattern from its reciprocal lattice, and we derived it in §8.4.3, in the simpler context of ordinary light diffraction. Here we extend this context to X-ray diffraction as conceived by Bragg (1914). Although individual atoms scatter X-rays, Bragg realized that a ray gets diffracted from a crystal when the scattered wavelets reinforce, and this happens when a series of parallel planes of atoms comes into the correct geometrical relation, shown in Fig. 13.1 (a). The rays scattered from successive planes reinforce when their path difference RPS is a multiple of wavelengths λ. This condition is embodied in the triangle PQR (b), which is similar to triangle AFB (c), four of which pack together to give a double-sized copy in (d). There, the right-angle at C implies that it is on a circle of diameter AE. Indeed, as Ewald showed, C is on the surface of a sphere (Fig. 13.2). If this imaginary sphere has a radius of $1/\lambda$, and the crystal's reciprocal lattice (on the same scale, Fig. 13.3) is imagined centered at A, then any reciprocal lattice point C lying on the sphere will cause a *real* diffracted ray to emerge in the direction of the line FC.

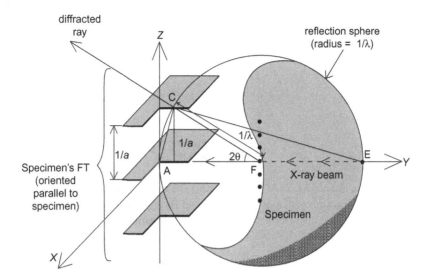

Figure 13.2: The Ewald sphere construction for finding a diffraction pattern from an FT. The specimen is supposed to be placed at the center of this sphere F, and irradiated from the right E by the X-ray (or light, etc.) beam. Where it leaves the sphere (A), we place the center of the specimen's FT, oriented parallel to the specimen and scaled correctly with respect to the sphere (radius $1/\lambda$). Everywhere the FT intersects the sphere, we get a diffracted ray radiating from the specimen.

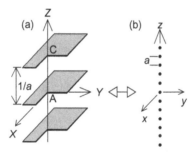

Figure 13.3: Line of equidistant peaks in real three-dimensional space. **(a)** Three-dimensional FT of **(b)** in reciprocal space.

Although we put the specimen's imaginary FT at the left edge of the sphere (Fig. 13.2), we put the actual specimen at the center F, from which all diffracted beams radiate. They leave the sphere at any point (e.g. C) where it intersects the FT, and this must include the FT center A, which is why the specimen's FT has to be separated from the specimen itself. The diffracted ray's brightness is proportional to the squared amplitude of the intersecting part of the FT. The curvature of the Ewald sphere causes the centrosymmetrical FT amplitude to produce a pattern of diffracted rays that is often not at all centrosymmetric (Fig. 13.4). Although the diagrams suggest that the reciprocal lattice points are just points, they actually

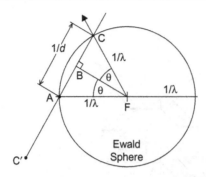

Figure 13.4: The specimen's FT, C'AC, is tilted and the diffraction angle 2θ is bisected by OB. Hence asymmetrical diffraction (OC but not OC'), and Bragg's equation: $\sin \theta = BC/CF = (\frac{1}{2}d)/(1/\lambda) = \lambda/2d$, giving $\lambda = 2d \sin \theta$.

have an effective size determined by the mosaic spread of the crystal (a quality parameter of the crystal, like resolution).

Macromolecular crystals can have very big real unit cells and therefore very small reciprocal unit cells; so the reciprocal-lattice planes can be tightly packed with points, which have a finite size determined by the crystal's mosaic spread. In such a case many points like A and C would be spread in a circle, center B (Fig. 13.4). If the next reciprocal-lattice plane parallel to C'ABC touched the sphere near B, the photograph might show tens of spots, in a clear lattice, filling a circle within the previous circle. Indeed, the recorded part of the reciprocal lattice could be intuitively interpreted in terms of small portions of successive 'layers' and with large viruses a film can show hundreds of spots in a photograph taken with no movement of the crystal (a 'still'). To ensure that no lattice-points are completely omitted, however, the crystal (and hence its reciprocal-lattice) is usually subjected to a small oscillation during each frame of the data collection. (A big oscillation would cause the different 'layers' to overlap.)

'Area detectors' efficiently record electronically all diffracted rays arriving simultaneously over a substantial area. Having collected (or while collecting) data, the reciprocal lattice parameters are determined. All the spots must be assigned reciprocal lattice coordinates (**indexed**), their intensities corrected for partial recording ('postprocessing'), and for some geometrical factors. The completeness of the data-set (% of available reflections measured) and the number of independent measures (called 'multiplicity') are estimated. Often a dataset involves combining records from different crystals; these must be carefully scaled together. These are tracked by 'quality control' statistical data, 'reliability factors' (R factors) which naturally measure the deviations, from each other, of measurements that should be identical, particularly when merging and through inherent symmetry. These reliability statistics are the start of a chain of quality assessments, so that the final structure has an objectively-known reliability. (This is an area where crystallography has inherent advantages over electron microscopy.)

13.2.4 Preliminary Analysis of the Data-Set

Just as ontogeny (sometimes) recapitulates phylogeny, so the study of each new protein repeats in some ways the history of protein crystallography. The protein must be first purified, then crystallized, and then a data-set must be collected. This puts us in the position of the first protein crystallographers who had no phases but only diffracted intensities, and so had to be content with studying their symmetries.

From the positions of spots on the film, one can use the Ewald sphere construction to calculate their position in reciprocal space. That gives the unit cell dimensions, and the cell shape sets some limits on the possible space groups. Much more information about the space group comes from the symmetry of the intensities of the spots in the reciprocal lattice, and especially if they have zero intensity 'spots' arranged systematically on certain planes or lines. Then the space group tells us how many asymmetric units there are per unit cell volume, i.e. the volume available per asymmetric unit (v_s). Knowing the molecular weight M of a protomer, we can estimate its volume. Perhaps v_s is not much bigger than this, or perhaps it is over twice as much, allowing the possibility of two protomers per asymmetric unit. Indeed, (v_s/M) is called the *Matthews volume* (measured in Å^3/Dalton), which has an expected range in protein crystals of a similar general type. If more than one chain is present per unit cell, it is possible that the chain forms symmetric aggregates whose symmetry may fit the space group (as a subgroup), or may not (non-crystallographic symmetry). Then crystallography can fix the precise point-group of a symmetrical aggregate. A classic example was *E. coli* aspartate transcarbamoylase, at first believed to have the composition c_4r_4 (c and r representing the catalytic and regulatory chains); but X-ray diffraction of different crystal forms established the composition to be c_6r_6 and the true symmetry to be D_3 (Wiley & Lipscomb, 1968).

After this, we know something about the packing of protomers (asymmetric units or chains) within the crystal and we are ready for the next stage of the recapitulation: phasing.

13.3 *Ab Initio* Phasing

13.3.1 The Phase Problem

We recall the Abbe theory of the microscope, as applied to a single magnifying lens (§9.1.1). About half-way between the lens and its image there is produced the object's diffraction pattern, the first of two FTs generated by the lens (Fig. 9.1). Then the waves continue, with a kind of momentum imparted by the lens, to generate the second FT. This is the FT of the object's FT, or the upside-down image (§9.1.1). It has long been lamented that, as there are no satisfactory X-ray lenses, we can get no further than the first FT. All we can do is to record it, and all records are generated by energy or the wave intensity, from which we can easily find the FT amplitudes. But the relative *phases* of the different diffracted rays – essential information for reconstructing the image – gets lost in the record. If we had that information, we would know the phasors at each point of the reciprocal lattice: we would know the crystal's FT, which we could invert to give an image of the crystal. A computer can do the work of a lens, provided it has all the necessary information. How to get the lost FT phases was the famous 'phase problem'.

That was the historical perspective on X-ray crystallography, but the long development of that subject and the related one of electron microscopy, where images *are* available, gives us a somewhat different view. Perhaps, when crystallography had its phases withheld by the evil fairy, the good fairy's gifts of data-condensation into sharp spots, of automatic averaging to minimize noise and (the most precious gift of all) of keys to unlock the phase-puzzle, more than compensated. How are we now enjoying the happy ending? We start by looking at how the original solution works, and then examine how it has since been complemented by other solutions.

We can't appreciate the value of Perutz's original solution unless we understand the magnitude of the problem. There was no intrinsic solution, no magic way to calculate the phases from the diffraction pattern; the problem was, strictly speaking, insoluble. Any conceivable phases would generate an image consistent with the experimental diffraction pattern and, if we knew nothing whatever about the structure of matter, we would have no grounds for choosing one 'image' from another. When decoding an enciphered message, it's essential to have some idea about likely messages. Similarly, the background of our solution was the development of chemistry for over a century, giving us enough information to distinguish a true solution from a 'fake' one. Nevertheless, this was not enough. Even now, when we know a great deal about the general principles of protein structure, there's no way to work backwards to find the required phases. And perhaps that's not so bad, as it provides many criteria for judging a supposed 'solution'.

But we were left with the problem, where to get the information that could solve the puzzle. That information must concern some feature that must be in the structure, when it has been solved. And, since *general* structural features have failed, it has to be a *new* feature of this particular structure. The only way to *know* it will be there, is to add it ourselves. So we shall need the diffraction pattern of the original protein, and also a second pattern after adding the new feature. Obviously, there has to be a difference between these patterns, and we have to be able to use that difference to get phase information. So, besides knowing the nature of the new feature (how to recognize it), we must be able to calculate its effects on the diffraction pattern. Our only hope of doing this is to make the new feature as simple as possible: an isolated peak in the image, i.e. a new atom that scatters strongly. The FT of such a peak is easily found once we know its position.

Thus our solution consists of four stages: (i) adding the atom(s) to make a measurable change, without producing any other change (this is the meaning of 'isomorphous'); (ii) finding the atoms' position(s); (iii) calculating the change to the overall diffraction pattern; and (iv) getting phase information from that change.

13.3.2 Adding a Heavy Atom and Finding its Position

Conceptually, the simplest way to introduce a new peak is to bind a new heavy (i.e. strongly-scattering) atom to the protein (giving the 'derivative' crystal). Proteins have several different side-chains, many at the surface and some also reactive (especially cysteine, with a reactive sulphydryl group). So a molecule containing a heavy atom could be chemically bonded to the protein before it is crystallized; but it needs to crystallize in *exactly* the same crystal form (i.e. *isomorphously*) as the unaltered protein (called the 'native' protein). Fortunately, that problem could usually be avoided by utilizing the special feature of protein crystals, that they contain a substantial fraction of solvent. This allows heavy-atom molecules to be diffused into the crystal, with much less likelihood of disturbing it. However, it is unusual to add just one heavy atom per unit cell, even if only because most space-groups contain rotation axes which would multiply everything in each unit cell. But it is also unusual for one binding site to be so much stronger than all others that only one atom gets attached; so derivatives often have several sites. Another problem is low **occupancy**, i.e. incomplete binding.

An unsatisfactory 'derivative' crystal will be evident from an unchanged, or an excessively changed, diffraction pattern. The next test is: can we find the heavy atom's position(s)? It might seem surprising that we could find anything at all in a diffraction pattern without knowing the phases. Recall that, without knowing any phases, we can find the autocorrelation function (ACF, also called the Patterson function) by simply setting all phases equal to zero (§3.4.6). This function gives a 'map' of all interatomic vectors, shifted so that their bases lie at the coordinate-frame origin. Now, although all the interatomic vectors of the protein form a hopelessly complicated bundle (and so of course do those of the 'derivative' protein), nevertheless most of the complexity is common to both forms. Thus the difference between these functions (the difference Patterson function) contains heavy atom–protein vectors, plus the heavy atom–heavy atom vectors that give us information about those atoms' positions. Obviously, this approach depends on fairly clear and strong heavy atoms, and also benefits if the crystal's space group allows projections with a 2-fold axis, making the FT real.

13.3.3 Getting the Phase Information

Every atom in the unit cell contributes its own phasor to every diffraction spot. At each spot, all phasors add like vectors, giving an overall phasor, and we need its amplitude and phase; but the recorded spot gives only its amplitude. However, having found the position of a heavy atom, we can calculate *its* phasor, F_H, (amplitude *and* phase) at each spot. Thus our data are, for each spot, its amplitude with the protein alone ($|F_P|$), and for protein plus the heavy atom ($|F_{PH}|$). Also, we have the calculated phasor contribution F_H of the heavy atom. So, for each diffraction spot, we know two amplitudes ($|F_P|$, $|F_{PH}|$) and one phasor (F_H). We would not expect to find *both* the remaining phases from only this information; but, surprisingly, we can nearly do that. Figure 13.5 (a) shows the situation. The three phasors, F_H, F_P and F_{PH}, are related by a vector diagram (b), since $F_P + F_H = F_{PH}$. We know

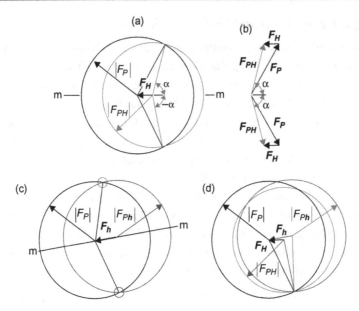

Figure 13.5: Harker diagrams for phasing a protein FT at one hypothetical reciprocal-lattice point. **(a)** Fitting native and derivative amplitudes with the phasor F_H. **(b)** Vector diagram showing the two possible solutions. **(c)** Fitting native and derivative amplitudes with the new phasor F_h. Two new possible solutions (circled). **(d)** Combining (a) and (c) should reveal the unique solution (here, near the bottom).

the phasor F_H and the amplitudes (radii) $|F_P|$ and $|F_{PH}|$. These three quantities fit the Harker diagram (a), where F_H is shown with zero phase for clarity (otherwise, just rotate the diagram appropriately). As for the other two phasors, not knowing the phase angle means that we know only the radius of the circle which the phasor must fit. Thus F_H tells us the displacement of the two circles of radius $|F_P|$ and $|F_{PH}|$. The entire diagram (a) has a mirror along F_H, so the circles intersect in two mirror-related points, giving two possible phase-angles that are equal and opposite. This information alone (all we get from a **single isomorphous replacement** or **SIR**) is thus insufficient to solve the phase problem, but only a little more information is needed. For example, one new heavy-atom derivative, obviously with the heavy atom at a new position, gives a similar diagram with two different intersection points (Fig. 13.5 (c) for this derivative *alone*). In principle, only one of these should agree with only one of the first two points (a), as shown in (d) (showing both derivatives together). In practice, the intersection points are very sensitive to errors in measuring the amplitudes (radii of the circles), so many derivatives are needed (**multiple isomorphous replacements** or **MIR**). Also, a statistical estimate of the reliability of each phase needs to be made (Blow & Crick, 1959), leading to the **figure of merit**, an important statistic for heavy-atom *ab initio* structure determination.

13.3.4 *Changing the X-Rays: Anomalous Scattering*

In the previous section, we concluded that the best way to solve a protein diffraction pattern would be by modifying the diffraction pattern through the addition of a simple peak(s). We then described how this is done by adding a heavy atom to the protein. But there is another way to achieve the same end: by changing the X-rays so that a specific atom, already part of the protein, scatters much more strongly, etc. The effects of this **anomalous scattering** have gradually grown in importance to the point where they now overshadow the effects of simple isomorphous derivatives. We concentrate on the principles in this section, and briefly survey major practical developments in the next.

Scattering much more strongly at certain wavelengths would be commonplace with ordinary light, whose absorption depends strikingly on wavelength. However, the X-rays normally used in crystallography show no wavelength-specific absorption by ordinary proteins, which they treat as uniformly 'grey'. Their 'colored' substances are heavier atoms, starting around iron in the Periodic Table. (This makes sense, as we noted that copper atoms emit crystallographic X-rays, and emission and absorption bands are close.) According to classical physics, X-rays set the outer electrons vibrating at the same frequency, and their vibrations emit the same X-rays, so they scatter X-rays. Atoms that absorb X-rays get into resonance with them (just as a radio receiver is tuned to get into resonance with a transmitter's carrier-wave). This resonance transmits energy to the absorber, causing it to vibrate more strongly and therefore to emit more strongly. Thus the first effect expected of 'tuning' our X-rays to match a medium-heavy atom, is an increase in its scattering power, in contrast to that of ordinary light elements and therefore 'anomalous'.

This first effect of *anomalous scattering* would make it essentially another way of changing a protein's diffraction pattern by adding a density-peak at a known position. We would still need to find some way of introducing a heavy atom into the typical protein (usually containing no atom heavier than sulphur), and into a specific site(s) with high occupancy. But, if the heavy atom's scattering can be changed, it is only important that the labeled protein should form good crystals. Then we can get our two diffraction patterns simply by using X-rays of two different wavelengths: one that is far from the absorption peak, and the other adjusted to show the maximum 'resonance' effect. The only difference would be that, since altering the X-ray wavelength cannot change the protein's crystal structure, the diffraction patterns are truly 'isomorphous'.

If that were the whole story, anomalous scattering would be much easier to understand but also much less useful. But changing the X-ray wavelength also changes the *phase* of the scattered wave. We saw in §8.5.1 that, at resonance, the absorber (here, the heavy anomalous scattering atom) lags 90° behind the driving oscillations (the incident X-rays). Consequently its scattered X-rays have a different phase from those scattered by light atoms. To represent the phase change, we start by taking the ordinary scattered X-rays as a standard, so *their*

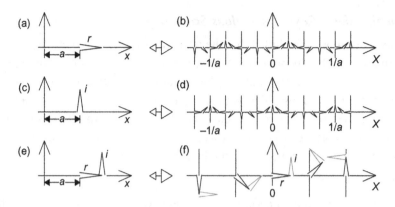

Figure 13.6: The 'real-anomalous' **(a)** and 'imaginary-anomalous' **(c)** parts of anomalous scattering, with their FTs **(b)** and **(d)**. The sum of (a) and (c) is shown in **(e)**, and the sum of (b) and (d) in **(f)**. Note how the complex anomalous scattering phasor's perpendicular components (e), with the left turn from the 'real-anomalous' to the 'imaginary-anomalous', remains as a unit throughout (e) and (f).

phase-angle is *defined* as 0° (so the phasor lies along the axis, as in Fig. 13.6 (a)). The X-rays scattered by the heavy anomalous scattering atom can now be represented as a phasor, with amplitude and phase-angle. That is the polar coordinate representation (Fig. 3.7), but crystallographers use the Cartesian representation, splitting the phasor into 'real' and 'imaginary' parts. The 'real' part (called[3] f_1) of the anomalously-scattered wave is just like ordinary scattered waves (essentially what we discussed in the previous paragraph) (see Fig. 13.6 (a)). The 'imaginary' part (called if_2) is like the phasor in Fig. 13.6 (c). So we view anomalous scattering as the sum of two effects: an in-phase wave (phase 0°) and a 'perpendicular' wave with phase 90°.

The phasors in Fig. 13.6 (a) and (c) are the simplest examples of the real and imaginary parts, respectively. The 'real' wave (a) gives an FT (b) with Friedel symmetry (§3.3.5). The 'imaginary' wave (c) gives an FT (d) that does *not* have Friedel symmetry, but mirror symmetry (relative to the origin as the mirror). Actual anomalous scattering has both the real and imaginary contributions, so its wave is like (e), and the FT will be the sum of (b) and (d). Notice that (e) has a real phasor (*r*) with an imaginary one (*i*) 'turning left' from its tip. Now we know that the FT of (e) must be the sum of (b) and (d); and that, at its origin, each of these FTs is the same as the phasor on the left side (as can be confirmed by looking at the origins of *x* and *X*). Consequently, when we add (a) and (c) to get (e), we also add the origin-copies of (a) and (c) to get an origin-copy of (e) (labelled *r* and *i* in (e)). And we also know that, with a phasor-wave, whatever phasor is at the origin simply rotates (here, anti-clockwise) to give all the other phasors. Thus we conclude that the FT of (e) is the same *r,i* phasor combination

---

[3] Alternate notations are: $f = f_a + \Delta f' = f' + if''$.

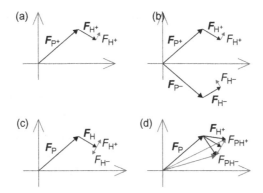

Figure 13.7: Phasors for protein (F_P) and for an anomalous scatterer (F_H for both parts; black for real and grey for imaginary); all shown at some specific X (+ for right of the origin; − for left of it). **(a)** shows all the vectors right of the origin, the real phasor from the heavy atom with its imaginary phasor forming a left-turn at the end. **(b)** shows the phasors at $+X$ (above) and $−X$ (below). The real vectors are upside-down through Friedel symmetry, but the imaginary vector is still a left-turn from it. If the lower part is reflected in the horizontal axis, we get **(c)**, and result of adding the heavy-atom vectors is shown in **(d)**.

at the origin, but rotating anti-clockwise elsewhere, as shown in (f). Thus all the phasors in the FT of (e) consist of an ordinary phasor (black) with a left-turn phasor (grey) at its tip. All anomalous contributions consist of this combination, though with different rotation angles and different amplitudes (lengths) of the real and imaginary phasors.

It is important to distinguish between two different uses of 'real' and 'imaginary' in the context of Fig. 13.6, where we were considering the two components of an anomalous scattering factor: 'real' like the scattering of any ordinary atom in the same position, and 'imaginary' at right angles (in the same position). Thus the total anomalous scattering phasor consists of the two components shown in (e) and (f), the 'real' behaving like that of any other atom, the 'imaginary' with a phase-shift relative to the real. Perhaps it would be clearer if we relabeled these terms 'real-anomalous' and 'imaginary-anomalous', to distinguish them from the previous use of 'real' and 'imaginary', meaning having a phase-angle of $0°$ and $90°$, respectively. To help the distinction, in this section we shall relabel those terms as '$0°$-real' and '$90°$-imaginary', within this section.

In Fig. 13.7 we explore how these composite anomalous phasors can be used for finding protein phase angles. In (a) we see all the phasors contributing to one reflection on one side (positive) of the origin. Here, F_{P^+} is the phasor of a protein (ordinary scattering) on the right (positive) side of the axis. On the other side, because of Friedel symmetry (which means the same '$0°$-real' component, but an opposite '$90°$-imaginary' component), the corresponding phasor would be reflected in the horizontal axis (see F_{P^-} in (b)). Attached to the protein phasor are the two anomalous heavy-atom phasors, both labeled F_{H^+}. The 'real-anomalous' phasor of the anomalous heavy atom is black, and its 'imaginary-anomalous' phasor (grey) is always related to it by a 90° left-turn.

In Fig. 13.7 (b) we show those same three phasors on the positive side, plus the corresponding ones on the negative side (i.e. at the two positions X and $-X$). (Normally, both those reflections have the same intensity, because of Friedel symmetry, so their intensities are averaged; but we shall find a difference when we include anomalous phasors.) We see F_{P^-} reflected in the axis by Friedel symmetry, and the same happens to the 'real-anomalous' phasor of the anomalous heavy atom (black). However, the 'imaginary-anomalous' phasor (grey) is *always* related to it by a 90° left-turn, so it is *not* reflected in the axis. Because of this, if we joined the origin to the tip of the grey 'imaginary-anomalous' phasor, the line would be longer on the plus (upper) side than on the minus (lower) side. Thus the reflections on opposite sides no longer have the same intensity when anomalous scattering is involved; Friedel symmetry is broken.

In Fig. 13.7 (c) we show the lower (minus) three phasors reflected back onto the positive side, to join the three phasors already there. (This is done to make the changes in intensity clearer.) The two black phasors, which were Friedel pairs, superpose exactly; but F_{H^-} ends up at 180° to F_{H^+}. Now we join the ends of those two phasors to the origin in (d), giving phasors labeled F_{PH^-} and F_{PH^+}, which clearly have different lengths $|F_{PH^-}|$ and $|F_{PH^+}|$.

We can see that $|F_{PH^-}|$ and $|F_{PH^+}|$ come from the measured intensities of the corresponding spots in the two positions X and $-X$. These intensities are different, so Friedel symmetry no longer applies when the imaginary contribution of anomalous scattering is significant. The lack of this symmetry means that the intensities $I(X)$ and $I(-X)$ differ, doubling the information available from the diffraction pattern, and raising the possibility that the diffraction patterns of one derivative might resolve the phase ambiguities.

Thus one anomalous heavy-atom derivative has given us, for each Friedel pair of reflections, two different intensities corresponding to $|F_{PH^-}|$ and $|F_{PH^+}|$. From these we find (as in the previous section) the heavy-atom position from which we can calculate the two phasors F_{H^-} and F_{H^+}. Besides these four quantities, we will have measured the same reflections at a wavelength where the heavy atom shows no anomalous effects (so they were a Friedel pair), giving us $|F_P|$.

This one derivative thus provides three amplitudes, $|F_P|$, $|F_{PH^-}|$ and $|F_{PH^+}|$, and two phasors, F_{H^-} and F_{H^+}, which must all be fitted into Harker diagrams and vector diagrams, as in Fig. 13.5 (a), (b). The two triangles in the center of Fig. 13.5 (a) form a mirror-symmetrical quadrilateral like two wings on each side of the heavy-atom phasor F_H. So the two heavy-atom phasors, F_{H^-} and F_{H^+}, in Fig. 13.7 (d) should generate two such quadrilaterals, OBAS and OLMS in Fig. 13.8 (a) with the solution point S (arrowed) in common. This is constructed from the more complicated Harker diagram (Fig. 13.8 (b)) since S is the unique common intersection point of three circles with radii $|F_P|$, $|F_{PH^-}|$ and $|F_{PH^+}|$.

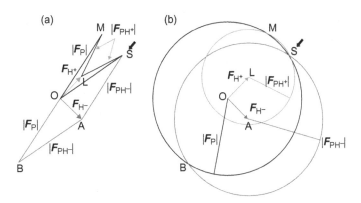

Figure 13.8: Finding the phase (phasor OS) from a single isomorphous replacement with anomalous scattering (SIRAS). Phasor notation as in Fig. 13.7. **(a)** Phasors connected in a vector diagram. **(b)** Harker diagram connecting experimental amplitudes (circles) with anomalous heavy atom phasors (grey arrows OA, OL). In both (a), (b) the goal is to find S (arrowed).

13.3.5 Applications of Isomorphous Replacement and Anomalous Scattering

Noise makes this **single isomorphous replacement with anomalous scattering (SIRAS)** usually unreliable, but when anomalous scattering is combined with several derivatives we get the powerful **multiple isomorphous replacement with anomalous scattering (MIRAS)** method. It is even more powerful if the anomalous scattering measurements are not confined to one wavelength, but are made at several different wavelengths close to the absorption/emission wavelength. This is called **multiple wavelength anomalous dispersion (MAD)**. A particularly important application of this uses proteins synthesized by bacteria requiring methionine, but which were supplied with selenomethionine (where the sulphur atom is replaced by selenium, which is very suitable for anomalous scattering). This introduces a relatively small number of anomalous scatterers that are part of the protein structure (and that also show the positions of methionines).

As mentioned, anomalous scattering occurs in the neighborhood of X-ray absorption bands, which involve the loss of photons that excite inner electrons. The most strongly-bound are the 1s or K electrons, which require the most energy (shortest wavelength) to excite electrons. After about bromine, their wavelengths become inconveniently short, so attention shifts to the next-most strongly bound electrons, from the L or $n = 2$ shell. There are three different X-ray energy-levels here, LI, LII and LIII (in increasing order of energy), and some of these bands ('white lines') in the heavier elements (lanthanides and later elements) are the strongest anomalous scatterers. With even heavier atoms, up to uranium, the M shell absorbances can also be used.

It is with MAD measurements with these anomalous scatterers (especially osmium LIII) that the heaviest asymmetric units have been phased (e.g. the ribosome and its subunits with molecular weights of the order of a million). It had been earlier thought that such structures, around a hundred times bigger than the first proteins solved, would require proportionately heavier masses for phasing. Large complexes containing 10–30 tungsten atoms have proved useful for MIR at lower resolution (up to 5 Å), but at higher resolution their extended structures weaken their contribution. On the other hand, although anomalous scatterers are much weaker (equivalent to only 20–30 electrons), the very low noise level and strict isomorphism compensate (Clemons et al., 2001).

Besides all these advantages of anomalous scattering, it allows us to determine the absolute hand of molecules. When Friedel symmetry is lacking, the basic data-set (i.e., each measured intensity attached to the appropriate point of reciprocal space) no longer has a center of symmetry. Instead, the data-set itself is a chiral object, so we can expect to calculate the correct chiral image (or density-map), from which the correct chiral structure can be built.

13.4 Other Phasing Methods

We now turn to other phasing methods, which make no use of heavy-atom derivatives. However, all phasing methods share the common principle of requiring some source of extra information. It can come from *outside*, from the structures of other proteins (or parts thereof); or it can come from *within* the crystals themselves (through the existence of redundancy). Obviously, this second (internal) information source is limited and often small. So it is insufficient for a complete phasing, though it can sometimes contribute enough information to make a previously inscrutable image interpretable. However, the first (external information) method provides much more information. There are two major types of application. Very general structural information is used in the *refinement* of protein structures, though it is essential to have an approximate solution as a starting point. That point is often reached from the other type of application, when the unknown protein structure has close relatives residing in the structural data-base. Then we might use the **molecular replacement** method, which comes in two different flavors, dependent upon whether this external structural information comes from another structure in the same crystal lattice, or from another structure in a different lattice(s). We start with the simpler case.

13.4.1 External Information, Same Lattice: The Difference Fourier

If the two proteins have exactly the same lattice, they must surely have very similar structures. So the structural differences are probably restricted to a few localized changes: bound substrates, substrate-analogs, drugs, or perhaps a very few amino-acid replacements from mutations. Two very similar proteins in identical lattices are the most favorable situation for structural comparison; indeed, the situation is very similar to isomorphous heavy-atom derivatives.

In this case, the analogy goes a little further, since the known protein can contribute approximate phases – the only information missing from the unknown protein. Combining the unknown protein's amplitudes with the known protein's phases would give the correct structure, if the unknown were the same as the known. So, if they were very similar, we would expect them to give an approximation to the unknown's structure. Then, to find the location of the changes, we would calculate the *difference* between that structure and the known protein's structure. However, we can find that in just one calculation, by finding the differences between the FT amplitudes (experimental minus calculated), combining *those* with the calculated phases of the 'known' protein, and then calculating the reverse FT. This calculation is known as a **difference Fourier**. It shows any new features as positive peaks, and any missing ones as negative troughs (though both are present to only half the correct density, as predicted by Luzzati, 1953, and there is also noise in the difference map since the phases being used are only approximate). It's the standard way to find the location of such minor changes as new heavy-atom sites or drug binding sites.

After making the original difference-Fourier calculation, we can interpret the peaks and troughs, make the structural changes they suggest, and use that improved structure as a basis for some new phases. These give an improved structure (though *not* an improved difference Fourier: Henderson & Moffat, 1971). Thus we can repeat the cycle, which is the basis of the original method for refining protein structures after the first *ab initio* phasing.

13.4.2 *External Information, Different Lattice: Molecular Replacement*

Now we consider a rather different situation. We have crystals of a protein U, of unknown structure, which is similar to another protein S whose structure has been solved. We want to use S's structure as a starting point for refining the structure of U. However, U is in a different lattice, and we don't even know exactly where U is in it. So we need to get into the same position we were in above, with both molecules oriented the same way in the same lattice. Then we can calculate the FT of S and compare that with the measured intensities of U. But, to put S into *the same place* which U occupies, we need to know where that place is. How do we find it?

We start by considering a much simpler problem. Suppose we have solved two protein structures that are very similar, for example because they are from related species; and we want to make a detailed comparison of these structures. However, the proteins crystallize in quite different lattices, so we have them arbitrarily oriented with respect to each other, and we shall need to re-align them. This is just the 3D version of the 2D problem (§12.2.1) of superposing two essentially identical copies of the same image. We can use the cross-correlation function (CCF) to find the translation which makes one copy fit exactly with the other copy, provided the copies are also oriented to be parallel. If they are not, then we must first rotate one of the images to lie parallel to the other; then we can undertake the translational CCF. The necessary rotation can be found with the help of a rotational CCF.

This is how alignment is done with isolated 2D images, but our 'simpler problem' was to match two 3D structures. The extra dimension is less trivial than it might seem. As explained in §12.2.4, 2D images require only *three* adjustments (2T + R), whereas 3D requires the full complement of *six* adjustments (3T + 3R). So, while a 2D alignment is in a total adjustment 'space' that is 3D, the corresponding 'space' for a 3D alignment is 6D. If we need to try ten different values for each parameter, the former requires a thousand different combinations, but the latter needs a million. So it is all the more important to split the search into two separate searches: rotational followed by translational.

With this background appreciation, we return to our original problem. Putting S into the correct place in the lattice of U is complicated in two ways: the structure (U) is unknown, and it is in a lattice. We focus on the first problem first. U is unknown, but its diffraction pattern is known, allowing us to calculate the Patterson function, the autocorrelation function (ACF) of U's crystal. This consists of all the inter-atomic vectors, shifted parallel so that their origins all coincide. We have to match U's Patterson function with S's Patterson function, by rotating the latter until it matches, the matching being tested by the CCF (or, better, by the correlation coefficient[4]). This 3-parameter rotation search (the **rotation function**) usually uses the Euler angle method for rotating structures (§12.2.3), even though the 'rotation formula' method is more intuitively comprehensible. Rotating the molecule's coordinates is a very heavy calculation which is considerably accelerated by the 'fast rotation function' (Crowther, 1972) that uses spherical harmonics (§16.3.4).

After orienting the molecules to be parallel (3R), we still need to translate them to coincide (3T). In rare cases, this is trivial, and we need only sample S's FT in U's reciprocal lattice. The ambiguity as to the phase-origin doesn't matter, unless the lattice has symmetry elements which define their own origin (in 2D or 3D). However, that complication is rarely avoidable, since almost all protein crystals have space-groups with rotation or screw axes (§10.4.3). These constitute a 'symmetry framework' defining their own origin, and they also generate extra copies of the protein with new orientations. This even interferes with the previous rotation search, by generating inter-molecular vectors that act as noise (so the search must be confined within a tight radius). The translation search must use Patterson functions, as well as being complicated by the symmetry-multiplied copies. One puts S successively into different places within the U lattice, and calculates and tests the Patterson vectors of all its symmetry-multiplied copies.

Although the broad principles are clear and fairly similar to those underlying image alignment in electron microscopy, in crystallography the technical details are complicated by space-group symmetries, combined with the need to use Patterson functions (autocorrelation functions).

13.4.3 Internal Information: Redundancy

An important aspect of a diffraction pattern is its information content, relative to the amount required to define the structure. If the structure is simple and the diffraction information

---

[4] This is the CCF scaled to normalise the lengths of the vectors being compared.

precise and extensive, we are in a much better position than if a complex structure has to be deduced from a noisy pattern with poor resolution. And, if we are in that better position, we should know how to exploit it. There are different ways in which the amount of structural information required may be smaller than might appear; we consider two of them.

Solvent Flattening

We saw that almost all protein crystals have a large fraction of solvent, which is essentially unstructured (on the average), and therefore has a uniform density. Now a region of uniform density needs virtually no information for its description, so there is in principle extra channel-capacity, if there were a way of utilizing it. Rescaling would convert a region of uniform density into one of zero density and (in 1D) we saw in §4.1.6 that such a region, referred to as 'zero-filling' or 'zero-padding', leads to a finer sampling of the FT.

How is that useful? In deducing the sampling theorem (§5.3.2), we saw that a rectangle of width D has an FT that's a sinc-function with zeroes $1/D$ apart. If the real lattice has the same spacing D (with reciprocal lattice spacing $1/D$), we have no redundancy. But zero-filling enlarges the lattice spacing and consequently shrinks the reciprocal lattice spacing, making the FT more finely sampled. If the FT is real, it must pass through zero to change sign, and a finely-sampled FT shows where it is likely to change sign (equivalent to phase). That gives us phase information in the form of useful constraints. However, if the FT is complex, the question is more difficult. But the even and odd parts of the density (parents of the real and imaginary parts of the FT) are each subject to the same restrictions, and the sum of their squares (i.e., the intensity) is also known. So zero-filling also provides some phase information in that way.

This 'solvent flattening' method is not powerful enough for *ab initio* phasing, but it can improve phases, especially when there is a high solvent content. However, there is a problem in assigning the precise boundary of the solvent, which should be kept small to avoid encroaching on the protein's space.

Non-Crystallographic Symmetry

Crystallographic symmetry (as described by the space-group, which applies to the entire crystal lattice) may simplify the collection of intensity data, but it is no help for phasing. However, *non-crystallographic* symmetry, which applies to only a portion of the unit cell, does provide a useful redundancy of structural information. We see something of this principle in 1D FTs (Chapter 4, §4.2.2), where *sub-periods* (internal repeats that fit within the box) provide 'non-crystallographic' information. The effect on the FT of sub-periods is *zero-interlacing*, where there are gaps between the peaks, shown in Fig. 4.9 (a). These gaps should be zeroes, and could be set to that (rather like zero-filling), improving the signal:noise ratio.

In practice, non-crystallographic symmetry usually takes the form of rotational symmetry, often a consequence of a symmetrical aggregate crystallizing in a space-group that does not (or even cannot) express its full symmetry. The classic cases have been icosahedral viruses (Chapter 16) whose symmetry group *I* contains 5-fold rotation axes that cannot apply to the entire crystal lattice (see Chapter 10, section §10.4.2). In the most favorable case, the entire 60-fold icosahedral group could be non-crystallographic, but usually some sub-group of this (which could be as much as tetrahedral, of order 12) is crystallographic, reducing the redundancy. But there is always a minimum of 5-fold redundancy, as that rotation must be non-crystallographic (§10.4.2). There is even a further gain from the quasi-equivalent symmetry of subunits: many small viruses have more than 60 subunits (§16.1.2) with basically the same structure.

Crystalline Sheets

Chapter Outline

This is the first of five chapters describing how structures are found at relatively high resolution by electron microscopy. So we start with a brief section outlining the main advantages and problems of electron microscopy. It is also the first of four chapters describing how symmetry aids this by permitting the use of convenient and powerful Fourier methods. However, the series of 1-, 2- and 3-D 'crystal' symmetry groups was surveyed in Chapter 10, together with their Fourier transforms (FTs).

14.1 Electrons Versus X-Rays

14.1.1 Differences

When we use electrons instead of X-rays to find molecular structures, many things are changed by two fundamental differences.

First, electrons can be focused to give images (the original reason for the microscope), and images preserve phase information. (This may seem an odd way to express the enormous advantages of getting a comprehensible picture instead of a puzzle; but the image, an end-point at low resolution, becomes at high resolution more a starting-point for extracting the full 3D FT, both amplitudes and phases.) This is the most fundamental difference. The single overwhelming phase-problem of X-ray diffraction is replaced by various problems of image-analysis, depending on the shape and organization of the specimen.

Structural Biology Using Electrons and X-Rays.

Second, electrons are much more strongly scattered than X-rays of the same wavelength, by a factor of about a hundred thousand (Henderson, 1995). Chapter 7 described the implications of this for the electron microscope (EM) itself. It also has major implications for the specimen, which can (indeed, has to) be much smaller, with some advantages but many serious disadvantages. These are aggravated by the need to have it in the high-vacuum environment of the microscope.

Besides these two great differences there are some relatively minor ones. Whereas X-rays are uncharged, electrons are both charged and very responsive to electric charges, a combination leading to specimen-charging effects. Electron wavelengths depend on the accelerating voltage, so they are easily adjustable and (to have sufficient energy to penetrate) much smaller than those of X-rays (which are more penetrating, as a consequence of the second difference).

We now summarize the main implications of these differences, starting with the specimen.

14.1.2 The Specimen

All structure analysis techniques have a serious initial problem: to obtain adequate pure samples of the molecule in a satisfactory form. Because of the strong electron scattering, microscopy requires the least material for data-collection and accepts it in a greater variety of alternative organized geometries: sheets, helices, polyhedra and even isolated molecules – almost any state except the 3D crystal, which X-ray analysis requires (except in special cases). However, both these techniques have the same inconvenient requirement: whatever may be its form of organization, the specimen must be *well-ordered*. (Obviously the isolated 'single particle' is here an exception, but at a very high cost; see Chapter 17.)

In a competition between well-ordered aggregates, a 3D crystal has the inherent mechanical advantage that extra layers help to strengthen each other. Electron microscope specimens rarely reach the precision of well-ordered crystals (resolution ≤ 2 Å) and, with one exception (aquaporin: Gonen et al., 2004 and Hite, 2010), the image resolution is further degraded and falls below this threshold. So the correction of residual disorder and noise is normally an essential aspect of high-resolution microscopy (§12.3). The EM grid is essential for keeping elongated structures flat, and even this is inadequate; the higher the translational symmetry of an EM specimen, the more extended it is, and hence the more liable to distortion.

Electron microscopes use such very small specimens, that an image inevitably exposes the specimen to more radiation damage than an ordinary crystal suffers for an X-ray diffraction pattern. Nevertheless, electrons have important advantages over X-rays (besides the possibility of focusing them). They are 10^5 times as strongly scattered, so the same signal is given by 2% of the volume needed by X-rays. Moreover, the damage per elastic scattering event is a thousand times less than with X-rays. If there were adequate lenses, an X-ray microscope would have major disadvantages relative to the electron microscope. Perhaps the main advantage of X-rays, for imaging purposes, is their ability to penetrate thick tissues; but that advantage is lost in microscopy.

14.1.3 Image Analysis

Almost all data-collection requires specimen tilting, to gain additional viewpoints which are eventually synthesized to give an all-round, in-depth picture (Chapter12). Because particles depend on the grid's support, tilting the particle usually means tilting the grid. Particles that attach in various orientations thus have an advantage which is however cancelled if we have to discover each orientation individually. Thus both the gathering and analysis of images in sufficient number and diversity depends on the specimen's symmetry type. Since images also differ in optical conditions, their analysis (essential for obtaining phases, and often amplitudes as well) involves the definition and refinement of more parameters, and the exploitation of more varied symmetries, than when solving X-ray patterns. Each major symmetry type therefore receives a separate chapter, and we start here with the most highly symmetrical EM specimen, the crystalline sheet.

Since an EM specimen must have at least one dimension of molecular size, the FT along that direction cannot be sampled like a 3D crystal, but must vary continuously. All EM specimens therefore have FTs with at least one continuous direction. Sheets have just one; helices have two; and polyhedra or single particles have three continuous directions. By contrast, the 3D crystals used for X-ray diffraction give lattice sampling in all directions and hence discrete spots.

Continuous directions of an FT can only be sampled, often at arbitrary positions controlled by the geometry. But for 3D reconstruction we need regular sampling at equal intervals, as required by the sampling theorem. This is what a crystal lattice does for us naturally, but we have a re-sampling problem in EM image analysis.

14.2 Electron Diffraction

14.2.1 Uses

When specimens have more than one continuous direction, we can only use their images, which we process to extract the best view of the 3D structure. However, a crystalline sheet gives a sharp, detailed electron diffraction pattern in two directions. How much use are such patterns?

The two main successful procedures of X-ray crystallography are unfortunately impractical. The isomorphous replacement method fails, since a heavy atom is weaker in electron diffraction (electron scattering is proportional to $Z^{3/4}$ instead of Z, where Z is the atomic number). Electron scattering depends on a kind of average electrostatic potential (§7.4.1) and heavy atoms squeeze their electrons close to the nucleus, so that in most places the positive potential is shielded. Thus the electron scattering power of atoms increases much more slowly with atomic number than their X-ray scattering power, and heavy atoms (that scatter X-rays strongly) are modest scatterers of electrons. The other important technique, anomalous scattering of X-rays, depends on scattering *by* electrons, not *of* them.

Thus (apart from molecular replacement), there are no good ways to solve the electron diffraction pattern, though it retains an importance in structure refinement. In this way we can hope to extract from the specimen a resolution beyond that seen in its images. There is another use for the diffraction pattern. It provides relatively accurate values of the FT amplitudes at reciprocal-lattice points. These supplement those obtainable from images. Thus electron diffraction was found useful in purple membrane (Baldwin & Henderson, 1984; Ceska & Henderson, 1990; Grigorieff et al., 1996), photosynthetic light-harvesting complex LHC-II (Kühlbrandt et al., 1994) and the αβ-tubulin dimer (Nogales et al., 1998). (See also Glaeser et al., 2007, Chapter 9.)

Perhaps a word might be added about the historical importance of electron diffraction. It was a useful technical halfway stage on the road to imaging cryo-preserved specimens. The electron diffraction pattern reveals immediately the resolution to which a crystalline specimen is preserved, over a substantial area, even with stage movement (which leaves diffraction patterns unaffected), and even shows whether the ice in the specimen is crystalline or vitreous (glassy). Achieving the latter was the important goal of rapid-quench cryo-preparation methods, developed by Dubochet et al., 1982 (see the classic review by Dubochet et al., 1988).

14.2.2 Tilt Geometry: The Missing Cone

Recall that, in X-ray crystallography, we need to record the *entire* (3D) diffraction pattern to solve the structure in 3D. Any given position (setting) of the crystal gives us a pattern with only a small part of that total FT. So we need to record many such patterns, each with a different crystal orientation, to get all the necessary data. In electron crystallography, each pattern comes from separate sheets from which we have to extract it and assemble it, arrayed properly on the crystal's 3D reciprocal lattice. We need many different orientations, and each orientation requires at least one diffraction pattern.

Note that, from the specimen viewpoint, recording a diffraction pattern corresponds to recording an image. It is only a question of the intermediate-lens setting: whether to image the specimen (for an image) or the back focal plane (for the corresponding diffraction pattern). The images are far more valuable, since they provide us with phases as well as amplitudes, which are all we get from diffraction patterns. However, the specimen geometry is exactly the same in both cases, so we discuss it for the diffraction pattern.

The basic geometrical problem is the same as in X-ray crystallography, but with some important differences. Most crystals are roughly isometric, but a flat sheet is extremely asymmetrical. As with needle-shaped crystals, some orientations are therefore much

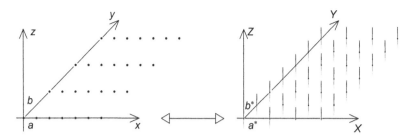

Figure 14.1: Main features of the 3D FT (right) of a crystalline sheet (left).

more convenient than others. This is aggravated by a orientations difficulty: the electron microscope objective-lens is an extremely tight space in which to rotate a specimen, so some desirable orientations are inaccessible[1], and there is always a maximum tilt angle. We now examine the implications of this.

The geometrical conditions for diffraction are the same as with X-rays (§8.4.3 and §13.2.3): the FT (oriented parallel to the specimen) must intersect the Ewald sphere, if a diffracted ray is to leave the specimen. But there are two differences. The electron wavelength is so short that the Ewald sphere (radius $1/\lambda$) is virtually flat; and the FT consists, not of isolated reciprocal-lattice points, but of parallel lines connected to the 2D reciprocal lattice (Fig. 14.1). The full three-dimensional FT of a thin crystalline sheet consists of a set of needles or lines, all parallel to Z and perpendicular to the (x,y) plane of the sheet but fitting its (X,Y) reciprocal lattice. We need to know the FT intensity along each of these lines, where the FT is not sampled but varies continuously.

The geometry of recording the patterns is shown in Fig. 14.2. Without tilting, we can record only the central X,Y-plane; with tilting, we start recording those 'needles' farthest from the tilt-axis (here along Y). But recording the central FT along Z would require a tilt of nearly $90°$; and, as we have seen, there is always a maximum tilt angle – in practice, around $60°$. Given this, how much of the FT can be recorded? The situation in Fig. 14.2 is repeated in Fig. 14.3 (a). But that represents only the result from one position, whereas we want to accumulate all the data in the same FT. So we rotate the FT into a standard position in (b), with its X,Y-plane horizontal, but the Ewald sphere intersection tilted. If the grid were tilted as in (a) but in the opposite direction, the standard position (c) shows that, if we were using the maximum tilt angle, we could record the part of the FT shown as shaded lines in (c).

Figure 14.3 shows that the portion of the FT recordable from tilts about one axis is a wedge. We can explore more of the FT if we change the axis position; axes at right angles give us two

---

[1] Orientations where the electron beam is nearly parallel to the grid surface, which cause the most practical problems, also reveal most acutely the crystalline sheet's departures from exact flatness. Of course, the specimen's effective thickness makes it almost opaque to electrons in these orientations.

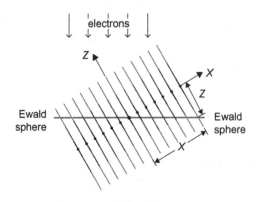

Figure 14.2: A tilted crystalline sheet's FT giving diffraction spots where it intersects the Ewald 'sphere' (here flat, seen in cross-section). The FT (center where *X*- and *Z*-axes would meet) consists of parallel 'needles' along *Z*, perpendicular to the reciprocal lattice in the *X,Y*-plane (shown black with thickenings). The rightmost intersection gives a spot with coordinates (*X,Z*), dependent on the tilt angle θ.

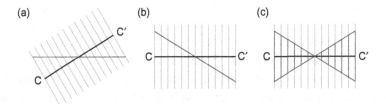

Figure 14.3: (a) as in Fig. 14.2. **(b)** FT 'rectified', rotated back so that *XY* plane horizontal. **(c)** Shaded lines show extent of FT recorded with tilts from 0° up to max tilt, in both directions.

intersecting wedges (Fig. 14.4 (a)). If we use axes in all possible orientations, all the various wedges intersect to give a cone (c), which shows the *inaccessible* part of the FT. This is called the *missing cone*, whose volume diminishes with the maximum tilt angle. Consequently, the loss for 60° tilts is only 13% (Amos et al., 1982).

14.2.3 Making and Processing the Records

As explained in Chapter 7 (section §7.2.1) we can record diffraction patterns in the electron microscope by adjusting the intermediate lens so that it focuses the back focal plane onto the detector. Adjusting the focus changes the spot size; small spots increase the signal:noise ratio, but result in fewer silver grains recording the intensity information (when film is used). The crystal must be thin, to avoid the many undesirable effects of thick specimens (e.g. multiple scattering); and, although diffraction patterns can be recorded at lower electron doses than images, low-dose precautions must still be taken.

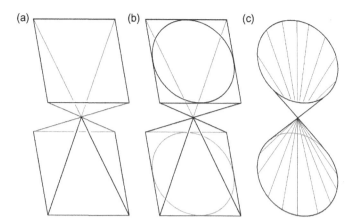

Figure 14.4: The excluded volume of reciprocal space when diffraction (or image) data are obtained by limited tilting about two axes **(a)**, **(b)** or many **(c)**.

For long the usual *detector* was film, which requires densitometry, but it is being replaced by electronic detectors that excite phosphors whose emitted light is recorded with a CCD sensor. In both cases, careful calibration is necessary for spatial and intensity distortion. (But, since the image being detected consists of (countable) high-energy electrons with no extraneous background, perhaps more efficient and linear detectors might be developed eventually.)

Each record consists of spots on a lattice (shape and dimensions depending on the recording geometry), but overlaid with a heavy continuous background of (especially) inelastically-scattered electrons (which could be reduced by energy-filtering), but also thermal-diffuse scatter (as with X-rays). Almost all this noise can in principle be removed, because the signal is restricted to small spots on a precise and predictable lattice; so all continuous scatter outside them can be interpolated and subtracted. Discrete spots can also be rejected if they differ from the main lattice, or down-weighted even if they belong to it but lack sufficient Friedel symmetry.

Different tilts contain mostly different information, so it needs collating and organizing. Thus the spots need to be indexed, i.e. assigned their reciprocal-lattice coordinates. These are integral in (X,Y) but continuous along Z. However, we ignore this for the moment and suppose that there were discrete spots along Z with different indices. Then each spot would have three integers defining its (X,Y,Z) positions, and indexing would be the same as for X-ray photographs. Spots with exactly the same indices should have the same intensity in all records, so we could assemble a set of index-defined intensities matching the experimental intensities on each record (after applying that record's scale-factor). To achieve this condition, we set up an overall error-function ('residual' or R-factor) that measures how far the current set deviates from the condition. We then need to change the assigned intensities so as to keep lowering the residual, using a process of iterative optimization, as outlined in §11.2.4 This process works, provided we start sufficiently near the optimum.

But we must return to the implications of the continuous nature of the FT along Z. In theory that requires an infinite number of Z-indices, implying an infinite data-set. This problem is avoided by the sampling theorem (§5.3.2), which allows us to interpolate to nearby Z-values. So the problem is soluble, provided we have sufficient records at different tilt angles θ. But here a final difficulty appears: these angles are hard to measure accurately. So we include the angles as additional adjustable parameters, along with the average intensities, when defining the residual R. And, again, we can iterate towards the optimum, provided we start near it.

After this harmonization, further conditions can be imposed to refine the assigned intensities. They should show Friedel symmetry (intensities equal for any two diametrically opposite spots with the origin midway between them). Furthermore, the sheet may have a 3-, 4- or 6-fold rotation axis, which should also apply to the intensities. These conditions can be expressed in a similar residual and imposed, in the same way, on the data-set.

14.3 Two-Dimensional Imaging

14.3.1 Feasibility

The crucial contribution of the electron microscope is providing images, equivalent to giving phases as well as amplitudes. The feasibility of this 'molecular microscopy' was revealed by Unwin & Henderson (1975) who combined several technical advances. They found that vacuum-dehydration could be prevented by replacing the water with glucose[2]; and they used the purple membrane of *Halobacterium halobium* (Stoeckenius & Kunau, 1968), a natural 2D crystal of a light-driven membrane proton pump (bacteriorhodopsin) driven by the same molecule (retinal) found in the visual pigments (Oesterhelt & Stoeckenius, 1971). It gave 0.35 nm resolution electron diffraction patterns, but the necessary weak exposure of ½ electron/Å^2 only yielded meaningless images of noise. However, optical diffraction dramatically converted these into precise peaks fitting the reciprocal lattice[3]. Even this was not sufficient to give the all-important phases. But the Thon rings could be identified from these amplitudes (compared with those from electron diffraction), and confirmed by optical diffraction from second, stronger exposures of the same sheet. Thus the phase contrast transfer function (PCTF) could be identified, the diffraction spots receiving reversed contrast could be corrected, and a reliable projection calculated at 0.7 nm resolution, revealing the bacteriorhodopsin trimers with an inner ring of α-helices. Crystalline sheets are still

---

[2] Negative stains also preserve structure (phosphotungstate: Ceska & Henderson, 1990; uranyl acetate: Trachtenberg et al., 1998), but they add a high-density background and probably disturb the structure.

[3] Lord Rayleigh, in his private laboratory, made diffraction gratings by photography, finding that they work even when (through under-exposure) the photograph shows nothing but noise.

useful for finding structures of membrane proteins, a most important group (Vinokumar & Henderson 2010).

14.3.2 Projections

Although 2D projections are generally uninformative, the projection onto a parallel plane of a thin crystalline sheet displays much of its symmetry and even some of its structure. So it is common to begin the study with this 'parallel projection'. Its FT fits the reciprocal lattice, and it has the same 2D point-group (though with the substitution of mirrors for 2-fold axes and glide-lines for screw axes). The analysis of crystalline sheets follows the crystallographic procedure slowly built up in studies of small molecules, where it became standard practice to commence a crystal structure analysis by finding the space-group, of which there are 230 (§10.4.3). But finding the plane-group of sheets is easier as there are only 17, though there is also less information. This includes the lattice shape (2-, 3-, 4-, or 6-fold); the intensity symmetry in the FT; the phases of the FT (having moved the phase origin to the rotation axis); and systematic absences, i.e. recognizable patterns of zero-intensity lattice points. The details are processed by the computer program ALLSPACE (Valpuesta et al., 1994), which fixes the projection plane-group.

We need to increase the signal:noise ratio by removing noise. This would be simple if the lattice were perfect, and only random intensity fluctuations needed to be removed. For then the signal from the repeating structure would be concentrated at reciprocal-lattice points, while the noise contributions would be spread everywhere and could therefore be almost entirely removed by accepting only the data from reciprocal-lattice points. Distortion unfortunately produces, in different places, different real lattices whose different reciprocal lattices superpose to produce extended 'lattice points'. The best solution is discussed in §18.2.2 on distortion, but this requires a good filtered image as a start.

14.4 Three-Dimensional Imaging

After Unwin & Henderson's original study of the parallel projection (the FT's basal plane), a 7Å resolution 3D study followed, for which Henderson & Unwin (1975) collected data from tilted specimens. In this they were helped by the plane group ($p3$) of the sheet which simplifies the task of data-collection. These two studies established the essential features of any 3D study of a crystalline sheet. (i) Images were taken of enough tilted specimens with the right orientations (here the 3-fold axis of the plane group made each image equivalent to three images). (ii) The data were extracted from diffraction spots in FTs of almost featureless images, as in the earlier study. (iii) The initial amplitudes and phases were corrected for tilt-induced defocus. Here, the directions of densitometer scans were chosen so that each entire line had a constant distance from the specimen axis, and hence a constant defocus. This allowed a simpler type of defocus correction to be applied.

(iv) Different images were relatively scaled like electron diffraction amplitudes (§14.2).

(v) The resulting 3D FT was inverted by taking its FT.

In the resulting image, the 'missing cone' produced an asymmetric resolution resembling astigmatism, though in this specimen that merely blurred α-helices along their lengths. Thus the 7-helix bundle was clear, but the helix connections were completely obscured. Although model-builders proved capable of rising to this challenge, the final verdict needed an atomic model based on high (3.5Å) resolution data. For this, several new techniques had to be developed[4].

14.4.1 Collecting and Assembling Data

These developments included experimental techniques, such as using cryo-specimens (in place of glucose embedding) and two new electron microscopes with a superconducting lens (Dietrich et al., 1977) or a spotscan illumination system (Henderson & Glaeser, 1985). However, we focus here on developments in image-processing.

The tilt-image data must be processed and assembled to give a file of FT amplitudes and phases. Three main corrections were necessary: for defocus and astigmatism (next section, §14.4.2); for distortion (Chapter 18, section §18.2.2); and an additional correction for beam tilt (Henderson et al., 1986; Baldwin et al., 1988).

Besides these corrections, the procedures outlined at the start of this section were followed. This led to the stage where all the image and diffraction data had been reduced to a list of amplitudes (image and diffraction) and phases (image) at different points along the Z-lines ('needles') of the FT, up to the edges of the missing cone. To calculate the 3D structure, these data were needed at equal sampling intervals along Z. (This problem was also faced with electron diffraction data, §14.2.) So the experimental data had to be interpolated (program LATLINE), making use of the sampling theorem, for which a good estimate was needed of the crystalline sheet thickness. That was provided by various X-ray diffraction studies of stacked purple membranes, which also indicated that only weak FT intensities were lost in the 'missing cone'.

14.4.2 Correcting for Contrast Transfer Function: The Tilt Transfer Function

The simplest view of defocus, in Chapter 9, section §9.1, considers it as a convolution of the image with a blur function (*point spread function* or PSF). Like many convolutions, this has clearer consequences in the FT formulation, where FT(PSF) is called the *contrast transfer function* or CTF. In electron cryo-microscopy, matters are complicated by the fact that cryo-specimens convey information almost exclusively through changing the *phases* of the elastically-scattered electrons. Thus visualizing the images depends on phase-contrast,

---

[4]   This account is mostly based on a detailed account of 3D image processing by Henderson et al. (1990).

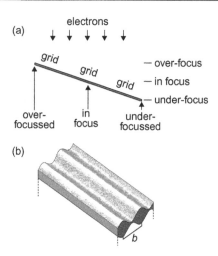

Figure 14.5: (a) Tilted grid, showing the effects of the tilt on defocus. **(b)** The tilted specimen (a weak sinusoidal phase object) on the grid and facing downhill. It transmits an electron wave with phase variations that are rendered visible by the tilt defocus.

obtained by changing the PCTF. This happens when the image is defocused, adding an aberration phase γ (equation (9.5)), whose sine is the PCTF (equation (9.3)). The specimen tilt inevitably defocuses the image. With progressively increasing defocus Δf, a given sinusoidal specimen (spatial frequency X) yields an image with alternately positive and negative phase-contrast (see Chapter 9, section §9.2.6). We start by considering a hypothetical specimen containing only one spatial frequency, a *weak sinusoidal phase object* (WSPO) shown in Fig. 14.5 (b). We suppose this object is on a grid positioned as in Fig. 14.5 (a) so that different parts of the grid are subject to varying defocus.

If the grid were level, then a single defocus would render the phase variations visible as alternating light and dark bands (horizontal in Fig. 14.6), stretching along the specimen's length[5]. However, since the tilt causes a progressive PCTF change from left to right, these bands must show a parallel change. So there is a (short wavelength) specimen-based alternation from top to bottom, and a (long wavelength) defocus variation from left to right. The combination is in Fig. 14.6 (a), showing a checker-board pattern produced by the interaction of the two variations. (Changing the focus would cause the pattern to move left or right.)

The simplest way to compensate for the PCTF's effects is to use the very same PCTF as a multiplication filter in reciprocal space (see section 9.2.7). After this, the correct image has been filtered with the PCTF's *square*. (Note that the PCTF varies only with x, so the sign-alternation along y is unaffected.) This is shown in Fig. 14.6 with (a) (one filtering) and

---

[5] Of course, the specimen will 'contain' Fourier components (WSPOs) pointing in many different directions, but only variation along y is portrayed as it is the most informative.

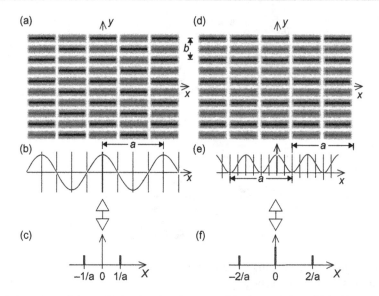

Figure 14.6: (a) Image of the WSPO wave in Fig. 14.5; black: positive; dark grey: negative; white: zero (nodal lines). The rapid specimen phase variations (the 'signal') run along y (spacing b); the slower CTF runs along x, spacing a (which depends on the tilt angle). **(b)** Plot of the PCTF in the left-right direction. **(c)** FT of (b) is a pair of peaks, i.e. the peak is split. **(d)** The squared (corrected) PCTF. **(e)** Plot of (d) like (b). **(f)** FT of (e) has (with a wider pair of peaks) a strong central (undivided) peak.

(d) (filtering with the squared PCTF). We see that the (horizontal) alternation of contrast has been replaced by continuity across the nodes.

This filtering, like all CTFs, occurs in reciprocal space. Its FT, in real space, is the PSF: the blur function that gets convoluted with the image (which, ideally, would be a sharp peak). The *uncorrected* image of the tilted specimen has the PCTF in (b) whose FT is shown in (c). Instead of a sharp peak, this is a double or split peak. The effect of the *correction* is to produce an image whose PCTF is shown in (e) with its FT in (f). Here the split peak has been replaced, not unfortunately by a sharp peak, but at least by a major central peak with only two weaker satellites. (These FTs have been previously deduced in Fig 5.4, but the FT in (f) can also be deduced as the ACF of that in (c).)

This treatment of the CTF of a tilted specimen originated with Henderson et al. (1990), who called the CTF in (a) or (b) the 'tilt transfer function' or TTF. Unfortunately for clarity of exposition, they found that the TTF and distortion corrections were interrelated; that is, one correction could not be performed entirely accurately until the other correction had also been carried out, so it was necessary to iterate them in combination.

Helices

Chapter Outline

Helical structures are particularly common in biology, both as polymers (protein or nucleic acid chains) and as macromolecules (especially in motile structures). One reason may be that helical symmetry is the most likely one to be generated, if identical molecules attach without blocking the attachment site, so that they bind consecutively in the same way. Therefore, one translation yields a convenient structure, without dominating the cell.

Despite the occurrence of helices in polymer chemistry, the subject of general helical symmetry was little explored by chemists or crystallographers before the development of molecular biology in the 1950s. An understanding of helical symmetry is fundamental to its applications in structure-determination, so we start with this topic. Here (as in other chapters) we consider only the symmetries of molecules that are chiral (i.e., present in only one 'hand'), so chirality-reversing symmetry-operations (like mirrors) are excluded.

Helices have a special advantage in providing all the data necessary for a full 3D reconstruction from only one image. Indeed, we already saw, in §12.1.3, that a single image of a helix of suitable symmetry can provide an excellent stereo-image. This advantage is probably why the first image to be reconstructed was a helical structure (the extended sheath of T4 bacteriophage: DeRosier & Klug, 1968). But the advantage, like so many, is part of a trade-off; and here the drawback is a relatively weak signal, for the following reason. Since any axial section of the FT of a helical wire contains the same data (Fig. 15.4), it follows that those data must be spread out in circles perpendicular to the axis (a feature we shall confirm in §15.2.1 below). This spreading stretches the FT intensity over the circles' perimeters, reducing both it and hence the helical signal, although the noise of course remains the same; so the signal:noise ratio is reduced. The effect is worse at large reciprocal radii, where the circles are bigger and the intensity more stretched; and this limits the resolution. So perhaps the 'one image reconstruction' benefit is mainly a feature of low-resolution pictures, where there is less stretching and where, moreover, layer-plane interference (§15.3.2) is usually absent. We should expect to have to combine many images in order to obtain high-resolution structures.

Helical structures and diffraction have been reviewed by Stewart (1988), Moody (1990) and DeRosier (2007). We adopt here a reciprocal-space (FT) analysis of helices, best for initial analysis to moderate resolution. A brief account of their real-space analysis and refinement is given in Chapter 18, section §18.3.

15.1 Helical Symmetry and Structure

We start with a section entirely in real space, which gives a background in helical symmetry that is relevant to understanding rod-shaped viruses, actin and other filamentary structures important for motility, etc, and which is also essential for real-space helical refinement (§18.3).

15.1.1 Helical Symmetry Groups

General helical symmetry can be viewed in two useful ways. (1) The formal symmetry viewpoint (Fig. 10.12, §10.2.3) apparently yields only a limited range of symmetries; but both components of the screw can vary continuously, giving many variations. (2) Helical groups operate in a cylinder which they leave unchanged. Although curved, a cylinder preserves all its geometrical relationships if it is opened flat (after cutting[1] parallel to the axis). This yields the *helical projection.* Then a screw becomes a translation with two components: h along the vertical coordinate z, and Ω (in appropriate units) along the horizontal coordinate x. That is one vector of a 2D lattice; the other, generated by the cut, is the *circumferential vector* which exists because every lattice point must repeat after one revolution. Thus we have two vectors, generating a 2D lattice on the cylinder, the *helical lattice.* (The limited circumference also imposes a minimal translation along z.) With a thick helix, if we mark the helical lattice at different radii on cylinders that are then opened flat, the lattice shape varies with radius; but the same helical lattice applies at all radii, provided we measure the horizontal coordinate as an angle (Fig. 15.1).

---

[1] However, this cut changes the *topology* of the cylinder, and our imagination must restore the connection between the left and right vertical edges.

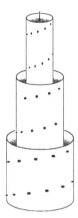

Figure 15.1: Three concentric cylinders cut from the same thick helical structure and consequently having the same helical lattice.

15.1.2 Helical Lattices Without Rotational Symmetry

Consider the simplest case of a helical lattice with an exact repeat[2], exemplified by the *ideal* α-helix, Fig. 15.2 (a). Here the screw has a rise distance h of 0.15 nm and a twist angle Ω *exactly* 100°, which represents a 'unit cell' φ-distance of the helical lattice. Then a full turn (360°) measures 3.6 'unit cell' distances of the helical lattice, so there are 3.6 *units per turn*, and $\Omega = 100° = 100°/360° = 1/3.6$ revolutions. So 18 units = 18/3.6 = 180/36 = 5 turns, an integer; thus 3.6 units per turn means 18 units in five turns. The number of units per repeat is called u, and the number of turns per repeat is called t. So we have:

$$\Omega(\text{revolutions}) = t/u \qquad (15.1)$$

After the rotations, we come to the translations. The distance per turn of the helix is its *pitch* P. Since a complete turn has 3.6 units = $1/\Omega$, and the rise per unit is h, a complete turn has a length of $3.6h = (1/\Omega)h = h/\Omega$. Thus the *pitch:*

$$P = h/\Omega \qquad (15.2)$$

If the twist angle Ω is a simple numerical ratio, there will be an exact repeat c. As it contains u units, and each takes a distance h, the repeat must be:

$$c = hu \qquad (15.3)$$

Unless external structures constrain a helix, an exact repeat along its axis is unimportant, and a helix-repeat might be called a purely geometrical fact. A *near* repeat is often used as a rough way to describe a helix, e.g. '18 units in 5 turns' for an α-helix, even when this is not quite exact. Such approximate repeats are also relevant to the helical Fourier transform (FT) (§15.3.2).

---

[2] We postpone discussion of its significance to below and section 15.3.2.

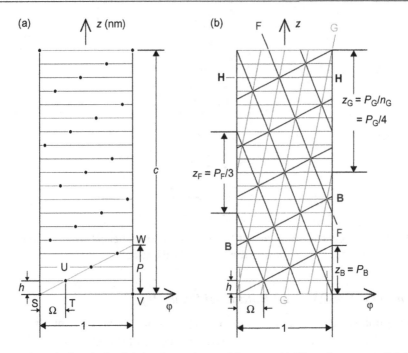

Figure 15.2: The 'ideal' α-helix. **(a)** Lattice points, with screw (h,Ω) and repeat c. **(b)** Subsidiary helices, numbered according to the number of helices cut by the equator. (H-'helices' are annuli perpendicular to z, and have the rise-distance h.) There is really only one B-helix ($n_B = 1$) which has pitch P_B;. There are three F-helices ($n_F = 3$) with pitch P_F; and there are four G-helices ($n_G = 4$) with pitch P_G. Helices intersect at lattice-points.

Any 2D lattice can be defined in terms of various pairs of lattice vectors. Each lattice vector of a helical lattice corresponds to a set of helices, as shown in Fig. 15.2 (b). Starting with a basic helix (shortest pitch and twist angle, labeled B), we call all other helix sets *subsidiary helices*. This diagram has four sets of lines (H, B, F, G), whose spacing is positive if they run from bottom-left to top-right (otherwise negative). The thin, horizontal lines (vertical spacing h) are labeled H. All other lines (labeled in Fig. 15.2 (b)) are counted (n) according to the number of intersections they make with each H-line (an intersection at the left or right *edge* counts as ½). *All* these lines intersect *each* lattice point, just as all the Fourier components (parallel lines) of a plane lattice intersect at a lattice point.

A set of subsidiary helices is defined by their pitch P and number n. (Their 'axial repeat' z is the distance between successive turns, $z = P/n$.) Here we shall notice mostly two subsidiary helix sets, F and G with numbers $n_F = 3$ and $n_G = 4$, respectively. We now examine how to calculate their pitches (given the basic helix parameters); later we look at their rise distances and twist angles.

In Fig. 15.2 (b), the basic set of helices (B) has only one helix ($n_B = 1$), so the equation $z = P/n$ gives us: $z_B = P_B = 3.6h$ (positive because right-handed). The F-set of subsidiary helices ($n_F = 3$) has $z_F = P_F/n_F = P_F/3$ which covers six of the intervals between horizontal lines (see diagram), so $z_3 = 6h$. Now, to make things concrete, we introduce the actual dimensions in the α-helix. As $h = 0.15$ nm, $z_F = 6 \times 0.15 = -0.9$ nm (negative because left-handed). But we already noted that $z_F = P_F/3$, so $P_F = 3z_F = -3 \times 0.9 = -2.7$ nm (negative because left-handed). The G-set of subsidiary helices has $n_G = +4$ (right-handed). The distance between two successive helices of this set is marked as $z_G = P_G/4$, and this distance covers nine horizontal lines (see diagram), so $z_G = 9h$. As $h = 0.15$ nm, $z_G = 9 \times 0.15 = 1.35$ nm (positive because right-handed).

It is easily shown that the vertical distance (in h-units) between successive lattice-points along a helix of any set is proportional to the number of helices in that set.

Finally, we consider the twist angles Ω. At the bottom of Fig. 15.2 (a), triangles STU and SVW are similar, so ST/UT = SV/VW or $\Omega/h = 1/P$. Thus the twist angle $\Omega = h/P$ (revolutions). This equation also applies to any subsidiary helix (from a set n in number) if the pitch P is replaced by the axial repeat z_n. (In general, $z_n = P_n/n$ but $n = 1$ for the basic helices where $P_1 = P$.) Thus the general equation is $\Omega_n = h/z_n$, and we apply this to the three sets of helix: B, basic helices where $\Omega_n = \Omega$; F, dark grey helices where $n_F = 3$; and G, light grey helices where $n_G = 4$. Thus we have $\Omega = h/P = 1/3.6 = 5/18$; $\Omega_3 = \Omega_F = h/z_F = -1/6 = -3/18$; and $\Omega_4 = \Omega_G = h/z_G = 1/9 = 2/18$ (see the penultimate paragraph about signs); and we now look for algebraic connections between them.

The lattices in Fig. 15.2 are ordinary 2D lattices, but with the x-axis labeled φ, and the y-axis labeled z which can be measured in multiples of h. However, since they are 2D lattices, their vectors can be added by adding their components. Thus the twist angles must satisfy simple additive equations, and indeed we find: $3\Omega = 3 \times 5/18 = 15/18 = 1 - 3/18 = 1 + \Omega_3$; and also $4\Omega = 4 \times 5/18 = 20/18 = 1 + 2/18 = 1 + \Omega_4$. So the helices in Fig. 15.2 satisfy the equations $\Omega_3 = 3\Omega - 1$ and $\Omega_4 = 4\Omega - 1$. The equations can be summarized as:

$$h/z_n = \Omega_n = n\Omega - 1 \tag{15.4}$$

15.1.3 Helical Lattices with Rotational Symmetry

The difference between cyclic (C_N, polar) and dihedral (D_N, non-polar) symmetry groups has no effect on the helical *lattice*. However, both point-groups have an N-fold rotation axis which enlarges the structure around the axis, promoting a central hole and creating N copies in the radial projection in the direction of the φ-axis, so all subsidiary helices come in multiples of N. A concrete example clarifies these changes.

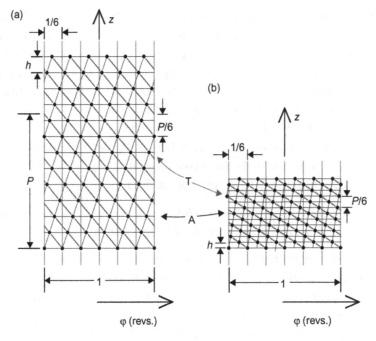

Figure 15.3: T4 (indeed, T-even) contractile sheath in the extended **(a)** and contracted **(b)** forms. All sixth-segments (present as the sheath has 6-fold symmetry) are identical. The double-headed arrows indicate subsidiary 'helices' that remain intact during contraction. T: 'transitional' helices, $\Omega = -0.1188$ revolutions in the extended sheath; A: annuli.

The T4 bacteriophage sheath is a primitive contractile structure that assembles, in its initial extended structure, around a DNA-transferring tail-tube. During infection, the extended structure contracts, but both forms have the point-group symmetry C_6. The extended sheath parameters are $h = 4.1$ nm, and $\Omega = -0.1189$ revolutions for an important set of basic helices. (Because of the C_6 symmetry there are six of these basic helices.) After sheath contraction[3], the sheath parameters become $h = 1.48$ nm (a contraction to nearly one-third) and the same twist now has $\Omega = -0.0756$ revolutions (Moody, 1973). The effect of rotational symmetry on the extended sheath is shown in Fig. 15.3 (a). Its C_6 point-group symmetry means that a diagram like Fig. 15.2 must fit into 1/6th revolution. It is one of the six vertical segments which, if there were no rotation axis, would constitute the entire helix.

15.1.4 Relation of Helices to Sub-Structures

Many helices have a central (axial) hole which prevents interaction between subunits across the center. If it is big enough, the helix can be described as a tube, and interactions are possible

---

[3] Contraction is driven by a spontaneous transition to a more stable protein structure, perhaps similar to prions.

only within the surface lattice. Quite often the hole has a function (e.g., bacterial flagella are a conduit for flagellin monomers). Perhaps others are hollow to gain mechanical strength.

When a tube is wide, its radius is less precisely determined by the assembly mechanism. It has the appearance of a rolled-up sheet of protein subunits linked in a 2D crystal that must fit one of the plane groups. When sheets form icosahedral viral capsids, that (local) plane group is *p*6; and this persists when aberrant 'heads' form long tubes by extending the middle (equatorial) part (e.g. polyoma/papilloma viruses, Kiselev & Klug, 1969, or T-even bacteriophage heads, Moody, 1965).

A similar set of tubes 'cut out' from the same 'sheet' is provided by the acetylcholine receptor tubes (Brisson & Unwin, 1984; Miyazawa et al., 1999), but the 'sheet' has only the symmetry *p*2 rather than *p*6 (though it consists of pentamers in an arrangement similar to *p*6). The 2-fold axes are preserved in the helices, with consequences for their FTs; see §15.2. Here the pentamers form protofilaments (Fig. 10.10, §10.2.2), which are 'subsidiary helices' that represent assembly stages. Other examples of protofilaments occur in polyoma-papilloma virus capsids (Kiselev & Klug, 1969; Baker et al., 1983) and in microtubules (Ray et al., 1993).

15.2 Helical Fourier Transforms

Before its application to electron micrographs, helical diffraction theory was developed for analysing X-ray fiber diffraction patterns (review: Finch & Holmes, 1967). This work culminated in the helical X-ray structure determination of TMV (Barrett et al., 1972; Holmes et al., 1972; Namba & Stubbs, 1986). However, even with the help of powerful magnetic fields, it is hard to make oriented sols with sufficiently parallel particles, the rules of helical diffraction allow layer-lines to overlap before high resolution is reached, and the diffraction intensity is far weaker than with crystals. Its main inheritance is a method for reconstructing from electron microscope images, using helical FTs.

We begin their study with the effects of the point-group, which are easily deduced in any 3D FT: the FT must have the same point-group symmetry as the helix – the same cyclic (C_N) or dihedral (D_N) symmetry. Thus it has N-fold rotational symmetry along the X-axis (reciprocal and parallel to the helix's z-axis). If the point-group is dihedral (a non-polar helix), the helix's FT also has a group of 2-fold axes in the perpendicular (X,Y)-plane, and each of these axes is perpendicular to a plane through the origin where the FT is completely real (Fig. 10.37, §10.5.2).

15.2.1 Cylindrical (φ,z) Fourier Transform and (n,Z) Diagram

In striking contrast are the complicated effects of the screw, both in real and reciprocal space. The same complexity was minimized with crystalline lattices, where the total FT is just the

FT of the unit cell contents, sampled by the reciprocal lattice. This is impossible with helices for, despite being doubly periodic like crystals, the angular periodicity *rotates* subunits so that the convolution theorem (based on translations; see §6.1.2) no longer applies. However, the linkage of translation with rotation provides some compensations.

For simplicity, start with a partial FT of a helical lattice, where the cylindrical coordinates (φ, z) are transformed, but the radial coordinate r is left unaltered at the lattice radius a. This hybrid FT, confined to one radius, consists of an invisible cylindrical support with attached lattice points. In the simplest case, the points are all related by one repeated screw operation. We find this lattice's FT in two stages: first, the FT of a thin helical wire (uniform helix) along z, and then the FT after the wire is intersected by equidistant planes perpendicular to z.

Stage One

The uniform helical wire (Fig. 15.4 (a)), has an exact 1D repeat (the pitch P) along the z-axis, so it is unchanged by convolution with a line of points a distance P apart. Thus its FT is unchanged by multiplication with a series of thin planes, perpendicular to Z and $1/P$, apart; so the FT is confined to these *layer-planes* (Fig. 15.4 (b)). They are numbered by an *order n*, so $Z = n/P$. (Thus, because the helix is periodic, it can be expressed as a *Fourier series*, so its FT exists only where the Fourier coefficients occur.)

Next consider the *angular* structure of the FT. Rotating the uniform helix one revolution leaves it unchanged, so the angular FT is also a *Fourier series* (making the lattice FT a

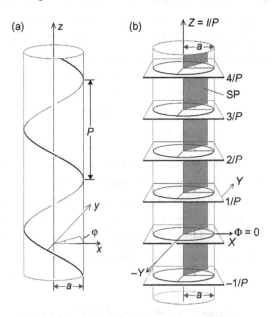

Figure 15.4: Every point of the wire has a 2-fold axis, perpendicular to the wire's axis, and the minus *y*-axis is one of the uniform helix's 2-fold axes. So the *xz*-projection has a 2-fold axis and its FT is real. (See flattened helix in Fig. 15.10 (a); the helix is left-handed so as to agree with the convention for Bessel-functions in Fig. 15.14.)

double Fourier series). Moreover, rotating the helix is equivalent to translating it; and this multiplies the FT by a unit phasor wave running along Z and so affecting bigger Z's more. Since the phase-changes cannot alter amplitudes, they must have *circular symmetry*. Thus the FT in Fig. 15.4 (b), when intersected by a section plane (SP) at an angle Φ, shows *zero* phase rotation on the plane $Z = 0$, some Φ-dependent phase rotation θ on the plane $Z = 1/P$, and a phase rotation 2θ on the plane $Z = 2/P$, etc.; also a phase rotation $-\theta$ on the plane $Z = -1/P$. (The sign of θ derives from the sign of Φ and the left-/right-handedness of the uniform helix.)

If we took the invisible cylindrical support in Fig. 15.4 (b), cut it parallel to Z at $180°$ and opened it up, we would have a long narrow rectangular sheet like that in Fig. 15.5 (seen from the outside). Each layer-plane, numbered according to its Z-coordinate $Z = \ell/P$, contains a phasor-wave that has a phase angle of $0°$ where $\Phi = 0°$. In the top half, where ℓ is positive, the phasor-waves are all positive (i.e. anti-clockwise) and they undergo ℓ revolutions in a complete circuit $0° \leq \Phi \leq 360°$.

The effect of N-fold rotational symmetry on this FT is to create a set of N parallel helices, so the repeat in Fig. 15.4 (a) changes to P/N. This convolutes (a) with a line of points along z (spacing P/N), which multiplies the FT (b) by a set of planes (spacing N/P). The FT now exist only on these layer-planes; the rotation axis eliminates all others.

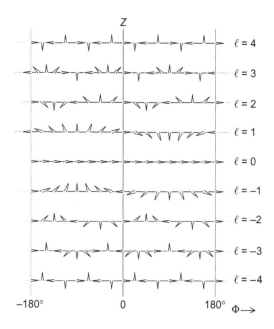

Figure 15.5: FT phases of a uniform helix (Fig. 15.4 (a)) on the cylinder (Fig. 15.4 (b)) which has been opened out. Layer-line orders marked on right, and phasors (with normalized amplitudes) show just the phases. Φ shows the polar angle; at $0°$ and $180°$ all phasors are real. (If a copy of this diagram is rolled into a cylinder, it will be seen that the phasors preserve Friedel symmetry).

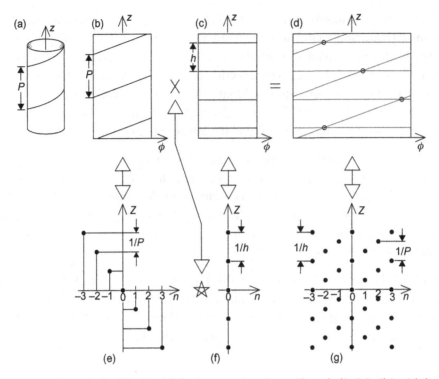

Figure 15.6: Forming a helical lattice **(d)** by intersecting the uniform helix **(a)**, **(b)** with horizontal lines **(c)**. Below we have the FTs, in the form of (n,Z) diagrams (see text).

Stage Two

The uniform helix, with successive turns P apart, embodies the screw's pitch P, but it still lacks the helical rise distance h. We incorporate this into the helical lattice through a set of parallel equatorial lines, a distance h apart (Cochran et al., 1952). In the top line of Fig. 15.6, part (a) shows the uniform helix, opened out in (b); (c) shows the same opened-out version of a cylinder with parallel equatorial lines; and (d) shows their intersection (i.e., their product), which is the helical lattice, with a rise distance h and a pitch P (This is clearer if $P \gg h$). The bottom line shows the FTs of the opened-flat cylinders above, (e) of the helices (d), and (f) of the equatorial lines (c). Their convolution is the reciprocal lattice (g).

The bottom line shows the FTs of the top, with Z the reciprocal of the z axis. In (b), (c) and (d), the horizontal axis is the polar coordinate angle φ, which measures the helices' angular spacing. The reciprocal of this is their number n, so the FTs in (e), (f) and (g) are in a reciprocal space called an (n,Z) diagram[4]. The convolution of (e) (where $Z = n/P$), with the points of (f) (where $Z = m/h$), gives the points in (g), where:

$$Z = m/h + n/P \qquad (15.5)$$

---

[4]   Its significance was first pointed out by Crick (1953).

with $m = ..., -2, 1, 0, 1, 2,...$ and $n = ..., -2N, -N, 0, N, 2N,...$ This is the most basic form of what is called the *helical selection rule*.

By multiplying Z with h, giving the dimensionless coordinate Zh, we get the more convenient equation:

$$Zh = m + n(h/P) = m + n\Omega \tag{15.6}$$

where Ω is the twist angle (measured in revolutions). This form of the helical selection rule connects the FT with the fundamental helical parameters: twist angle Ω and rise distance h. It corresponds to the real-space equation (15.4), $h/z_n = \pm n\Omega \pm 1$, when $m = 1$ and we put $1/z_n = Z$.

The third form of the helical selection rule is appropriate when the real helical lattice has an exact repeat c. Then the FT exists at values of $Z = \ell/c = n/P + m/h$, so $\ell = n(c/P) + m(c/h) = n(\text{turns/repeat}) + m(\text{units/repeat})$ or $\ell = tn + um$, where t is the number of turns per repeat and u is the number of units per repeat. The best way to understand this is to apply the selection rules to a simple helical lattice: the α-helix discussed in section §15.1.2.

To get familiar with how the 'selection rule' works, we use it to plot the reciprocal lattice. Like any 2D lattice, it needs only two basic vectors, i.e. just two points to plot, which we calculate by taking the *simplest* values of m and n. First we take $m = 1, n = 0$ so $Zh = m + n\Omega$ gives $Zh = m = 1$, and $Z = 1/h = 1/0.15\,\text{nm} = 6.66\,\text{nm}^{-1}$; and the first (n,Z) point is $(0, 6.66\,\text{nm}^{-1})$, the upper (thick) point in Fig. 15.7 (a). For a second (thick) point we choose $m = 0$ and $n = 1$ giving the second (n,Z) point $(1, 1.85\,\text{nm}^{-1})$. These two basic vectors allow us to plot the lattice Fig. 15.7 (a).

If we want to postpone finding the exact magnification scale of the micrograph, we can plot Zh instead of Z, as in Fig. 15.7 (b). (This diagram also left/right reverses the n-axis; see below.) But we could plot the lattice more easily by using a repeat $c = 18h = 18 \times 0.15\,\text{nm} = 2.7\,\text{nm}$. Now all points lie on horizontal lines at multiples of $1/c = 1/2.7\,\text{nm} = 0.37\,\text{nm}^{-1}$, as in Fig. 15.7 (c) (also with reversed n-axis). Then the first, axial or 'meridional' ($n = 0$) point, is at $Z = 1/h = 18/c$ on the 18th line on the ℓ-axis in (c). The second point, at $Z = 1.85\,\text{nm}^{-1}$, is on line number $1.85/0.37 = 5$. This plot uses the 'integral selection rule' $\ell = tn + um$, where t (the number of turns per repeat) is 5 and u (the number of units per repeat) is 18. Though less accurate than (b), this method is easier for sketching.

The selection rule equations also connect with subsidiary helices. To show this, the α-helix (n,Zh) plot from (b) in Fig. 15.7, and the diagram of its subsidiary helices from (b) in Fig. 15.2, are brought together in Fig. 15.8. The fine divisions (horizontal lines) in (a) are in (dimensionless) units of $1/18 = 0.0555...$, so the Zh-coordinate of the upper square point H is 1. If (a) is turned through $\pm 90°$ it fits (b), so points in (a) correspond to helices in (b). Thus the two points B in (a) lie on the B helices in (b); and similarly with F and G points in (a) and F and G helices in (b).

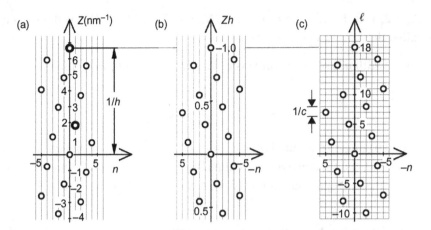

Figure 15.7: Three different plots of essentially the same α-helix layer-line diagram. **(a)** (n,Z) plot, showing layer-line order n corresponding to layer-line reciprocal coordinate (Z) in reciprocal length (nm^{-1}). This is appropriate when not simplifying the data to fit a repeat. **(b)** (n,Zh) plot, the same as the (n,Z) plot except for plotting Z in units of h, so a point at $Z = 1/h$ in plot (a) now has $Z = 1$. **(c)** (n,ℓ) plot, when applying the simplification of a repeat. The points now fit the integer equation
$$\ell = tn + um.$$

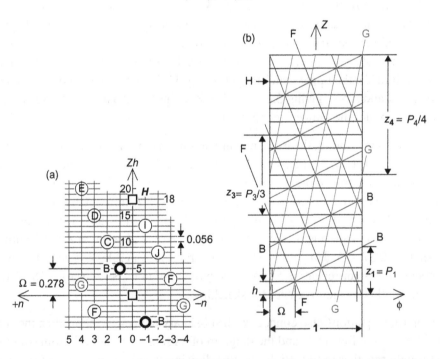

Figure 15.8: The 'ideal' α-helix. **(a)** The (n,ℓ) and (n,Zh) plots from Fig. 15.7 (b)and (c). (Integers are ℓ; to get Zh divide the ℓ-coordinate by 18.) **(b)** The diagram of its subsidiary helices from (b) in Fig. 15.2, but with the subscript letters (B, F, G) replaced by the corresponding helix numbers (1, 3, 4), as explained in section §15.1.2 and shown in equation (15.4).

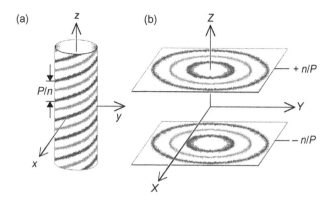

Figure 15.9: Each pair of layer-planes **(b)** relates to a set of helical density-waves **(a)**. If the layer-planes indices are (n,Z), the helices are n in number and have a pitch $P = n/Z$.

Now the *number* of helices in a set corresponds to the n-coordinate in (a). Thus, ignoring signs for the moment, we find from (a) that the B helices have $n = 1$; the F: $n = 3$; the G: $n = 4$. (To find the relative *handedness* of the helices we need the relative[5] signs of the n-coordinates.)

We can also find the *pitches* of the helices, if we know the value of the rise-distance h (here 0.15 nm). We divide 0.15 into the value of the Zh-coordinate to get Z, the reciprocal of the axial repeat z, which we multiply by the number of helices to get their pitch P. Thus, for the F helices Fig. 15.8(a) gives $\ell = 3$, the Zh-coordinate $= 3/18 = 0.167$ so $Z = 1/z = (0.167)/0.15 = 1.111 \, \text{nm}^{-1}$, or $z = 0.9 \, \text{nm}$. The F points in (a) always have n-coordinate ± 3, so $P = \pm 3 \times 0.9 = \pm 2.7 \, \text{nm}$. We have already shown that the correct sign is minus, so $P_3 = -2.7 \, \text{nm}$ and the helices are left-handed.

Finally, we can find the helices' *twist angles*. These angles, measured horizontally in (b), are measured vertically in (a). Thus, in Fig. 15.8 (b), the basic (B) helices have a Zh coordinate equalling the basic twist $\Omega = 0.278$ revolutions, and the F helices have the Zh coordinate $= 0.167$ revolutions.

Density Waves

If the image of a helix is blurred, the FT could have only one point pair for each helix. So the real helical lattice is then formed by the intersection of a few sinusoidal *density-waves*, like those in (a) of Fig. 15.9.

---

[5] *Relative* handedness is essential for indexing; *absolute* handedness (discussed in §15.3.4) can be postponed until after structure determination.

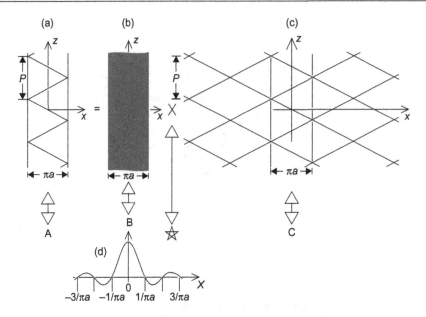

Figure 15.10: (a) Flattened helix equals... **(b)** ...a uniform rectangle of the same width times...
(c) ...a lattice of intersecting straight lines. Below, the FTs of (a), (b) and (c) are A, B and C. The
last two will be deduced later, and their convolution (the FT of multiplication) will give us A.

15.2.2 Radial Fourier Transform: Amplitudes

We can get useful insights into the radial part of a helix's FT by studying a *flattened* helix ((a)
of Fig. 15.10), which we choose to have a projection where the FT is real. Very close to the
z-axis, this flat helix is identical to a cylindrical helix projection; and it is sufficiently similar
overall that it should provide an approximation to its FT. In Fig. 15.10, the flattened helix (a)
is viewed as something cut from an infinite 2D repeat pattern (c), where the cutting process is
achieved by multiplication with a 2D rectangle function (b). So the FT of the flattened helix
(A) is the convolution of the two FTs (B) and (C); therefore we examine (B) and (C) in turn.
As shown in (d), (B) is simply the FT of an infinitely extended one-dimensional rectangle
function: a sinc-function (Fig 5.1).

(C) is the FT of the two-dimensional repeat pattern (c), consisting of two intersecting sets
of oblique parallel lines. As in Chapter 6, the FT of a set of parallel lines is effectively a 1D
comb. Here the two sets differ in their relation to the origin: the set running from top-left to
bottom-right intersect it, whereas the other set have the origin midway between two lines. The
equivalent 1D combs are shown in Fig. 15.11, where (a) + (b) = (c), so (b) = (c) − (a). Thus
we see that the set of lines intersecting the origin give all-positive signs, whereas the other set
give alternating signs, as shown in the middle frames of Fig. 15.11. Thus (C) of Fig. 15.10 is
found in (c), (f) of Fig. 15.12.

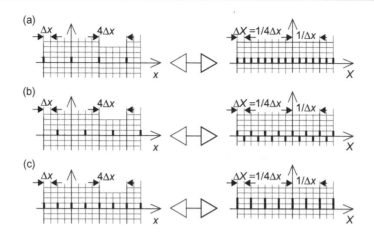

Figure 15.11: The one-dimensional FT (combs) of various combs (left) equivalent to the sets of parallel lines in Fig. 15.10. Note that **(a)** + **(b)** = **(c)**, whose FT is clearly correct: halving the spacing of the real pattern gives twice the spacing in the reciprocal one.

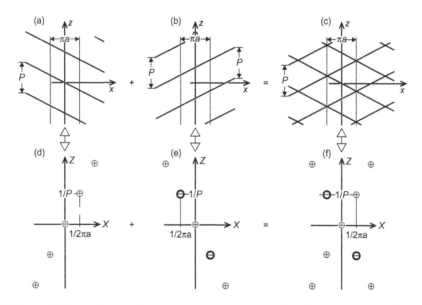

Figure 15.12: (a) Oblique lines, whose projections correspond to Fig. 15.11 (a), left. **(b)** Oblique lines, whose projections correspond to Fig. 15.11 (b), left. **(c)** Sum of (a) and (b). **(d)** FT of (a). **(e)** FT of (b). **(f)** FT of (c) = (d) + (e). The pair of lines at (f) contains the essentials of the 'helical cross' found in diffraction patterns of helical structures. (The vertical band of width πa is for reference when this is used in Fig. 15.10.)

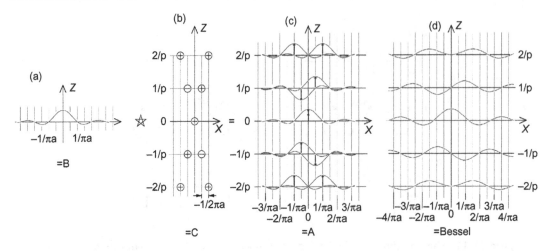

Figure 15.13: Finding the FT of the flattened helix. On the left is the FT (B), the sinc-function from Fig. 15.10, in the middle is the FT (C) from Fig. 15.12, and on the right their convolution, which is the FT of the flattened helix.

We can now find the flattened helix (A) from Fig. 15.10 by convoluting (B) with (C). This is shown in Fig. 15.13 (c), where the curves are combinations of sinc-functions. The main features of the flattened FT (c) are present in the *un*-flattened helix projection plotted in (d). There the curves are Bessel functions[6], which we discussed in §6.4.2 and §6.4.3. Many features of Bessel functions receive an intuitive explanation through comparison of (c) with (d) in Fig. 15.13. Thus the Bessel peaks (d) are seen to correspond to diffraction from the projected helical lines, shown flattened in Fig. 15.10 (a). The positions of these peaks, found in Fig. 15.12, are shown in Fig. 15.13; they are found where X is (approximately) a multiple (n) of $1/2\pi$(helix radius a). We use $R_{max}(n)$ to refer to the X-value (R in polar coordinates) of the first Bessel peak on layer-line number n, shown arrowed in Fig. 15.14; then $R_{max}(n) \approx n/2\pi a$, so $2\pi a R_{max}(n) \approx n$. This is the first term of an excellent approximation[7] given by Abramowitz & Stegun (1964).

Note the effect this has on the diffraction pattern of a continuous helix. There is a barren cone, mostly empty of any diffraction data, centered on the origin and whose axis is the Z-axis. The highest order Bessel-functions only start to become significant at the surface of this cone, where R is very roughly proportional to n. Thus the FT intensity associated with the first Bessel peak gets spread around a radius that is very roughly proportional to n. Thus

---

[6] Bessel functions are discussed in any book on mathematical physics, programs are provided in Press et al. (2007), tables with the principal formulae are in Abramowitz & Stegun (1964); and, for a comprehensive mathematical treatise, see Watson (1958).

[7] $2\pi a R_{max}(n) \equiv j'_{n,1} = n + 0.8086165 n^{1/3} + 0.07249/n^{1/3} - 0.05097/n + 0.0094/(n.n^{2/3})$.

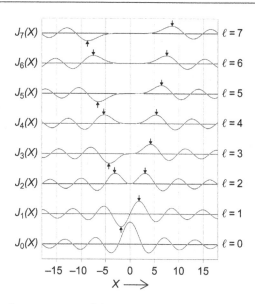

Figure 15.14: Plot of the first seven Bessel functions. Note that the even order functions are even (like cosines), the odd orders are odd (like sines). So the odd orders must be zero at the origin, but so also are the even orders, beyond the zero order. The central minimum increases with the order. It is flanked by the highest peaks (or lowest troughs), both arrowed.

the actual intensity is, on average, roughly proportional to $1/n$, and the FT amplitude (height of the first Bessel peak) is the square root of this. (See Fig. 12.6 of DeRosier, 2007.) This contributes to the difficulty of collecting high-resolution data from helices.

15.2.3 Radial Fourier Transform: Phases

Bessel functions describe the *amplitudes* of layer-planes of a uniform helix's FT. As discussed above (§15.2.1), their *phases* rotate when making a circuit around the Z-axis but keeping within a layer-plane. The phases make n complete revolutions at layer-plane n. But, since the line-helix is chiral (left-handed in our diagrams), its FT must also be chiral. For the upper part of the FT (*positive Z*), if we circuit a layer-plane in the sense $X \rightarrow Y \rightarrow -X \rightarrow -Y$, the phase rotates *clockwise* (for a left-handed helix, and according to our FT convention, §3.4.2). However, it rotates *anti*-clockwise in the layer-planes with *negative Z* ($\ell = -1/P$ etc.). This can be deduced from Fig. 15.5, and is a consequence of the FT of a real structure having Friedel symmetry.

15.2.4 Overall Fourier Transform: Atoms

Although we now have the full (R,Φ,Z) FT of an isolated uniform helical wire, we have only the cylindrical part (Φ,Z) of the FT of a helical lattice. However, that cylindrical part is the (n,Z) diagram, where each point represents a layer-plane, with a Bessel function of order n

on a plane at position Z. The same helical lattice applies at all radii (Fig. 15.1), so the (n,Z) diagram summarizes the FTs of sets of uniform helical density-waves, all at the same radius, which intersect to give the helical lattice. (Their intersection involves multiplication, whose FT is convolution, creating the (n,Z) diagram: see Fig. 15.15 for two helices.)

In the simplest (indeed, hypothetical) case, where the helical lattice has no exact repeat, the layer-planes all have different Z-coordinates, so they cannot interfere, and the FT of the complete helical *lattice* is simply their sum. This simplification often applies at low resolution, so we ignore its limitations until we have discussed the complete FT of the *structure*, to which we now turn.

Although we know the FT of a helical lattice, that is just copies of one 'lattice point' P, equivalent to only one atom from each asymmetric unit of the real helix. To get the *structure*'s FT, we need *all* its atoms added together. So we take in turn each atom (A, B, C, …) from the group, and find the adjustment required to bring the lattice point P (coordinates $r = 1$, $\varphi = 0$, $z = 0$) into coincidence $n\varphi$ with it. If we are considering atom A, with polar coordinates (r_A, φ_A, z_A), we shall need to adjust P's radius to equal r_A, to rotate it by φ_A, and to translate it by z_A. The first operation involves radial expansion (or contraction), the second operation rotates the already modified FT by φ_A, and the third operation multiplies the twice-modified FT by a unit phasor-wave along Z that repeats after $1/z_A$. After undergoing these three adjustments, the lattice FT must be multiplied by the scattering amplitude of the atom A. Then we add the modified lattice FT into a file that will build up the structure's FT, as the contribution of each atom is included.

None of the modifications to the lattice FT changes the layer-plane positions or the distribution of Bessel orders. As each atom's contribution is included, the addition processes are confined to these layer-planes and involve adding Bessels of the same order, but differing only in their radial expansions, their starting phases and their amplitudes. Thus the final helical structure's FT has exactly the same overall pattern as the FT of the helical lattice.

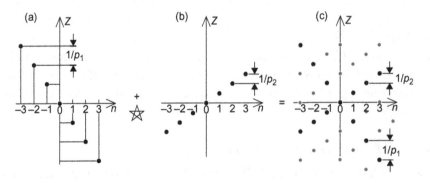

Figure 15.15: A helical lattice's (n,Z) diagram **(c)** viewed as the convolution (grey) or sum (black) of the (n,Z) diagrams of a pair of its helical lines **(a)** and **(b)**.

(For interpreting optical or X-ray diffraction patterns, we need to compare with the FT *intensity*, which is the squared amplitude of the phasors.)

15.2.5 G-Functions and Radial Inversion

We have seen how all the simplicity of the helical FT lies in the cylindrical helical lattice and its 'FT', the (n,Z) diagram. So a regular FT can be applied to all the atoms (or voxels) of an asymmetric unit of the helix, provided we keep the radius r constant. That is, we can calculate a partial FT that doesn't apply to r. A special terminology has therefore been developed, which is commonly used in helical diffraction theory. This partial FT is called $g_n(r,Z)$, though this usual way of writing it obscures the fact that n and Z go together (as 'reciprocal space' coordinates), whereas it is r that hasn't been changed and remains in 'real space'; so we shall write $g_r(n,Z)$ here. Thus $g_r(n,Z)$ is an ordinary FT of the height z and angle φ of each atom (or voxel), but still preserving its original radius r. To get the full 3D FT, we must (first) do a *radial* FT of each $g_r(n,Z)$ and add them all together, getting a new function $G_R(n,Z)$; and then (second) convert the n into Φ by making use of the n-fold layer-plane phase-rotation of phasors per rotation of Φ.

The functions $g_r(n,Z)$ and $G_R(n,Z)$ (Klug et al., 1958) are related by the radial (r,R) transform. Although R is orthogonal to Φ and Z, it isn't at a uniform scale, because the radius vectors aren't parallel, but intersect at the axis. Towards infinity they become more parallel (and r becomes more uniform like x), but towards zero the proximity of their intersection makes them more *non-linear*. So we must replace sinusoids by Bessel functions and also use *non-uniform scaling*, as we shall now see.

Consider the simplest example of this problem: diffraction from a uniform cylinder of infinite length. As it is long and uniform, its FT is confined to a perpendicular plane, where it is a zero-order Bessel function (Fig. 6.25, § 6.4.2). That plane reduces the problem from 3D to 2D, where the FT's circular symmetry reduces it further to 1D (Fig. 15.17). Suppose we now want to extend this thin cylinder to a thick one with 'salami-slices' which are concentric cylinders. Consider just two of these (Fig. 15.16 (a)). They have the same density, but the

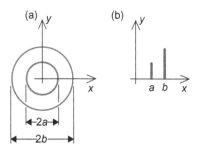

Figure 15.16: (a) Projection of two concentric uniform cylinders of radius a and b. **(b)** Radial plot of the relative areas of the two cylinders.

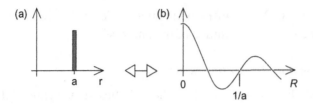

Figure 15.17: The Fourier–Bessel transform **(b)** of a peak **(a)** at $r = a$.

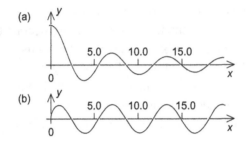

Figure 15.18: Adjusting the zero-order Bessel function so it has the orthogonality features of sinusoids. **(a)** Plot of the equation $y = J_0(X)$. **(b)** Plot of the equation $y = (\sqrt{X})J_0(X)$. The factor $(\sqrt{X})$ has made it similar to a sine-curve. (The same factor is also used with other Bessel orders.)

outer ring, with its bigger radius, makes a bigger contribution to the FT. To allow for this, we multiply the density of each ring by its radius. This correction is the key to using Bessel functions in cylindrical geometries.

We recall from Chapter 3 that sinusoids of different frequencies give a zero CC and are thus said to be 'orthogonal'. But $J_0(r)$ functions of different frequencies couldn't be orthogonal because of the large positive contributions near $R = 0$ (Fig. 15.17 (b)).

However, we need to multiply the density of each ring by its radius. When we find the CC of $J_0(r)$ functions of different frequencies, this is equivalent to multiplying *each* Bessel by $\sqrt{R}$. The effect of this scaling is shown in Fig. 15.18 (a), (b). The curve (b), after scaling, is remarkably similar to a sine-curve. (Other J_n functions are similar, though without the problem at the origin, where they are all zero; and the near-zero region expands with increasing n.)

15.3 Getting a Structure from Helical Diffraction Data

15.3.1 Indexing

Getting a 3D structure from helical diffraction data must obviously depend on assigning each datum to its correct place in the reconstruction. That place is defined by the (n,Z) plot, so the first roadblock is constructing this plot, a process called *indexing*.

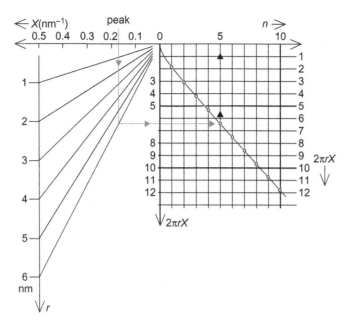

Figure 15.19: Assigning numbers of helices to a layer-line using an (X,r,n) diagram. The X-coordinate of the diffraction peak is entered at the top left, the radius r is chosen from the appropriate 'radius line' on the left edge, and the order n is read off at the top right. The grey arrowed line shows an example, running from a peak at $X = 0.18\,\text{nm}^{-1}$, via $r = 6\,\text{nm}$, through $2\pi rX = 6.4$ to $n = 5$ at the top. The oblique line on the right is plotted from the approximation for $2\pi aR_{\text{max}}(n)$ given in §15.2.2.

Figure 15.8 shows (n,Z) plots next to the helical lattice. Both are lattices that relate to one surface of the helix. But the helical diffraction pattern derives from *both* surfaces, and thus from *two* identical helical lattices, and thus *two* identical reciprocal lattices, related by an axial mirror. And each reciprocal lattice derives from a *curved* surface.

We start by tackling the curvature. We convert the (X,Z)-diffraction spots into provisional (n,Z)-diffraction spots, using the (X,r,n) diagram (Fig. 15.19) (or its equivalent). Here the experimental datum was the X-coordinate of a diffraction spot's peak ($0.18\,\text{nm}^{-1}$) and the effective radius of the helix ($r = 6\,\text{nm}$). Of course, this procedure rarely yields an exact integer, as in the example; so this method usually gives two alternative n-values, plus further variation through uncertainty about the effective radius r. Thus the provisional (n,Z)-diffraction diagram has layer-lines at the correct Z-coordinates, but with a range of n-values.

We next try to refine these n-values. One useful method compares the phases at $+X$ and $-X$ to discover the parity of n: if n is even, the phases are equal; if n is odd, they differ by 180°. (On each layer-plane, the phases rotate n times in a complete circuit; so, if n is even, the phases on opposite sides are the same; otherwise, as in Fig. 15.20 (b), they are 180° out of

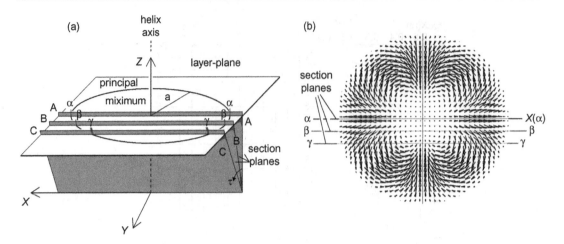

Figure 15.20: (a) View from above of the central maximum of the layer-line plane for $n = 3$, shown in **(b)** being intersected with section planes at different angles. Section α: no tilt (X-axis); β and γ at increasing angles. Note the progressive downward turn of the phasors in β and γ. **(b)** View from above of the ring in the partial (r,Φ,Z) [or (φ,z)] transform. Here the different positions (α, β, γ) of the tilted section plane are slightly closer.

phase.) A third method for refining n-values uses tilt experiments to get additional data (i.e., to find the helical arrangement by tomography); but this is only used as a last resort.

Now suppose that we have assigned the n-values to each layer-line. To draw a provisional (n,Z)-plot, we recall that every layer-line has (like the original diffraction pattern) mirror-symmetry, so each group of n-values on the left must be mirrored on the right. Our next task is to choose just one of these groups. They arise because each surface of the helix gives rise to its own reciprocal lattice. These are superposed together, so we have to disentangle them, a 'game' subject to strict rules. (i) Each layer-line contributes just one point to the chosen lattice, and one point to the discarded lattice. (ii) These two points are mirror images (reflected in Z and also, because of Friedel's law, in X), so they come from opposite sides of the layer-line. Thus some layer-lines contribute the left point, and others contribute the right point, to the chosen lattice. (iii) The chosen lattice must be a proper lattice, with two defining lattice-vectors, with coordinates (n_a,Z_a) and (n_b,Z_b); and any layer-line with coordinates (n,Z) must satisfy the lattice conditions $n = i.n_a, + j.n_b$, and $Z = i.Z_a, + j.Z_b$.

Uncertainties as to the n-value of each layer-line adds difficulty to this 'game', and it may not be soluble. There may be no lattice satisfying the conditions, and then we must allow some extra flexibility to the assignment of n-values. Or there may be alternative solutions, when we need to reduce the flexibility to leave only one. The real challenge is to steer a route that plausibly avoids both problems, if possible without resorting to tilt experiments.

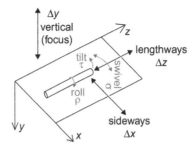

Figure 15.21: Orientation parameters for a helix examined in the electron microscope. There are three translational parameters (black), Δx, Δy and Δz, and three rotational parameters (grey), *roll* ρ, sideways *swivel* (or turn) σ and *tilt* τ (sometimes given the symbol ω). (However, σ corresponds to ω in polar coordinates.)

Efforts have been made to develop computer programs for indexing. Toyoshima (2000) developed one requiring operator assistance. Another (Ward et al., 2003) tested different layer-lines as possible base-vectors to generate the (n,Z) plot. An alternative, when the helical lattice is completely defined by just one screw with components (h,Ω), would scan over Ω from $0°$ to $180°$ and, at each scan point, adjust h to find the best cross-correlation function (CCF). But, whatever indexing method is used, it appears that there are some situations where the available data yield at best a 'most likely' indexing, and more certainty requires new data.

15.3.2 Interfering Layer-Lines

It simplifies matters if layer-planes keep separate, each with one order of Bessel function amplitude with circular symmetry, and a phase that rotates n times per circuit. But layer-planes have a minimum width w of 1/(minimum particle width in that direction). And no (n,Z) lattice can always avoid having lattice-points with the same Z ($\pm w$), as n increases endlessly. So layer-planes inevitably collide at higher resolution.

Such a collision leads to interfering Bessel orders, so the amplitudes lose their circular symmetry and the phases their uniform rotations. Except in the very simplest cases, further analysis requires extra data. We need to sample layer-planes at different orientations to map out the interference. Such sampling requires new section planes of the FT, for which we require tilted images (preferably with a rotation that changes ρ in Fig. 15.21). This can give enough equations to find the extra unknowns. The calculation (Crowther et al., 1970, 1985; Jeng et al., 1989) is similar to the problem of 'reverse interpolation' discussed in section §11.1.7 and uses similar least-squares methods (Chapter 20). Besides the answer, we also get error estimates.

This becomes very complicated with a large helix giving data at high resolution, and it has been replaced by an approximate method (Miyazawa et al., 1999; Unwin, 2005). If the layer-line intensities are averaged over many different values of ρ, interference disappears and different Bessel orders add intensities like incoherent waves (Franklin & Klug, 1955; Waser, 1955; Finch & Holmes, 1967, p.410, section 3.D.2). However (as mentioned at the start of this chapter), layer-plane interference concentrates the signal, improving its detectability. Increasing the signal:noise ratio is probably more valuable than merely minimizing the number of different images needing processing.

15.3.3 Tilt Correction

The indexing process should include the parity test for n (explained above). Although we cannot expect measured phase differences to be *exactly* 0° or 180°, a systematic deviation probably indicates that the helix axis is not perpendicular to the beam. Any deviation is described as tilt τ (Fig. 15.21) and its qualitative effects are easily predicted. Thus, when n is *odd* the phase-difference is maximal (180°), and tilting can only *diminish* it. But when n is *even* the difference is minimal (0°), so tilting can only *increase* it. The quantitative effects (DeRosier and Moore, 1970) are illustrated in Fig. 15.20 (a) and (b). When n is bigger, the phase changes faster around the Bessel maximum, so the same tilt (τ) has a bigger effect. But the effect is smaller if the radius a (measured by the X-coordinate) of the principal maximum increases. However, a bigger Z-coordinate amplifies the effect of τ. Clearly, (n/X) measures the spacing of phase changes, while (τZ) measures the shift of the section plane. Thus we expect that, at small tilt angles τ, the phase angle θ is changed by[8]:

$$\Delta\theta \approx \tau\left(nZ/X\right) \tag{15.7}$$

This can be used to correct the phases in the FT of a tilted helix. For a given (n,Z) plot, which fixes all n-values, we can measure the deviations, get their variance and then calculate the tilt τ that will minimize this.

15.3.4 Handedness

The process of tilt-correction also provides an opportunity to get data concerning the helix's chirality, which determines in which direction the phases rotate. So each chirality alternative leads to a different prediction for the direction of the actual tilt of the helix. If these don't point to one answer, we might be able to give the helix a known tilt (at least, of known direction) and measure the effect on the calculated τ. (Any way to determine the handedness of any set of helices – such as metal-shadowed preparations – fixes the chirality of the entire lattice.)

15.3.5 Averaging Data from Different Particles

Correcting the tilt angle τ (§15.3.3) finds one of the 2R out-of-plane rotations. The other is the roll angle ρ (Fig. 15.21), which can be found absolutely if the symmetry framework contains

---

[8]   This approximation holds for tilt angles less than about 10°, the most common situation; when τ is bigger, the change is $n\tan^{-1}[(Z\sin\tau)/X]$. This is similar to the equation for diffraction spacings, §9.4.4 (see footnote).

a symmetry-marker. That is the case if the helix has the point-group symmetry D_N, when 2-fold axes are the important extra markers. Otherwise, ρ's can only be compared between particles, and such comparisons are also needed for the (2T + R) in-plane adjustments (Δx, Δz and σ in Fig. 15.21). Every helix has a unique z-axis, which is a symmetry-marker for a perpendicular translation and for in-plane rotation (T + R). This leaves ρ and the in-plane translation along the axis (Δz) as the parameters lacking symmetry-markers in the case of a polar helix (point-group symmetry C_N). Then the comparison of particles depends on their images, using the CCF. Of course, the ordinary translational CCFs serve for Δz, but there is also a rotational (angular) CCF for ρ (§6.4.3).

Icosahedral Particles

Chapter Outline

Having surveyed all particles with translational symmetry, we are left with those having only rotational (if any) symmetry. This chapter surveys those with high rotational symmetry: particles that are intrinsically bounded, unlike the sheets and rods that grow until they meet a boundary[1].

The rotation groups were discussed in Chapter 10, section §10.4.2, illustrated in Figs 10.32, 10.33 and their connections are displayed in Fig. 10.34. The groups yielding the most compact assemblies are T, O and especially I, since spherical viruses were found to have an icosahedral shape. The first insight into its biological significance came from Crick & Watson (1956) who proposed that, through shortage of nucleic acid, a small virus would need to construct its capsid from many copies of one protein, which (to maximize capacity) would form the assemblage with highest symmetry: icosahedral with sixty units. But even that is still inadequate, and viruses have found ingenious ways of enlarging these cramped quarters with more complicated polyhedra (*deltahedra*), which combine high quasi-symmetry, curvature and intrinsic strength which maintains an ordered structure capable of preserving detail.

With polyhedral structures, analysis loses its most powerful tool, the Fourier analysis of translations. But it can still use Fourier analysis for *one* rotation, and this is much better than nothing.

---

[1]   The electron microscope analysis of icosahedral viruses has been reviewed by Chiu (2007).

Structural Biology Using Electrons and X-Rays.
© 2011 Michael F. Moody. Published by Elsevier Ltd. All rights reserved.

16.1 Deltahedra

16.1.1 'Platonic' Polyhedra

Each of the three 'Platonic' rotation groups has its archetypical polyhedron whose faces are equilateral triangles (called **deltahedra** by analogy with the Greek 'delta' Δ). Each can be visualized as derived from a symmetrical hexagonal net from which 60° segments have been removed. Each 'missing' segment leaves an angular deficit of 60°; and a closed surface requires a total angular deficit of 720° (a theorem due to Maxwell, 1854). This can be made up from four 180° segments (4 × 180° = 720°), giving a tetrahedron, or from six 120° segments (6 × 120° = 720°), giving an octahedron, or from twelve 60° segments (12 × 60° = 720°), giving an icosahedron. Of these, by far the most important is the icosahedron, which can be visualized as removing twelve 60° segments in two stages. First, we remove three 60° segments (and three 180° segments) to give a hemispherical 'cup' or 'half-icosahedron', as shown in Fig. 16.1 (a), (b). Second, we combine this with a second identical 'cup' that is added on top to give the icosahedron, as in (c).

We shall often need to refer to the icosahedron in this chapter, so we start by counting its faces, vertices and edges.

  (i) *Faces.* Figure 16.1 (c) shows that the icosahedron has two pentagonal caps (e.g., 1, 2, 5, 9, 6, 3; each with five triangles) separated by an equatorial region of 10 triangles (five pointing up and five pointing down). Thus there are 5 + 10 + 5 = 20 triangular faces.
 (ii) *Vertices.* Each pentagonal cap has one central, and five peripheral, vertices: a total of six vertices. Thus the two pentagonal caps make up a total of 6 + 6 = 12 vertices, and this is all, since the equatorial region has no vertices exclusively its own.
(iii) *Edges.* Each pentagonal cap has five radial edges and five tangential ones; so the two caps have (5 + 5) + (5 + 5) = 20 edges. Adding the 10 edges in the equatorial region (that do not belong to either pentagonal cap), the total of edges is 20 + 10 = 30.
(iv) *Summary.* Thus the icosahedron has 20 faces (F = 20), 12 vertices (V = 12) and 30 edges (E = 30), fitting Euler's formula for a polyhedron: V + F = E + 2.

We can see from Fig. 16.1 that the vertices occur in opposite pairs, like 5 and 7 in (b). Similarly, the triangular faces also occur in pairs, like the topmost and bottommost faces in (c). The edges also occur in opposite pairs. All this fits with the fact that the numbers of faces, vertices and edges are even, and so is the constant (2) in Euler's formula. This pairing allows us to count the rotation axes of the icosahedron (see below).

If we assemble these polyhedra, not from a hexagonal net but from a sheet with the *p*6 plane-group, we reveal features of their corresponding point-groups. Thus Fig. 16.2 (a) shows how each icosahedral vertex symmetry axis arises by removing one 60° segment from a *p*6 sheet. This converts the original 6-fold axis into a 5-fold axis, and we have another polyhedron in which the highest-order axes run through the vertices. We find the order of the icosahedral

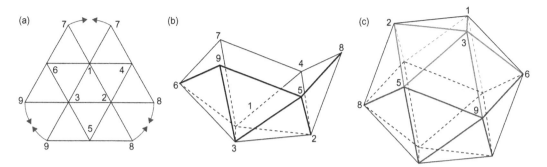

Figure 16.1: (a) Half an icosahedron is formed by folding a hexagonal net. The folds, along sides that join two triangles, bring together edges at an angle of 120°. (Only 10 of the 20 faces are drawn.) **(b)** Folded icosahedral 'cup'. **(c)** Two 'cups' put together to give an icosahedron with its upper vertices marked.

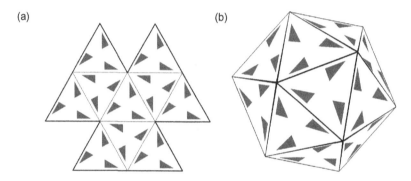

Figure 16.2: (a) Ten triangles from a $p6$ sheet, where the three vertices near the center have already lost a 60° segment. These assemble into a 'half icosahedral cup', as in Fig. 16.1 (b). **(b)** The fully assembled icosahedron, with 5-fold as well as 3- and 2-fold axes.

group I by counting the subunits (grey triangles). Each face has three subunits, and we have just seen that the icosahedron has 20 faces ('eikosi' is Greek for 20). So the number of subunits is $20 \times 3 = 60$.

We shall also need to know the number of symmetry axes in the icosahedron. As a start, the rotational symmetry axes of the icosahedron are made explicit in Fig. 16.3 (b). The asymmetric units of icosahedral symmetry are shown in (c). These are the portions of the surface which, by repetitive copying through the action of the rotation axes, cover the surface of the icosahedron. Now we count the rotation axes. Each 5-fold axis connects an opposite pair of vertices and, since there are twelve vertices, there must be six 5-fold axes. Similarly, each 3-fold axis connects an opposite pair of faces and, with twenty faces, there must be ten 3-fold axes. Finally, each 2-fold axis connects an opposite pair of edges, so thirty edges imply fifteen 2-fold axes.

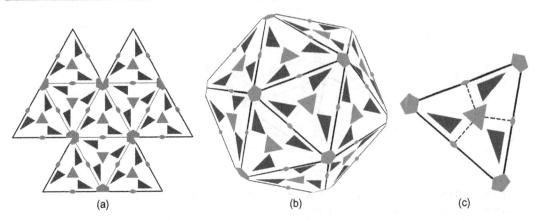

(a)   (b)   (c)

Figure 16.3: (a), **(b)** A version of Fig. 16.2 where the symmetry axes are made explicit. **(c)** Three asymmetric units of icosahedral symmetry.

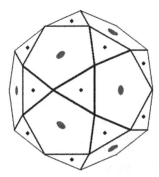

Figure 16.4: Effects of clustering of the asymmetric unit. View of an icosidodecahedron nearly down a 3-fold axis. Its 30 vertices occur at the positions of an icosahedron's 2-fold axes. The triangles have 3-fold axes at their centers, the pentagons have 5-fold axes at their centers.

The asymmetric unit, symbolized as a grey right-angled triangle in Fig. 16.3 (b), (c) is actually an asymmetrical three-dimensional structure adjacent to 5-, 3- and 2-fold axes. Depending on its shape, it is likely to protrude more in the direction of one or other of these axes, so its adjacent copies will generate a cluster around that axis which gives the particle a characteristic appearance. Thus a cluster around the 5-fold axis produces 'pentamer clustering' with protrusions around the vertices of an icosahedron; 3-fold or trimer clusters cause protrusions around the vertices of a dodecahedron; and 2-fold or dimer clusters protrude around the vertices of an icosidodecahedron (Fig. 16.4).

The basic dimensions of the icosahedron are shown in Fig. 16.5 (a). It can be viewed (Coxeter, 1989) as an assembly of three perpendicular identical rectangles with side-lengths 2 and 2τ, where $\tau = 1.618...$, the 'golden ratio'. (The shorter sides are bisected by the x,y,z axes.) The rectangles are assembled with D_2 symmetry and the Cartesian coordinates oriented

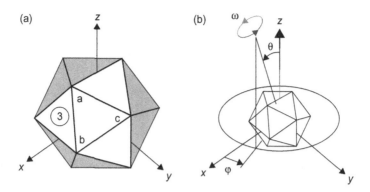

Figure 16.5: (a) The icosahedron, with the (x,y,z) axes passing through three of its perpendicular 2-fold axes. The coordinates are 'a' = $(1,0,\tau)$, 'b' = $(\tau,1,0)$ and 'c' = $(0,\tau,1)$, all coordinates taken in order from the sequence $..\tau,1,0,\tau,1,0...$ **(b)** Polar coordinate system applied to the (x,y,z) axes. This system can be used either with the 'rotation formula' or with Euler's angles (see §12.2.3).

along the 2-fold axes of the rectangles. The polar coordinate system fitting those axes is shown in (b). The points 'a' and circled-3, or (3), in (a) have polar angles $\varphi = 0°$ (for both) and $\theta = 31.7°$ (for 'a') or $\theta = 69.1°$ (for (3)), which marks the division of three asymmetric units, Fig. 16.3 (b), (c). The spherical triangle with corners at 'a', 'b' and (3) represents an asymmetric unit of the icosahedron: we only need to know that section of an object with icosahedral symmetry, in order to fill in the rest[2].

16.1.2 Viral Capsids

Icosahedral symmetry is most commonly found in the shells or **capsids** of spherical viruses. As mentioned above, these usually have more than the 60 subunits expected if each asymmetric unit were a chemical subunit. So it must be a multiple of 60 (for the shell to have icosahedral symmetry), and a simple geometrical theory for this was provided by the **quasi-equivalence** theory of Caspar & Klug (1962).

The capsid still has icosahedral symmetry with its 60 exact copies of triangular (notional) 'faces', but the 'faces' need to be larger. In the simplest case (Fig. 16.3 (a)), the 'face' cut out from a $p6$ sheet was the minimal triangle. To get the bigger face we need, we cut from the same sheet an equilateral-triangular face whose second vertex is at a non-adjacent lattice point, as in Fig. 16.6. Many of the triangular faces bigger than (b) come in two enantiomorphic forms (the triangular frames are mirror images, not the contents). Figure 16.3 (c) and (d) are the simplest examples of this. The preferred enantiomorph is preserved over a wide group of viruses (e.g. T = 7*dextro* in the papovaviruses: Belnap et al., 1996).

---

[2] In the case of an icosahedral FT, even less is needed, since Friedel symmetry fills in everything from a hemisphere.

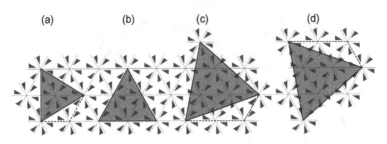

Figure 16.6: Three larger equilateral triangles constructed from the lattice points of a hexagonal lattice with small grey triangles arranged in a *p6* plane-group. **(a)** Next-nearest lattice points give a triangular facet that is three times bigger than the simplest (T = 3). **(b)** The next permitted one has four triangles, which can be counted (T = 4). **(c)** The next bigger has seven small triangles (T = 7*laevo*). **(d)** The same size as (c), but with the enantiomorphic triangle (T = 7*dextro*).

The second lattice point can be defined by *h* unit vectors along one lattice direction, followed by *k* unit vectors along another. The integers (*h,k*) fix the area, and hence the number of subunits, of each face and therefore of the entire polyhedron. This is T = $h^2 + hk + k^2$ (Caspar & Klug, 1962), where *h* and *k* are positive integers, so T equals 1, 3, 4, 7, 9, 12, 13, …, but not 2, 5, 6, 8, 10, 11, … (see Appendix §16.4).

16.1.3 Curvature of Deltahedra

We have seen how to get an icosahedron from a *p6* sheet by removing 60° segments. Removing one such segment converts a flat sheet into a cone, the apex of which is a 5-fold vertex (Fig. 16.7). To get an icosahedron, which has twelve vertices, we need to remove twelve 60° segments.

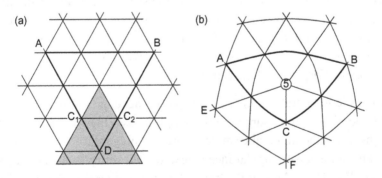

Figure 16.7: (a) Flat sheet of equilateral triangles, with a larger triangle ABD marked with a darker line. The 60° shaded segment is to be cut out. **(b)** After removing this segment, the segment edges fuse, bringing C_1 and C_2 together as C. This increases by 60° the angle at the bottom edges of the triangle. Thus, whereas triangle ABD in (a) had a total internal angle of 180°, triangle ABC in (b) has a total internal angle of 240°. This results from the point of positive Gaussian curvature at the indicated 5-fold vertex.

We earlier met a different reason why we need to remove twelve 60° segments. Each 'missing' segment leaves an 'angular deficit' of 60°, and a closed surface requires a total 'angular deficit' of 720° (by Maxwell's theorem), and twelve missing segments leaves a deficit of $12 \times 60° = 720°$.

The concept of an 'angular deficit' is connected with the concept of **Gaussian curvature**. This generalizes the intuitive concept of spherical curvature: a smaller sphere is more strongly curved (e.g. possessing more surface-tension energy, which small droplets can diminish by fusing to give bigger droplets). 'Gaussian curvature' finds the surface's bend (measured by its deviation from a tangent plane) in two perpendicular directions, oriented so that the two bends differ as much as possible. For example, in a cylinder, one of the two directions lies parallel to the z-axis, where there is no deviation from a tangent plane (so no curvature); and the perpendicular direction (xy-plane) sees a circle, whose curvature is measured. One measure of overall curvature would average these two bends, but 'Gaussian curvature' multiplies them, getting zero (because of no bend along z). This may seem counter-intuitive: surely a cylinder isn't the same as a flat plane? But it *is* the same in one important respect: a map drawn on a flat sheet still applies after that sheet has been rolled up to give a cylinder. And it turns out that two surfaces can use the same map, without needing any distortion, provided they have the same Gaussian curvature. If their Gaussian curvatures differ, however, maps on their surfaces must differ. The classic example is the earth, whose map can be represented accurately on a globe (also positive[3] Gaussian curvature) but not on a flat sheet in an atlas (zero Gaussian curvature).

This mapping problem is relevant to the structural analysis of polyhedral aggregates. We saw that Fourier analysis can be applied equally well to translations and to a rotation; so we can have a double Fourier transform (FT) either for flat structures (two translations) or cylindrical (helical) structures (one translation and one rotation). Polyhedral structures have no translational symmetry, but two rotations, so it might have been expected that these would allow a double FT. However, two different rotations cause the surface to be curved with a positive Gaussian curvature, whereas all our examples of double FTs applied to surfaces with zero Gaussian curvature. So we can only have one FT with polyhedral structures, adding unwelcome complexities to their analysis.

Another interesting property of Gaussian curvature is its connection with the shape of a triangle. Any triangle drawn on a flat surface has its shape fixed by the three angles of its vertices, which add up to 180° (a straight line). But that is not true for a triangle drawn on a curved surface. Of course, there is the little problem that such a triangle cannot strictly have 'straight' sides, but we can extend the definition of a 'straight line' as the shortest

[3] Positive Gaussian curvature gives a closed shell like a sphere; negative Gaussian curvature, where the curvatures work in opposite directions, gives an 'unclosed' surface, like the inner surface of a torus or tyre, where the edges get further away from each other at bigger distances.

path between two points. So, on a sphere like the earth, a 'straight line' (**geodesic**) is a 'great circle' like a circle of longitude or the equator. And a triangle whose vertices are the North pole and any two points on the equator has two right angles (90°) at the equator, plus some other angle at the pole, giving more than 180° for the sum of its angles. When we remove a 60° segment, we change the corner-angles of any triangle that surrounds the 5-fold vertex (triangle ABC in Fig. 16.7). Now, if the corner-angle of a triangle is enlarged to 180° is becomes a straight line; and a 'triangle' with three such corner-angles is a circular loop. Such a 'triangle' needs enlargement of 120° per corner-angle (starting with 60°). Therefore we get a circular loop after adding two 60° portions per corner, i.e. six 60° portions, which require six internal 5-fold vertices. That gives us a triangle like a circular loop, appropriate to a cylinder; so six 5-fold vertices give a cylinder, and thus twelve 5-fold vertices give a closed shell.

A deltahedron made like a paper model has the positive Gaussian curvature concentrated at the vertices. Elsewhere it is zero (like the paper in a model), and the inter-subunit bonds (the equivalent of the triangle sides) have equal lengths. Like a paper model, it also looks 'polyhedral', like the mature T4 bacteriophage head which presents the profile of a lengthened hexagon. By contrast, immature viruses (procapsids) are often round, a shape requiring slightly unequal bond lengths with >60 subunits (an aspect of the impossibility of mapping a sphere onto a plane). So the procapsid is relatively weak, but the Gaussian curvature is uniformly distributed (as in a sphere), which should allow the head size to be determined by 'built-in curvature' (see the review by Moody, 1999). Then the head transformation both enlarges and toughens the newly assembled procapsids, equalizing the bond and thereby 'angularizing' the head.

16.2 Projections

We deal first with the image analysis of isolated single projections (postponing to Appendix §16.4 their combination to give solid structures). We saw in Chapter 12 how images with two-dimensional translational symmetry can be processed to reveal it and also filtered to extract the symmetrical content. Here we discuss how similar operations can be performed on images with approximate rotational symmetry. Indeed, the first image analysis (Markham et al., 1963) was of this type. As discussed in §6.4, we need polar coordinates so that rotation can be represented as a convolution in φ. As discussed in the last chapter, there is usually no repeating structure in r, so the analysis is one-dimensional; though there is the complication that the center of rotation must first be found. However, we postpone this to get a clear initial view of the method, due to Crowther & Amos (1971).

Given the rotation center, we start by dividing up the micrograph into concentric narrow annuli and analyzing each annulus separately. An annulus is essentially a one-dimensional image where density is a function of distance along the periphery (Fig. 16.8 (c), (d)). Its Fourier analysis is shown in Fig. 16.9. Because the density distribution repeats over the strip

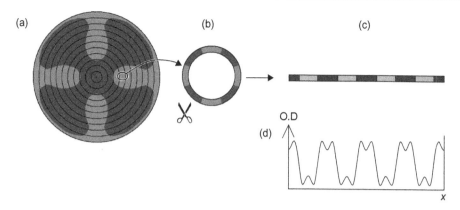

Figure 16.8: The Fourier method for finding rotational symmetry. The image **(a)** is divided into annuli, of which one **(b)** is cut and opened to give the strip **(c)**. Its density distribution is plotted in **(d)**.

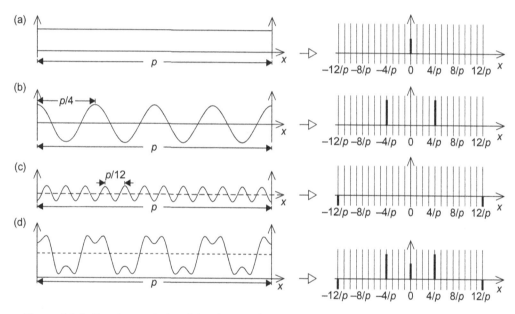

Figure 16.9: Fourier analysis of the density distribution in Fig. 16.8 (d). The three Fourier components are shown in **(a)**, **(b)** and **(c)**, with their FT on the right. The corresponding sums are shown in **(d)**.

length p, we get a Fourier *series*, as shown in (a)–(d). The plot in (d) and Fig. 16.10 (b) shows the contributions of different Fourier components, plotted as amplitudes of complete waves in $360°$. The power spectrum (d) plots the squares of the quantities in Fig. 16.10 (c).

A complete analysis of the image in Fig. 16.8 (a) would add the power spectra of each annulus. The total 4-fold power spectrum is apparently the sum of the contributions

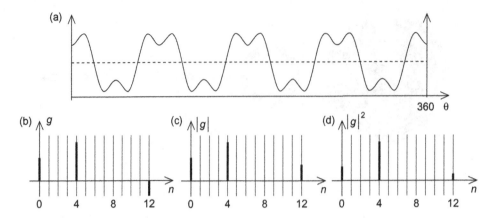

Figure 16.10: The density distribution **(a)** of Fig. 16.8 and Fig. 16.9 (d) shown as a function of angle θ and represented as a Fourier series **(b)**. The amplitudes of the components are plotted in **(c)**, and the power spectrum of the annulus, which plots the squares of the amplitudes, is shown in **(d)**.

for 4, 8, 12,... However, that procedure would make the total 2-fold power spectrum even bigger than that for 4-fold symmetry, simply because more terms contribute to 2-fold symmetry. So some form of order-scaling would be useful.

Obviously, the power spectrum depends on the choice of rotation center. It could be chosen as the center that optimizes circular symmetry (though this might bias the analysis to favor high-order rotation axes), or different centers might be chosen to optimize the overall 2-fold, 3-fold etc. powers. Then the final comparison of power spectra would give each rotational symmetry its best rotation center.

So far, the analysis in terms of power spectra parallels optical diffraction analysis of a noisy image to find the translational symmetry. But that is only the prelude to an optically-filtered image, so we might expect a similar 'rotational filtering'. Indeed the method of Markham et al. (1963) produced rotationally filtered (i.e., averaged) images and chose the one with best contrast. However, the averaging can only be undertaken as a function of angle; the radial direction has no symmetry which restricts pure rotational filtering. A more recent method of projection analysis was MSA (§17.3.4). The rather similar procedure of rotational orientation of particles is discussed in section §16.3.3.

16.3 Three-Dimensional Reconstruction

Previous chapters dealt with structures having 3, 2 and 1 translation that contributed many repeated images to get good-quality averages for reducing noise. Now we consider for the first time particles lacking any translational symmetry, so the number of asymmetric units is relatively small (60 at most). We need to compensate for the lack of intra-particle averaging by much more inter-particle averaging; but that demands an exact knowledge of the particles' alignment parameters (§12.2).

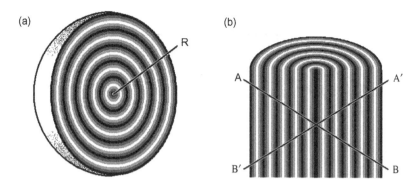

Figure 16.11: Self-common-lines in highly symmetrical density-distributions with **(a)** spherical and **(b)** cylindrical symmetries.

We find these alignment parameters from inter-particle comparisons, which usually depend on a reliable symmetry-framework. Now, although structures with high rotational symmetry have symmetry-frameworks adequate for reference, they remain 'noisy' without internal translations to enhance their signal:noise ratio. So small particles, even with rotational symmetry, tend to have 'noisy' symmetry-frameworks, leading to unreliable alignment parameters from inter-particle comparison. We get the best alignment parameters when one of the comparison partners is a reliable 'average particle', but we cannot calculate that average without reliable alignment parameters; and, to get those, we need that same 'average particle': a vicious circle.

Before addressing this problem (p. 352), however, we must look at how to find the relative alignment parameters of different particles.

16.3.1 Common Lines

Suppose we calculate from an image the FT, which is a central section of the particle's 3D FT. If the particle has only rotational symmetries, does this section contain any information about its orientation? A highly symmetrical object should leave traces of its symmetry on at least some of its FT cross-sections (sections passing through its center).

To take an extreme example, consider the central section of an object with spherical symmetry, so each cross-section has circular symmetry (Fig. 16.11 (a)) and any radius R passing through the center has the same density-distribution. Thus all the lines are identical or common; all are **common lines**. The next most symmetrical density-distribution is cylindrical (b), and any two lines AB and A'B' have the same density-distribution, provided they intersect at the center and have opposite gradients. Such lines, with identical density-distributions in the same image, are also common lines. (However, these extremely symmetrical objects actually have common *planes*; those are absent in molecular structures which can only have common *lines*.)

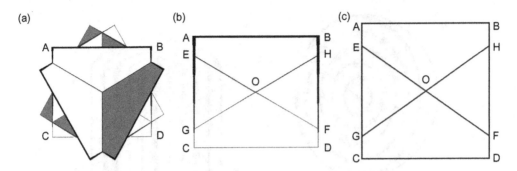

Figure 16.12: The 'common lines' method for finding a particle's orientation through that of its rotation axes. **(a)** The FT viewed down a 3-fold axis. The plane ABCD intersects the two symmetry-related planes in two lines. **(b)** The plane ABCD tilted relative to the paper, exactly as in (a). The line EOF comes from one plane's intersection, and HOG comes from the other (shaded). **(c)** The plane ABCD made parallel to the plane of the paper. E has the same phasor as H, and G the same as F, etc.

We might expect similar (though more restricted) common lines associated with *any* rotational symmetry, approximating (a) with several rotation axes and (b) with just one. Consider the simple case of one 3-fold rotation axis in both the particle and its FT. Suppose we have an image of the particle in some random direction (relative to the 3-fold axis). That image's FT is a section of the particle's 3D FT (Fig. 16.12 (a)). The FT's 3-fold symmetry generates two equivalent plane sections, which intersect at three lines. As the intersection of two identical planes, each line must have the same distribution of phasors; so all three lines are the same: they are common lines. Two of those common lines will lie within each plane section and could, in principle, be identified by scanning that section. (See EOF and HOG in (b) and (c).)

To find the orientation of the particle producing any given image, we first choose the origin at the best center of the particle (as explained in section §3.4.8). Then we investigate every possible orientation (i.e. of the particle relative to the 3-fold axis, equivalent to every orientation of the axis relative to the particle, a hemisphere). For each possible orientation, we find where the common lines would be and then calculate an error-function measuring the overall FT differences between corresponding points on them.

Obviously, the error-function will have a large 'random' value if we have mistaken the 3-fold axis's position. But consider the case where we have hit on exactly the correct orientation. Then the FT values are *theoretically* identical at all pairs of corresponding positions along EOF and HOG, so the error-function should be zero between all pairs. In practice, of course, noise will increase the error-function, but its contribution varies along the pair of lines. The point O, the origin of the FT (where the resolution is zero), is common to all line-pairs, so there is no contribution to the error-function there. At the other extreme, points E and H lie near the outer boundary of the FT, where the resolution is highest and the effects of distortion

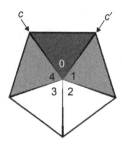

Figure 16.13: How a 5-fold axis generates two pairs of common lines. The central point is the FT origin, lying on a 5-fold axis that coincides with the direction of view. That axis meets plane 0 at an angle, and generates the symmetry-related planes 1–4 from it. (The five planes form an inverted pyramid standing on its vertex which is at the FT origin.) All the planes are oblique to each other, and must therefore intersect. The intersection of 0 with 1 and 4 can be seen in the V meeting at the origin; that intersection generates the first pair of common lines, c and c'. (The intersection of 0 with 2 and with 3, which generates the second pair, is not shown in this diagram.)

etc. are greatest[4]. Thus the error-function will vary with resolution, starting at zero (at O) and gradually rising to the typical 'random' value as the boundary is approached. The resolution where it reaches that value measures the expected resolution of the particle's data, based on its state of structure-preservation.

This indicates the essentials of the *common lines method* (Crowther et al., 1970; Crowther, 1971), the first method for finding the orientation of an isolated icosahedral particle. Of course, icosahedral particles have many symmetry axes, each of which generates common lines within a central section plane of the FT. We have just seen how a 3-fold axis generates one pair of them. A 5-fold axis generates two pairs, as shown in Fig. 16.13. Even a 2-fold axis generates a pair, but the lines coincide while running in opposite directions (Fig. 16.14).

We saw, in section §16.1.1, that there are six 5-fold axes, ten 3-fold axes and fifteen 2-fold axes. Since the 5-fold axes generate two pairs of common lines, they contribute $6 \times 2 = 12$ pairs. Each 3-fold axis generates one pair, so these axes contribute $10 \times 1 = 10$ pairs. Finally, the fifteen 2-fold axes, each providing one (coincident) pair, contribute fifteen pairs. This makes a grand total of $12 + 10 + 15 = 37$ pairs of common-lines, of which fifteen are coincident.

All this intrinsic orientation information is contained in the FT of just one icosahedral particle's image, under favorable circumstances. In a perfect world, this would be enough to find each icosahedral particle's *absolute* orientation. In practice, it is very useful or even essential to compare different particles to get their *relative* orientations. This has led to a

---

[4] However, there is a tendency for the FT of an icosahedral virus to have higher intensity along the direction of rotation axes (Caspar, 1956), perhaps because of its polyhedral shape. Polyhedra have FTs with intensity 'spikes' perpendicular to the faces, similar to those around polygons (§6.3.3), shown by the 'Laue transform': Hosemann & Bagchi (1962).

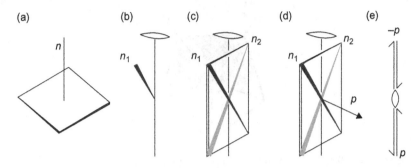

Figure 16.14: Common lines from a 2-fold axis. **(a)** The orientation of a plane is conveniently specified by its perpendicular or *normal, n*. **(b)** So, when a 2-fold axis (vertical) intersects an arbitrary plane, that plane can be represented by its (arbitrary) normal n_1. **(c)** The 2-fold axis generates from it a second normal n_2. **(d)** At their intersection point, the corresponding planes intersect to generate a common line. Since each plane is perpendicular to its normal, their common line must be perpendicular to both normals, i.e. the line *p*. **(e)** A view down the 2-fold axis showing the perpendicular *p* and its extension *–p* on the opposite side. Thus the two common lines here coincide and, because they run in opposite directions, the FT of a real particle is real along them.

refinement of both methodology and terminology. If the particle is considered in isolation, it has *self-common-lines* within its own FT. But, when two different particles of the same type are compared, we look for *cross-common-lines* connecting the two FTs[5].

These lines give information about the *relative* orientations of different particles. They connect the projection planes of the two particles. Figure 16.15 shows an example of just one such line, but icosahedral symmetry multiplies them. Cross-correlations are particularly useful to establish the *scale factors* between particles (for radial dimensions and amplitude). Dimensional scaling is particularly important if particle images are to be combined into an average with good resolution. Also, cross-correlations can ensure that the *handedness* of different particles is consistent. For it is important to preserve the *relative* chirality of different particles, especially if the T-number is itself chiral (§16.1.2). (The *absolute* chirality can be ignored or postponed, unless the resolution allows interpretation in terms of known secondary structure elements.)

16.3.2 Cryo-Image Problems

The common-lines method was introduced in the 1970s when all high-resolution specimens were still negatively-stained, and it gave reconstructions as good as 2.5 nm with favorable specimens (bushy stunt virus: Crowther et al., 1970; Crowther, 1971), but of course this gave no details of protein structure.

---

[5] The cross-common lines method, when applied to unsymmetrical particles (as in 'single particle' methods, §17.4.2) is called 'angular reconstitution'.

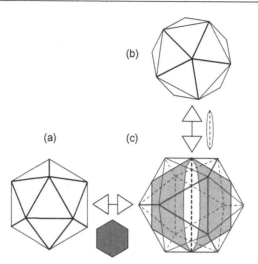

Figure 16.15: Examples of cross common-lines that connect different projections of the same particle-type (here an icosahedron). Projections are shown down a 3-fold axis **(a)** and down a 5-fold axis **(b)**. **(c)** The FT of each projection specifies a central plane section of the 3-dimensional FT, which has icosahedral symmetry. Projection (a)'s FT is perpendicular to its 3-fold axis (perpendicular to the paper) and projection (b)'s FT is perpendicular to its 5-fold axis (nearly parallel to the paper). These two FT sections intersect in a common line (vertical broken line in (c)).

That became a conceivable possibility after the advent of cryo-specimens in the 1980s, but they also posed the challenge of much fainter images that reduce the signal-to-noise ratio. So accurate alignment parameters became harder to get, just when they were more desirable. This need led to new developments in finding orientations, which we consider in turn.

(i)   To determine the orientations of projections of cryo-preserved particles, we need to use all the meager information available. But the common lines method utilizes only the information in the immediate neighborhood of the common lines, and discards that in the gaps between them. Because all the lines radiate from the transform center (origin), these gaps, though minor at low resolution, comprise much of the FT at high resolution. Matters are even worse when an image's projection direction nearly coincides with a rotation axis. The underlying geometry was shown in Fig. 16.14. We represent (a) the projection plane by one of its normals n, and then choose (b) a normal that intersects the rotation axis under consideration. Call this normal n_1. The rotation axis generates copies of n_1, so we consider both n_1 and a copy, i.e. n_1 and n_2. The direction of the common line must be perpendicular to the plane containing n_1 and n_2, so it must resemble the line p in (d). But this construction fails if the normal n_1 is either parallel or perpendicular to the rotation axis. If it is parallel, it cannot intersect that axis. If it is perpendicular to it, then so too must be the symmetry-generated normal n_2; so the plane (d) containing n_1 and n_2 is now *perpendicular* to the rotation axis, and the common line p (perpendicular to both n_1 and n_2) coincides with that axis. Thus all the common lines generated by the rotation axis coincide if the normal is perpendicular to it, i.e. if the projection plane is parallel to it. This means that there are fewer common lines when an image's projection direction lies very close to one of the particle's rotation axes. Unfortunately, this happens fairly frequently with an icosahedral particle, which has 31 rotation axes.

Other problems reported with the common-lines method are the difficulty of finding lines when images are well-focused (and therefore have low contrast), and the heavy computational demands with large data-sets. But some of the problems of common-lines methods have been mitigated (Fuller, 1987; Fuller et al., 1996). Also, more recently, the method has been used to give a 0.74 nm resolution reconstruction[6] of hepatitis B virus core protein from 6400 particle images (Böttcher et al., 1997), and it is still frequently used.

(ii) The use of model-based methods relates to a familiar theme. Noisy or 'weak' data are hard to compare with each other, so we need to start the analysis by comparing the noisy data with inherently 'stronger' data. Of course, later in the analysis these 'stronger' data are an average structure derived from previous particles; but that does not exist at the beginning. So we have a 'starting problem': what appropriate 'strong' data could replace a comparison of 'weak' with 'weak'? In the case of cryo-preserved icosahedral particles, an appropriate 'strong' comparison might be a low-resolution structure, derived from either negatively-stained specimens or strongly under-focused cryo-specimens. Thus a common procedure is to start with the common-lines method, which avoids bias towards any particular model, but to confine its initial application to high-contrast under-focused or even negatively-stained specimens. Once a reliable low-resolution model has been obtained, this can be refined with the model-dependent PFT method (next section).

16.3.3 *Polar Fourier Transforms*

The most widely-used alternative to common-lines for finding particle orientations has been the **polar Fourier transform** (PFT or EMPFT) method of Baker & Cheng (1996). It requires an initial model that can form a good starting-point for refinement. However, it is not this initial model that will be refined, but the orientation angles assigned to the experimental images. This is done by comparing each image with a 'projection library' consisting of all the essentially different projections of the model. Such a 'library' needs to be quite large (e.g. 382 projections, if they are taken at 1° intervals), and many experimental images are needed to achieve high resolution structures (usually a hundred initially, and at least a thousand to get the final structure). So there need to be very many image comparisons, and each comparison uses the entire content of both images. The comparisons are facilitated by a geometrical transformation of the images.

---

[6] The starting point was a 3.5 nm model by Crowther et al. (1994). A parallel paper on the same virus (Conway et al., 1997) used the PFT method. The virus structure was later determined at high resolution by X-ray crystallography (Wynne et al., 1999). See the review by Crowther (2008).

Both images are roughly circular, so much of the comparison involves a rotational adjustment, based on rotational cross-correlation. We have seen that a single rotation is formally like a translation, the rotation angle playing the role of a translation distance (§6.4). Also, §3.4.6 explains how a 1D CCF is more quickly calculated with FTs. Thus we can do angular correlations by taking angular 'FTs', multiplying them and then doing the inverse FT. For this, we must first divide the particle image into annuli, as in section 16.2 of this chapter. We recall that each annulus was effectively 'straightened out' (Fig. 16.8). Similarly, in the PFT method, the annuli are straightened on a 'polar grid' (Fig. 16.16 (c), of which the geometry is explained in Fig. 16.17). Note that, between Fig. 16.16 (c) and (d), the straightened

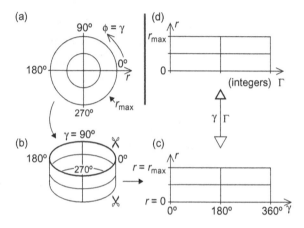

Figure 16.16: 'Polar projections' and their FTs. **(a)** The polar coordinate system appropriate for a rotationally-symmetric particle like (a) in Fig. 16.8. The polar angle φ is called γ here. **(b)** The diagram (a) after the transformation illustrated in Fig. 16.17. It is being cut vertically along the 0° line at the right. **(c)** After cutting, the cylindrical diagram is opened out with the cut line on the left. This is the coordinate system for the 'polar projection'. **(d)** The 'polar projection' is Fourier-transformed along only the angular coordinate γ to give a 'hybrid' FT (like in section §6.1.1) that has a *real* vertical coordinate r and a *reciprocal* horizontal coordinate Γ. This is the 'polar FT' or PFT.

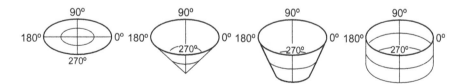

Figure 16.17: The transformation of (a) to (b) in Fig. 14.30. **(a)** Corresponds to (a) in Fig. 16.16. The maximum radius is shown as a thicker line, but the minimum is a mere point. **(b)** The center becomes the vertex of a cone. **(c)** The vertex is enlarged into a circle, so the diagram is now a cone's frustrum. **(d)** Finally, the lower circle is made as big as the upper one, and the diagram ends as a cylinder. Thus we have the reverse of the historical transformation of Edison's 1877 cylinder into Berliner's 1894 disc.

and stretched annuli (horizontal) are Fourier-transformed, whereas the radial coordinate r cannot be so transformed without introducing Bessel-functions and losing the computational advantages. Thus a 'hybrid' FT is taken, with the angular parameter γ becoming Γ while r remains unchanged.

Baker & Cheng's (1996) paper includes a flow-diagram of their method. Besides a loop that improves particle origins and orientations, there is an outer loop in which improved data yield an improved model (i.e. best current structure), giving a new 'projection library' for the inner loop.

16.3.4 Other Methods for Finding Particle Orientations

Of course, different specimens pose different problems, and an armory of alternative methods is needed to deal with the more 'refractory' particle types. However, other methods (such as the variance technique of Cantele et al., 2003) mostly use methods akin to those employed for asymmetric 'single particles' (Chapter 17). The 'random-conical' data collection method is used in single-particle work as a better way to get particle orientations (see §12.1.4).

Mention should also be made of the method of **icosahedral harmonics** (also called 'icosahedral symmetry-adapted functions' or ISAF). Although relatively old, it has never become widely-used, apparently mostly because of its computing demands. It attempts to remedy the problem of doing 2D FTs on the surface of a sphere (§16.1.3). For this, the two-parameter angular part uses 'spherical harmonics' while the radial part uses 'spherical Bessel functions'. 'Spherical harmonics' are special functions that have become familiar through their use for atomic s, p, d,... wave-functions. Combinations of them can be devised that have icosahedral symmetry (Cohan, 1958, and unpublished contemporary work by Klug, outlined in Finch & Holmes, 1967). The method was revived and applied by Provencher & Vogel (1988a,b), and later resuscitated by Navaza (2003) and Liu et al. (2008) with an application by Cheng et al. (2008). Despite this chequered career, the method might still offer advantages, if it could be made computationally practical. Perhaps it could apply to spherical particles some of the advantages of sinusoids in analyzing weak, noisy images of structures with translational symmetry.

16.3.5 Fourier Transforms: Interpolation and Inversion

Once particle orientations can be worked out reliably, the structure analysis enters the second stage. As more and more particle orientations are found, it is useful to plot them as a function of (φ, θ) within the asymmetric region of an icosahedron (Fig. 16.5). This plot reveals if the particles have a preferred orientation. Thus, virus/antibody complexes show a specific interaction at the air/water interface (Thouvenin et al., 1997), and preferred particle orientations (e.g. Fig. 11 of Bernal et al., 2003) can lead to gaps or holes in the data. Gaps or holes increase, and thus become more serious, further from the center (origin) of reciprocal space, so they determine the resolution of the reconstruction. To restore resolution, supplementary data from tilted grids are needed.

When there is an adequate number and distribution of views, we are ready to calculate the 3D FT. Of course, each image contributes all its icosahedral symmetry-related central sections to the FT; this is where the high symmetry of icosahedral particles makes a critical contribution. However, we only know the FT on these irregularly-oriented sections; we know only (true FT) × ('section-function'), the latter being 1 on each section and zero outside. If we were to invert this by taking the FT, we should get: FT(true FT × 'section-function') = FT(true FT)★FT('section-function') = (particle structure)★FT('section-function'). That is, we should get the desired particle structure distorted by being convoluted with FT('section-function'). It is de-convoluted by Crowther's method (see §11.1.7) which was originally developed for this very purpose. The example in §11.1.7 was simplified as far as possible, dealing only with a real FT defined by only two unknown values in one dimension. The sampling theorem applies to equally-spaced samples, for which the 3D generalization is a cubic lattice of sampling points. Ideally, one would calculate the FT on such points by '*reverse interpolation*'. However, at the time (1969) when icosahedral particle reconstructions were being developed, this would have rendered them impractical with contemporary computers (IBM 360 series). Therefore the calculation was very much reduced (Crowther et al. 1970; Crowther, 1971) by an adaptation of the coordinate system and the associated Fourier–Bessel representation of FTs used in helical reconstruction (§15.2).

It will be recalled that these cylindrical coordinates have equally-spaced concentric cylinders intersected with equally-spaced planes parallel to (*X*,*Y*) (Fig. 16.18). Where

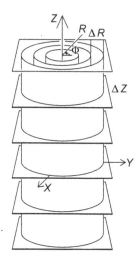

Figure 16.18: Coordinate system for the Fourier–Bessel interpolation of an icosahedral FT. Cartesian coordinates are (*X*,*Y*,*Z*), cylindrical polar coordinates are (*R*,Φ,*Z*). Sheets parallel to the (*X*,*Y*) plane are separated by a distance ΔZ. On these, annuli are separated by ΔR and, on each annulus, the functions have an angular dependence on Φ. (ΔZ and ΔR are approximately the reciprocal of the particle diameter.)

the cylinders meet the planes there are *annuli* (visible at the top of Fig. 16.18) on which the FT is interpolated. With helical diffraction, each plane was a layer-plane with its own order of Bessel function given by the (n,Z) plot. But here they are merely sampling-planes, like the planes perpendicular to Z in a Cartesian sampling-grid; so each plane has, in principle, *all* orders of Bessel functions. However, some Bessel functions may be excluded for other reasons. Figure 15.14 shows that there is a central 'gap', starting at J_2 and increasing with the Bessel's order. This means that, at low resolution (inner annuli), only a few Bessel orders matter, because the others (higher orders) have a 'gap' that fills reciprocal space within the resolution limit. (Of course, if we make our reconstruction at higher resolution, we use more annuli and fewer orders have a 'gap' on them, so more orders are needed.)

Besides the sampling of the FT along Z and along R (the different annuli), we also need to sample along Φ. Here the phases of Bessel functions rotate in the same way that we studied in Chapter 15: the phase rotates n times per revolution for a function of order n. Thus we only need to find the (complex!) amplitude of each different order Bessel, on each annulus, at $\Phi = 0°$. Thus, on each annulus, the number of equations needed to solve is only the number of Bessel orders relevant to that particular annulus. This is far less than the number of points in a three-dimensional Cartesian lattice. The least-squares calculation of the necessary sample values supplies the errors (as explained above) and, when these indicate a reliable interpolation, the FT is inverted just as with a helical FT. Of course, this general method is even more efficient if one chooses the Z-axis to lie along a 5-fold axis of the particle's FT (so Bessel orders must be multiples of 5).

This simplified Fourier–Bessel method is still used, even though computer power has increased considerably in the last forty years. However, during the same period, the advent of cryo-specimens has increased the potential resolution of reconstructions, thereby increasing the number of equations to be solved by the least-squares method[7].

16.4 Appendix: Calculation of T-numbers

In Fig. 16.19, the big triangle has a fixed point B and a variable one A, defined by h and k unit distances (respectively). The area of the big triangle (side AB) is proportional to the square of AB, i.e. proportional to the 'triangulation number[8]' $T = t^2$. So the relative area of the big triangle is (using Pythagoras' theorem from Fig. 19.1):

---

[7] Recent improvements in data-processing are reviewed by Baker et al., 1999; Thuman-Commike & Chiu, 2000; Zhou & Chiu, 2002 and Jiang & Chiu, 2006.

[8] We use T for the triangulation number and I for the tetrahedral group since we make little use of the tetrahedral group.

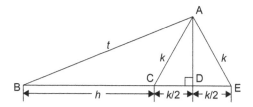

Figure 16.19: Geometry for calculating $T = t^2$. ACE is an equilateral triangle (as angle $A\hat{C}E = 60°$; see Fig. 16.6) with all sides equal to k. (See text for the simple derivation.)

$T = t^2 = AB^2 = BD^2 + AD^2 = (h + CD)^2 + AD^2 = h^2 + 2h(CD) + CD^2 + AD^2$.
Now $2(CD) = CE = k$, so $2h(CD) = hk$. Also, triangle ACD is right-angled, so
$CD^2 + AD^2 = AC^2 = k^2$. Thus:

$$T = t^2 = h^2 + hk + k^2 \qquad (16.1)$$

(h and k positive integers).

Note Added in Proof

A recent 3.6Å structure of adenovirus from Zhou's laboratory (Lin, H., Jin, L., Koh, S.B.S., Atanasov, I., Schein, S., Wu, L. & Zhou, Z.H. (2010). Atomic structure of human adenovirus by cryo-EM reveals interactions among protein networks. Science 329, 1038–1043) allows polypeptide chains to be traced, revealing how the major capsid proteins are held in position by minor proteins.

CHAPTER 17

Unsymmetrical ('Single') Particles

Chapter Outline

17.1 Introduction

After exhausting all the particle symmetries, we discuss, in these last two chapters of Part IV, particles with *inadequate* or *deficient* symmetry, which can take two forms: *defective* symmetry, treated in the next chapter, and *insufficient* symmetry, i.e. low or no symmetry, which we discuss here. Members of this very numerous group are usually termed 'single particles', perhaps in contrast to those held within the symmetry framework of a sheet, helix or polyhedron (although icosahedral particles are sometimes also called 'single particles').

This was the last area of electron microscope image-analysis to appear. Various factors delayed its development. It is probably the most dependent on improved technology: computer facilities, image-scanners, and cryo-specimen techniques. It is also inherently the most difficult, and was thus postponed until symmetry-based methods had, first developed, and then 'cherry-picked' the best symmetrical fruit. And, despite the existence of many specimens with insufficient symmetry, not too many are big enough (§12.3.1) to provide the necessary data for finding the particles' alignment parameters with sufficient accuracy. Each area of image analysis benefits from an 'ideal' specimen, and the role played by the purple membrane in developing 'electron crystallographic' methods (§14.3) was played by the ribosome (a completely unsymmetrical particle or pair of particles) in developing single-particle methods (especially by the groups of Frank and of van Heel). Indeed, these methods might well have been the only route to solving its structure, but for the pioneering work of Yonath to initiate serious work on obtaining and improving 3D crystals of ribosomal subunits suitable for X-ray crystallography. As it was, the results of electron microscopy made important contributions to the final structures. But, if suitable particles of *no* symmetry are uncommon, they are supplemented by a plentiful supply of those (e.g. hemocyanins, proteasomes, chaperonins and membrane protein complexes) with low or inadequate symmetry. Of course, there is a borderline between low-symmetrical and polyhedral particles, and consequently the interesting question of how best to combine their characteristic image-analysis methods.

17.1.1 The Reasons for Single-Particle Methods

It helps us to understand the characteristic single-particle methods if we look at their central problem in more detail. Insufficient symmetry implies no satisfactory symmetry framework. In symmetrical specimens, this framework is the basis of both the *intra*-particle and *inter*-particle averaging and tilt data (features classically combined in the helix (§15.3.5)) that are the basis of 3D structure. Therefore the lack of such a framework has important consequences.

 (i) Without intra-particle data, structure-determination becomes dependent on *inter-particle averaging* and *multi-particle tilt* data. The finding of very many accurate alignment parameters thus becomes particularly important.

 (ii) The disappearance of the symmetry framework, i.e. the loss of rotational and translational symmetry, means that Fourier transforms (FTs) lose most of their value. This most valuable theoretical support must be substituted.

 (iii) The symmetry framework also contributes to *particle recognition*. (Thus mirror-symmetry makes faces more easily recognizable, so passport photographs are full-face and not side-face images that lack any symmetry. Also the high symmetry of helical or icosahedral particles is a great help in their recognition.) So the lack of a symmetry framework led to a search for alternative methods of particle recognition.

 (iv) The dependence on data from many different particles increases the importance of *particle homogeneity*. Whereas, for example, a micrograph showing a mixture of various particles including a few long helices might be a suitable start for helix analysis, a similar convenience does not apply to a mixture of different asymmetrical particles. Obviously, all specimens are first made chemically homogeneous, but that might not suffice. Single particles might have a natural *flexibility*. Although this might be detectable, many forms of particle *distortion* would be more difficult to find without a symmetry framework.

As a consequence of these factors, single particle methods tend to use a characteristic constellation of techniques. They focus on image recognition problems. They perform calculations in real space rather than reciprocal space, and use sinograms rather than common lines, cross-correlation functions (CCFs) rather than phase residuals, back-projection rather than Fourier reconstruction. Fourier transforms are reduced to the status of numerical algorithms (for calculating CCFs through the fast Fourier transforms (FFT)). All this is understandable with structures lacking translational or rotational symmetry (the feature that makes reciprocal space preferred with icosahedral structures). But perhaps the rejection of Fourier methods occasionally goes further than is strictly desirable, especially as many 'single particles' have *some* rotational symmetry.

The absence of a symmetry framework means that there are some analogies between 'single particles' and NMR. Both deal with unsymmetrical macromolecules in free solution, so that conformational variants and ligand binding are free to occur without the constraints of a crystal lattice. Moreover, both lack the absolute spatial framework of symmetry, so distances must be measured relative to some recognizable feature of the particle itself.

We shall survey this important new methodology under a series of headings placed in the order of their probable application.

17.2 Alignment

As we saw in Chapter 12 (§12.1), building a 3D structure from images requires either a 'well-covered viewing-sphere', or a 'well-filled FT'. In either case, we need the alignment parameters of different particle images to put them (or their FTs) in the right places. How we get these alignment parameters depends on the type of specimen. Symmetrical particles have a framework of symmetry elements which act like an absolute reference. When this is lacking, only relative alignment parameters are possible between specific particle pairs. But, as in outer space, large-scale averages can act as quasi-absolute references for some purposes. We shall look into how such 'average particles' ('reference based' as opposed to the relative or 'reference free' comparisons) are constructed. But first we need to examine the basic methodology of aligning particles from a random field.

17.2.1 Reference-Free Alignment of Particles Differing by (2T+R)

The basic tool for finding alignments is the cross-correlation function (CCF) (§2.1) between particle images. But its naïve application is hindered by the low signal:noise ratio and the orientational disorder of most particle fields.

We therefore start by considering what might be theoretically possible with a field of particles differing only in their in-plane translations, or in (2T) (in the notation of §12.2.2–§12.2.4). Even a very short exposure would allow a 2D crystal to give its diffraction pattern, from which much could be deduced. But could anything be done, at that noise level, with our less ordered array? If we took its diffraction pattern, the disorder would make a big difference. A disordered array is a random (i.e. extremely complex: §11.1.1–§11.1.2) 'lattice'. This samples the FT amplitude but garbles its phases (equivalent to an incoherent sum: §8.1.2). Nevertheless, the autocorrelation function (ACF) could be calculated from the FT amplitude, and this information seems available even at very low exposures.

Of course, the supposed parallel arrangement would be most likely to occur after partial processing of a less ordered state, most probably where the particles differ by (2T+R) (§12.2.4, which is a conceivable state, if they adhered by just one surface). At extremely low electron doses (where individual particle comparisons would be impossible), the total diffraction pattern would show an incoherent sum of copies of the particle FTs, each showing this same projection but rotated randomly about an axis perpendicular to the grid plane. Therefore we would get the circularly-averaged FT intensity, as if the FT were rotated about its center. Obviously, this rotation smears much of the detail, but some valuable information remains. We could find a kind of circularly-averaged ACF, a series of rings surrounding a central peak. The rotational disorder (R) destroys information more efficiently than did the previous two translational disorders (2T).

Therefore, in an actual specimen with (2T+R) disorder, the first priority is to correct for (R), the in-plane rotations. Unfortunately, we have to find the translations to get the correct rotation axes, but it is hard to get an accurate translation between two images with different orientations. To clarify these adjustments, we can introduce a kind of artificial 'symmetry framework' from each image's ACF, which always has a 2-fold axis in 2D (§6.3.1). There is a (2D) translational ACF and a (one parameter) rotational ACF, both easily calculated with the aid of the FFT (§3.4.6). The translational ACF has a central peak, and the rotational ACF has (at each ring into which a particle is divided, as in Fig. 16.8 (a)) peaks at 0° and 180° rotation. (Of course, there has eventually to be a decision between these alternatives, based on the original image.) It is possible to start with the translational ACF (e.g. Langer et al., 1970) and to use the ACFs to reorient the particles; or to start with rotation (as in crystallographic searches which have to use a Patterson function). A third option is to use a double ACF (both translational and rotational): Schatz & van Heel (1990).

17.2.2 *Alignment of Particles Differing by (2T+3R)*

This is the least ordered particle arrangement; the only order is imposed by the grid, which keeps their translations T_z (along the optic axis) fixed. Now the only information at extremely low electron doses (i.e., the field's diffraction pattern) is the incoherent sum of copies of the particle's FT in all possible 3D orientations. This is the information we get in X-ray solution scattering, where the scattering patterns allow us to calculate the particle's spherically-averaged vector-map (Patterson function or 3D ACF). This function (the 'pair distribution function' or 'distance function' since it's a map of all intra-particle vectors, not preserving orientations but with origins coincident) is spherically symmetrical and consists of blurred spherical shells, but it does give some information about the particle's size and even a little about its shape.

However, averaging moderately noisy images from particle fields with (2T+3R) disorder poses a special problem. Through out-of-plane rotations of the particles, most of the images present different projections that cannot be simply averaged. They need to be sorted out, i.e. to be positioned correctly around the 'viewing sphere' (Fig. 12.1). This is relatively simple if the particle is highly symmetrical, when the symmetry framework indicates the orientation (see Chapters 14–16 for the different symmetries). Without this framework, we have the problem that has been best tackled by multivariate data analysis and classification methods (Section §17.3).

17.2.3 The Reference Particle Problem

The pairwise mutual alignment of particle images works satisfactorily with the high-contrast, fairly high-exposure images from most negatively-stained specimens. But this procedure tends to fail with the images of low-dose, unstained cryo-specimens, which have very low contrast, especially if well-focused. In such cases one needs a 'reference' particle for averaging. Though mentioned before, it is worth recalling the simple argument. The CCF $\odot$ of two noisy images, each consisting of an image and a noise component (I and N, respectively) is $(I_1 + N)$ $\odot(I_2 + N) = (I_1 \odot I_2) + (I_1 \odot N) + (N \odot I_2) + (N \odot N)$. If the noise N is at least as big as either image component (I_1 or I_2), then the 'signal' term $(I_1 \odot I_2)$ is swamped by the other three noise terms. In that case, it is much better to calculate an average image $<I>$ (called the 'reference') and calculate the CC with it: $(I + N) \odot <I> = (I \odot <I>) + (N \odot <I>)$, where the two terms are now of comparable weight. Having found their CCF, we can align I with $<I>$ and combine the two into a stronger $<I>$ that makes an even better partner in the next CCF. Thus the reference itself grows during the averaging process, a process of positive feedback which amplifies the initial choice of reference. But this amplification carries risks; the initial choice of images for averaging 'steers' the original 'reference' in their direction, rather like the first speakers at a meeting. The 'reference' average retains a 'memory' of the original sequence which guided the alignment of later images. The crucial 'memory' lies in these relative alignment parameters, from which the average is the same, in whatever order they may be taken. (Indeed, the alignment parameters could be used to align all the images on a square lattice, and the average found by optical-filtering which takes them all simultaneously.)

Attempts have been made to mitigate this 'steering' effect (often called 'model bias') when the signal:noise ratio isn't too small. One of these is the 'iterative alignment method' of Penczek et al. (1992). Imagine you have a gallery of images all aligned on a square lattice and ready for filtering to give the average. Then you want to adjust the alignment of each individual image to improve the average. So, in sequence, you take each image out, filter the rest to get *their* average, and align the removed image to this. You then replace it in its new position, and proceed to remove the next image. This method removes each 'individual voice' but it leaves the 'general will' in control. So it leaves a memory of the most influential voices from the earliest phase of 'steering'. Perhaps the only way to eliminate 'steering' is to make many independent starts (using different images during the critical early phase). The different final averages can then be compared statistically. A similar 'steering' problem occurs when calculating 3D structures from NMR data. A random-coil structure is chosen, the NMR data-restraints are applied, and they guide the folding process. But a different starting structure can lead to a somewhat different final one.

17.3 Multivariate Statistical Analysis

17.3.1 The Need for Statistics

Micrographs showing many different views of an identical novel particle present us with a confusing variety of projections. We are faced with the puzzling challenge of finding a single 3D structure to explain all of them. In the early days of negative staining, the most successful analyses were of symmetrical particles, where the symmetry framework could be exploited. Such analyses were later replaced by more objective methods using FTs, as explained in Section §12.4. However, unsymmetrical particles presented a more difficult challenge. The first step was to accumulate galleries of characteristic views, each of known orientation. To make any further progress, it was necessary to reduce the resolution and to consider only simplified models of uniform density bounded with a smooth surface. Thus the structures could, in principle, be obtained from the boundaries as 'shadows' (appropriate for negative-staining, and soluble in principle by back-projection for convex surfaces). However, even these simplifications left plenty of room for disputes, especially over that important but asymmetrical organelle, the ribosome. The most objective data came, not from model-building, but from antibody labeling (e.g. Lake, 1976).

This kind of sorting of different projections (images) is objectively done by statistics. We saw in §11.1.6 how the distribution of objects defined by two measured parameters can be reassigned to two new parameters, chosen because of their significance to the distribution. The most important parameter produces the greatest variation in the objects, while unimportant parameters of course produce no variation at all. We need a method of identifying the parameters, in decreasing order of importance, so that we can sort the objects into groups that resemble each other in the most important parameters. The method (or, rather, group of allied methods) chosen are all closely related to *principal component analysis* (sometimes abbreviated to PCA), described in §11.1.6. One of these is *correspondence analysis*, and they all go under the umbrella name of *multivariate statistical analysis* (MSA) (or MDA, replacing 'statistical' by 'data')[1].

---

[1]   There are good accounts of these methods in the review of van Heel et al. (2000), the book by Frank (2006) and
      his summary chapter in Glaeser et al. (2007); though probably the clearest account is in van Heel et al., 2009.

17.3.2 Multivariate Statistical Analysis Pre-Processing

For pre-processing, the particle images must be clear enough to recognize, orient, match and 'box' within an appropriately-shaped **binary mask**. If the particle images are collected from more than one micrograph, it is important to correct for defocus by multiplying each image by the phase contrast transfer function (PCTF) (§9.2.7). Then, by the methods discussed in §17.2, we find, for each of the I images, the alignment parameters that will bring it to a standard position. Consequently, after each pre-processed particle image has been scanned, the 1st, 2nd,..., Pth pixels of different images all correspond as nearly as possible[2]. Thus each of the I images gives a string of P pixels, such that the corresponding pixels of different strings refer to corresponding parts of the particle images. We shall suppose that $P = I$, i.e. the number of pixels in an image equals the total number of images.

At the end of the pre-processing stage, all our data are collected into a rectangular (square, if $P = I$) array of pixel densities. Each row is the scanned image of a different particle, and has P pixel densities; there are I rows, where I is the number of scanned images (i.e. particles). Each of the P columns has I pixel densities, the set of corresponding pixels from the same part of each particle. This rectangular array is scaled by division by the total pixel density, and the average particle image (average row) is subtracted from each row.

17.3.3 Multivariate Statistical Analysis Processing

The real processing stage can now begin. The above rectangular matrix (square, if $P = I$, which we suppose for simplicity) is subjected to a process very similar to least-squares analysis (§11.1.5), in which the first step is to multiply it by itself (strictly, its transpose §20.2.4) to get a square *symmetric* matrix with I rows and I columns. Next, instead of matrix inversion, we find the eigenvalue-eigenvector spectrum (a related operation; see §20.4). As explained in §11.2.1, the eigenvectors constitute orthogonal axes of a new coordinate system, and the importance of an axis is indicated by the size of its corresponding eigenvalue.

We now have a series of I eigenvalues, each associated with its eigenvector (P pixels in length, since $P = I$). Each eigenvector is effectively an image, called an **eigen-image**. The size of the eigenvalue gives the relative contribution of its eigen-image to the total set of particle images. Out of the I eigen-images, only a relatively small number are needed to contain the real information in the original I particle images §3.2.1. The situation is analogous to the Fourier analysis of a noisy repeating function. There, the Fourier component's size (corresponding to the eigenvalue) indicates the relative contribution of its corresponding density-wave (corresponding to an eigen-image) in the original data. When the Fourier component's size falls below the noise level, nothing is gained by including the corresponding

---

[2]   If there are uncertainties in aligning the particle images, one could replace each image, in the MSA, by its double ACF (DACF, Schatz & van Heel, 1990), which is unchanged by shifting or rotation. After establishment of a factor map, the DACFs should be replaced by the original images.

density-wave in the filtered image. Thus we exclude all Fourier components beyond a certain spatial frequency (i.e. beyond a certain resolution). Similarly, in MSA, eigenvalues below a certain cut-off value are ignored[3].

We therefore arrange our I eigenvalues in order of declining value, and plot this series as an **eigenvalue histogram** (or spectrum). Their values will decline rapidly at first, but soon level off. Only the values that are declining contain useful information, and the 'level' values can be ignored, since they describe pure noise. This amounts to a considerable *data compression* of the original data-set. Suppose that we retain just the first eight eigenvalues (although eight is quite arbitrary, it is clearer to have a concrete number). Then their corresponding eight eigen-images contain all the real image information in this data-set. Thus each of the I images can be satisfactorily approximated by a suitable combination of those eight eigen-images, and the 'suitable combination' is given by eight parameters or coefficients.

This is analogous to deciding that a data-set only needs eight Fourier components, and then expressing each particular curve of the set in terms of those eight components (by doing a Fourier analysis and terminating it after the first eight components). There is a further analogy. We saw (§3.1) that Fourier components having a different spatial frequency are orthogonal. The eigenvectors (or eigen-images) of a symmetric matrix are also orthogonal (§11.2.1 and §20.3.2), and the matrix we calculated was indeed symmetric. Thus the eight eigen-images are *orthogonal*, like the three coordinate axes of ordinary space. (Viewed in this way, the eigenvectors are called **factors**.) So, when each image is represented by its eight coefficients, these eight numbers can be used as coordinates in eight-dimensional space to plot the image as a point. If the number of images were much more than eight, we would now have a kind of 'data-cloud' distributed in 8D space, called **factor space**[4].

17.3.4 Multivariate Statistical Analysis: Interpretation of Factor Space

In order to interpret the distribution of data-clouds in factor space, it is most useful to be able to calculate a 'typical' particle image from a chosen point in this space. In Fourier analysis, we can reassemble a curve by multiplying its Fourier components with the corresponding sinusoids, and adding the products. Exactly the same applies to any image's eight coefficients, which can be multiplied by their corresponding eigen-images and added (together with the

---

[3] It is interesting that the study of symmetrical particles was made objective by the introduction of Fourier analysis, and the study of isolated particles was made objective by the introduction of MSA, a matrix method with strong analogies to Fourier analysis. Other applications of eigenvalue methods have been to the 'reverse interpolation' of icosahedral FTs (§16.3.5) and the analysis of muscle sections. Another major application of these methods is to the analysis of components in mixtures, using 'singular value decomposition' (SVD).

[4] Because of the analogies between eigen-images and Fourier components, *factor space* also has *some* analogies to *reciprocal space*, but also the big difference that the Fourier components in reciprocal space do not each need their own dimension! Unfortunately, eigen-images cannot (like sinusoids) be simply classified by a single number like spatial frequency, though that is probably their most important property.

average particle image, equivalent to the constant in Fourier analysis) to give a 'noise-free' version of the image. This process, termed **reconstitution**, has several uses. The significance of different factors is made clearer by reconstituting images from the extreme ends of data-clouds, where the image is dominated by one factor. These are called **explanatory images**. Another 'explanatory tool' is the **local average**, the average of several images occupying a small but important region in factor space; it can reveal the archetypical particle of that region.

We can now return to the interpretation of factor space. Of course, we cannot show or comprehend 8D space, so we have to be content with projections (usually 2D projections) of it. Such projections show the distribution of all particle images (as points forming a 'data cloud') with respect to the two chosen factors. If the 'cloud' shows two separate concentrations in the projection, this indicates that there are two distinct classes of particle. The local averages typical of each cluster are calculated and examined. Two such images might, for example, be nearly mirror-images; this would suggest that the particles are attaching by opposite surfaces. On the other hand, the 'cloud' may form a continuous band without any local concentrations; this would indicate that particle images can vary continuously with respect to these factors, perhaps as a consequence of different orientations about some preferred axis; again, this could be confirmed with local averages. The local concentrations of points – clusters – may be connected with each other in ways that relate the (random) particle images to the 3D structure of the particle.

Such analysis can reveal possible deficiencies in the pre-processing of particle images. If the alignment of particles failed (so that in-plane rotational disorder remained), then there would be a continuous band of points whose images would vary in that rotation angle. The failure of pre-processing would complicate the interpretation of factor space but not invalidate it. If particles from micrographs with different defocus were included without CTF correction, the CTF could also become a major factor[5].

Finally, an overlap between MSA and Fourier analysis occurs in the analysis of rotationally-symmetric images by Fourier methods (§16.2) and by MSA (e.g. bacteriophage portal proteins, Dube et al., 1993), where the rotational symmetry shows up in the eigen-images. A particularly interesting example of this is the study of *Lumbricus* hemoglobin by van Heel et al. (2009) (see next section). These particles, evidently 6-fold when well-preserved, gave a series of interesting eigen-images after MSA of particles aligned translationally but random in orientation. The first seven of these eigen-images were rotationally symmetric: the first circularly symmetric, the next two 6-fold, the following two 12-fold, and the following two 18-fold. These results mimic exactly the rotational Fourier analysis of the particle, except that here each 'Fourier component' (actually eigen-image) comes as a pair. And the members of each pair are related exactly as sine and cosine functions are related: by a ¼-repeat shift. Thus MSA is here giving us the full set of Fourier components needed for an unaligned periodic object with 6-fold rotational

---

[5]   Also, it would complicate use of the correspondence analysis version of MSA, which requires positive input data.

symmetry. (Perhaps, if the radial densities also had sufficient variability, the eigen-images might reconstitute Bessel functions.)

17.3.5 The Uses of Multivariate Statistical Analysis

MSA seems to have three major uses. First, it is a kind of initial sorting, to reveal potential problems in the particle data. Do they contain hidden sources of variation (like defocus, in-plane rotations, or sample inhomogeneity) that will interfere with other, more important factors? Perhaps the most serious of these 'interferences' is inhomogeneity, which might be divided into residual chemical impurities, inherent particle flexibility and conformational states (e.g. an equilibrium between allosteric 'T' and 'R' states), and distortions (because we need to assume that rotation affects the particles as rigid bodies). Although an early awareness of such problems is useful, their solution can involve lengthy modifications of many stages (see e.g. White et al., 2004, for size inhomogeneity).

A second use is removing the in-plane rotational variability. Van Heel et al., 2009 tested this on *Lumbricus* hemoglobin particles. They originally made the translational alignment by cross-correlation (CC) with a rotationally-symmetric ring pattern (presumably chosen to match the average radial distribution), but then avoided the (somewhat arbitrary) choice of the rotational origins. Instead, these were assigned arbitrarily (i.e., randomly), and the eigenimages calculated by MSA. The first of these to lack circular symmetry (number 2) was then used as the reference for assigning the rotational origins by CCF. These newly-aligned particles were subjected to a new MSA, and this time the first eigen-image was 6-fold (like the particles), and it now took twice as much of the total variance, besides having a distinct resemblance to the best-preserved particles. The authors believe that reference bias might be avoided by procedures like this, since new programs allow larger numbers of particles to be processed (7300 were used in this study).

The third use of MSA is also as another step towards assigning particle images to their correct position in the (notional) 'well-covered viewing sphere' (section §12.1.1). For (to use the shorthand notation of section §12.2.2) if the original particle data differed in (3T+3R), we removed (2T+R) when aligning the images during pre-processing, and then removed the remaining T_z when correcting for the CTF. Thus the remaining variation was, hopefully, confined to the out-of-plane rotations (2R), which can change a particle's projection profoundly. These last variations define each image's position/orientation in the 'viewing sphere' (equivalent to its plane FT's orientation in the 3D FT). If we can be sure that we have removed other problems, we are ready to use the factor map to start this last sorting operation. For this, we need a final operation: classification.

17.3.6 Classification

If the factor map has yielded a discrete set of well-separated data-clouds, each with a distinct local average, then MSA will have already classified the particle images into separate types ready for finding their alignment parameters so as to position them in the notional 'viewing sphere'. In the absence of this ideal situation, we need to classify the particle images by new procedures.

There are many classification techniques (Frank, 1990), but the one apparently most satisfactory for particle images is the **hierarchical ascendant classification** (HAC, apparently due to Ward, 1963). This works by minimizing the variance between members of a class, while at the same time maximizing it between different classes. It is hierarchical in the following sense. We start with each image in a class by itself, and view these as the 'twig-ends' at the edge of a large 'tree'. As we follow these twigs back to the tree, they join in pairs (i.e. the corresponding classes get merged) to form 'branches' which join to form thicker ones. Eventually all the twigs have become joined to the same 'tree-trunk', but the important matter is the *sequence* in which they got connected. At each step, the question is, which class should be merged with which other class? The procedure chooses the pair whose merging adds the minimum intra-class variance, as judged by Ward's formula.

The classes are assigned by a stepwise procedure in which the sequence of assignments is important. Consequently, wrong assignments can become prematurely 'frozen' in place. van Heel has therefore introduced the use of a post-processor to allow class-membership to be reviewed later (see review by van Heel et al., 2000). Indeed, the situation is reminiscent of the computational 'folding' of a polypeptide chain (§11.3.2), where similar premature assignments were subject to revision through the method of simulated annealing.

Now consider the resulting 'tree', called a **dendrogram**. Of course, the 'tree-trunk' is simply the total merging of all classes; when applied to the particle-images, it is their total average. This is a single vertical line placed at the top, when the dendrogram is printed out. As we follow the line down, it splits, and then each branch splits, etc. Most interesting are the branches fairly near the unmerged twigs at the bottom. There we find separate classes, each composed of several 'twigs', but each distinctly different from the others. This (rather arbitrary) division represents our choice of distinct views or projections. Each of these averages (a **class average**) represents a distinct view that we now wish to attach to the 'viewing sphere'. To do that, however, we must find its appropriate polygonal window in the sphere (cf. Fig. 12.1), i.e. its out-of-plane rotations.

17.4 Reconstruction

After a large collection of images (projections) of randomly-oriented particles has been subjected to MSA and classification, we have compressed the original unwieldy mass of data into a much smaller and more reliable gallery of projections. Obviously our goal is to reconstruct the particle's 3D structure; and for this we need a well-filled 3D FT or its equivalent, a well-covered 'viewing sphere' (§12.1.1) (where only an equatorial ring of images is strictly necessary). So we now need to assign each image from our projection-gallery to a specific place on the viewing sphere.

This assignment can be done at two distinct levels: topological, concerned only with neighbors, and geometrical (i.e. metrical), concerned with angles. Obviously, we shall ultimately need all the angles and, if we had them, we could find which projections are neighbors. However, the assignment of neighbors is in principle easier, and could at least provide a check on angles, if not help with the angle-assignment of projections on the basis of its neighbors' angles.

17.4.1 Data Assembly: Topology

The classes of projections are distinct, relatively noise-free, and contain some indications as to their similarities. But the classification tree or dendrogram does not relate in any simple way to the topology of the viewing-sphere's surface. If it were a deltahedron, each face would have three neighbors, so it would be best if the classification found the three closest neighbors to each projection-image (strictly, each class-average). For this we need some kind of (at least qualitative) distance-measure, and some such measure is available in the **generalized Euclidean distance** between any two images (see Frank, 2006, 2007). Consider two images that have been scanned, in the manner of a television image without 'interlacing', in exactly the same way. If they are similar and oriented in the same way, then corresponding pixels should mostly relate to corresponding features. The similarity of any two corresponding pixels is measured by their density-differences, rendered positive by squaring. When that is expanded[6], we get a constant, related to the images' variances, *minus* the CC between the images. Thus, the bigger the CC, the shorter the Euclidean distance. Unfortunately, these distances have not so far proved helpful, though a related method of 'self-organized maps', which uses neural networks, does produce maps with closely-related images appearing as neighbors. Such a map, with spherical or circular topology, might prove helpful in filling the viewing-sphere.

17.4.2 Data Assembly: Geometry

To fit the images geometrically into the viewing-sphere, we need to find their orientations, i.e. three angles (Euler or polar) relative to some coordinate frame. If we can find the relative orientation of any two images, then we can make one of those images the reference and try to find how all the others are oriented relative to it. Unless there is some symmetry in the aggregate, any of the images makes as good a reference as any other. (Of course, if the particle has *any* symmetry, even just a rotation axis, it would be advantageous to build the coordinate system around that axis.) So the essential problem is to find the relative orientation

---

[6] The corresponding pixel-densities are $f(p)$ and $g(p)$, where p is the pixel number. Then the squared differences are $(f(p) - g(p))^2 = (f(p)^2 + g(p)^2) - 2f(p) \cdot g(p)$. This equation has to be summed over the pixels, $p = 1,\dots,P$. Then the first term in ( ) brackets is unaffected by the image similarities, and the last term gives the cross-correlation between the two images.

parameters (the only relevant parts of the alignment parameters) between each image and the chosen reference. There are two methods generally used for this:

(a) Angular Reconstitution
Angular reconstitution (AR: see van Heel et al.'s review, 2000) is an adaptation of Crowther's (1971) common lines principle to systems with low symmetry. van Heel's group apply it in real space, using 'sinograms' and 'sinogram correlation functions'. A 'sinogram' represents all the projections of a 2D image, as a function of projection angle θ, where the projections are shown as marks on a horizontal line and θ becomes the vertical coordinate connecting the marks[7]. And 'sinogram correlation functions' (van Heel, 1987) relate each line of one sinogram to all the lines of another.

However, it is simpler to follow common lines methods in Fourier space, as used in Chapter 16. Any two projections of a 3D particle correspond to two central sections of the same FT, and those section planes must intersect in a line. If that common line can be identified in the two section-planes, we know of a common 'axis' joining them, but not the angle between the planes. To get that, we need to find common lines between each of these planes with a third. If these three section-planes happened to be related by a 3-fold axis, we would get the situation portrayed in Fig. 16.12. One of the planes (say, plane 1) would be ABCD in (a) and, in (b), the line EOF would be its common line with plane 2, and the line HOG would be its common line with plane 3. In (a), the three distances corresponding to EH would form the (equal) sides of a triangle, which would consequently be rigid. Thus the entire system of planes in (a) would be rigid, and that would also be true for the (less symmetrical) diagram that would usually apply to three section-planes all intersecting at the origin of the FT of an unsymmetrical object. Since the system is rigid, we can calculate the angles between all the planes, and hence of two of the planes to the one reference plane.

The crucial problem is to find reliably the common line between any two section-planes. This was found to raise problems even with some icosahedral particles, where there are many more common lines to facilitate the search (see Chapter 16). However, van Heel et al. (2000) find that it works satisfactorily in their real-space implementation. (Of course, any rotational symmetry in the particle helps by multiplying the number of common lines.) Like all analyses without tilting, extra information is required to decide on the handedness of the particles.

(b) Random-Conical Data Collection
Random-conical data collection has already been mentioned in Chapter 12. Figure 12.8 shows the geometry in the case where particles attach to the grid in parallel (i.e., when the out-of-plane (2R) are fixed[8] and only the in-plane (2T+R) can vary). It will be recalled that

---

[7] This is also called the 3D 'Radon transform', named after the very first person to suggest 3D reconstruction (Radon, 1917).
[8] Of course, in practice the (2R) will also change, giving richer tilt data.

each particle is made to give two exposures, an initial high-resolution tilted image and then a low-resolution untilted image with more exposure and consequently a clearer image at lower resolution. The untilted set simply gives the type of data assumed hitherto in this chapter. When analyzed by MSA and classification, they yield the images of different class-averages (differing in the (2R) out-of-plane rotations) with different, known in-plane rotation angles, (R) from (2T+R). Of course, corresponding particles are identified in both tilted and untilted sets, and the tilt angle is known. Consequently, knowing the in-plane rotation angle of a particle allows us to calculate its orientation in the tilted set. And once these orientations are known, we have the information needed for a 3D reconstruction.

17.4.3 Reconstruction Methods

As we accumulate 'classified' projection-images of known orientation, we plot their angular distribution. Are there big gaps that need to be filled before starting reconstruction? Or are the gaps unavoidable, in which case what can we do about them? Are there, on the contrary, regions where projections are excessively crowded? The ability to handle a very non-uniform distribution of orientations is one of the requirements of a good reconstruction algorithm. Another most important requirement is speed, since we shall need to find many reconstructions in the course of the later refinement phase (§17.4.4).

Reconstruction methods (Natterer, 1986) are important not only in electron microscopy (both for particle images and in electron microscope tomography) but in many other areas, especially medical imaging (especially CAT – computerized axial tomography, MRI – magnetic resonance imaging, and PET – positron-emission tomography). The principles of reconstruction were discussed in section §12.4, where the principal methods were described: the Fourier transform and back-projection methods. Both of these related methods face similar problems of unevenly-distributed data as a function of radius. The FT sections and back-projection rays both fan out from the center (of the FT or the reconstructed image, respectively). We saw how this problem was corrected (for arbitrarily-distributed FT sections) by 'reverse interpolation'. For uniformly-distributed back-projection rays, it can be corrected by 'R-weighting'. The greater speed of the latter method favored its use for getting the structures of single particles. However, it needs development for the case when the projection directions are unevenly distributed. R-weighting is a form of filtering in reciprocal space. When there are serious inequalities in the angular distribution of projections, this too can be minimized by exact-filter back-projection (Harauz & van Heel, 1986; Radermacher, 1988), where a filter is tailor-made for the distribution. More difficult is the treatment of any 'unfillable' gaps; are there any alternatives to zeroes? Perhaps slightly better would be the tail of the solution-scattering curve of the particle, corrected for the known scattering from assigned particles.

However, there is a third class of reconstruction method, if the real-space reconstruction equations are soluble. The standard way of solving such a set would use matrices

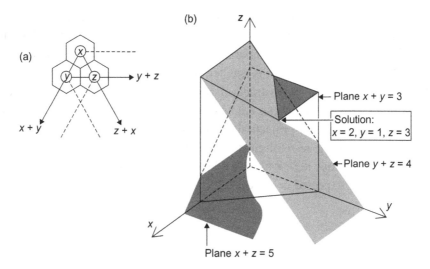

Figure 17.1: A simplified representation of the real-space reconstruction problem (§12.4.1). **(a)** The problem reduced to three pixels with densities x, y, and z, with their density-sums $(x + y)$, $(y + z)$ and $(z + x)$. **(b)** A geometrical representation of the problem. Each of the density-sum equations corresponds to an oblique plane (two of which are shaded), and the solution occurs where the planes meet. Starting at an arbitrary point, ART moves successively along each plane independently.

(see Chapter 20), but this is too slow to be practical (though it was used in the first CAT-scanners). Therefore a quicker *iterative* algebraic reconstruction method ('ART', another rediscovery of methods used in other fields) was suggested by Gordon et al. (1970). This compares the sum of the calculated densities along a line of pixels, with the sum-density at the line's end. Both sums are supposed to be equal, and a simple correction is applied to the line of densities to make them so. After that, another density-line is corrected in the same way. In the modified method (SIRT) of Gilbert (1972), all density-lines are corrected simultaneously (see Herman's (1980) book). A geometrical representation of this class of method is shown in Fig. 17.1. ART, and its later improved version SIRT, are relatively fast, and also allow the incorporation of constraints (i.e. extra knowledge about the 3D structure), but they are non-linear. (Perhaps it is also to their credit that they are iterative, since there seems an inherent preference for iterative processes in many single-particle methods; and of course they also fit well with the needs of computer algorithms, getting many calculations from the same code.) Yet another method used to solve the real-space reconstructions is the use of the conjugate-gradient minimization routine (§11.3.1).

17.4.4 Refinement

Once a preliminary reconstruction is available, the process of refinement can begin. Whereas the projection-images for the original reconstruction were oriented by reference to each other, we now have a source of much less noisy and much more self-consistent projections. Because they are less noisy (since they derive from a structure containing the averages of all measured images), we have an even 'stronger' comparison for CC with the 'weak' images (§2.3.2). Thus the noise contribution to misalignment is smaller, though there will remain some misalignment resulting from distortions in the reconstruction. So the process of iteration needs to continue. The newly re-aligned image-projections are again reconstructed to give a second version of the particle, from which a new set of projections is calculated. These projections are again used for aligning images, and these re-re-aligned images are used for yet another reconstruction, etc. These computer-intensive iterations are ended only when successive reconstructions prove to differ by less than the expected noise level.

CHAPTER 18

Distortion Correction

Chapter Outline

18.1 Introduction

18.1.1 Earlier Distortion Studies

We have seen, through most of Part IV, the important role that symmetry often plays in helping us to get different views and to find their orientations. But macromolecule symmetry is imperfect and difficult to preserve, so we end by giving some consideration to these natural or artifactual imperfections or distortions.

Some big natural distortions have an evident function, like the curved bacterial flagella whose rotation propels bacteria. These have been studied for their own sake, leading to a two-state theory (Calladine, 1978). Minor natural symmetry deficiencies are common, but of uncertain function, so their elucidation requires their study.

The grossest artifactual distortions were generated by the need to dehydrate the specimen before (if not during) insertion into the electron microscope. The principal distortion was flattening, minimized by freeze-drying or Anderson's critical point dehydration (transferring into liquid CO_2 that vaporized rapidly when pressure was released at the critical point). This dehydration problem was significantly reduced by negative staining, devised to enhance contrast but equally important for preserving symmetry. It is still used in preliminary studies, and it helps by replacing water with a heavy-metal salt solution that, on drying, becomes a glass that mostly resists the electron beam and concealed, for some time, its destructive

Figure 18.1: (a) T4 contracted sheath attached to a grid. **(b)** After drying and radiation exposure of negative stain, the upper part shrinks whereas the bottom, in contact with the grid, preserves its original dimensions. **(c)** The same distortion's effect on a helix lying flat on the grid (shown in cross-section). (All copied from Moody, 1967).

effects. But the beam can cause a severe contraction observed in unsupported layers over holes, and in small drops surrounding particles. These can shrink to a more conical shape (Fig. 18.1 (a), (b)), and helices lying flat can get their upper half contracted and flattened (c), producing a stretching of that part of the diffraction pattern (Moody, 1967) and thereby providing a chirality clue. Negative staining has limited resolution (perhaps 2 nm at best), so even a badly bent structure has less to lose by correction. The major distortion in helices is a curved axis, for which a correction was developed by Egelman, 1986, while Bluemke et al. (1988) developed ways of correcting irregularities in the twist angle.

The development of reliable cryo-microscopy (Dubochet et al., 1982, 1988) gives much better shape preservation and potentially much higher resolution. But, if there is less distortion to correct, there is also much less information to guide its correction. Distortion can be classified into short-range and long-range, i.e. by the level of resolution where it applies, and hence by the signal:noise ratio over the region corrected. This sets a minimum size to anything capable of correction, and *very* short-range distortion (i.e., what might be better called 'disorder') demands too much information from too small an area. Longer-range distortion, that is well-ordered on the very small scale, requires fewer parameters whose estimation is therefore more reliable.

18.1.2 Overview of Correction Process

This brief survey has now reached the topic that has been the subject of most work in recent decades: improving image resolution through the compensation of *small* distortions. For, if distortions were big, their effects would probably be (largely) irreversible, besides being hard to measure with the necessary accuracy; so only small distortions have been corrected. Also, these have been chosen from structures with *translational symmetry*; for this gives the biggest symmetry frameworks, providing an essential part of the correction information. Thus corrections have been developed for slightly distorted crystalline sheets (two translations) or helices (one translation). The dimensionality of the correction is 2D for sheets, and also for the surfaces of helices that are wide tubes. However, the overall shape of helices is of course 3D whereas their axes are only 1D.

Distortion curves structures, and any correction must start by analyzing the curvature. This follows essentially the path of differential calculus: the curve is cut into small 'increments', regions that are effectively straight. (Crystalline sheets are cut into unit cells; bent helices into short segments.) Then the segments are analysed for orientation, i.e. aligned with the original direction instead of following the bend. Here there is a choice for the alignment calculation: FTs or CCFs? After that, the segments are perhaps themselves corrected (if they are big enough to have been distorted) and then re-assembled into an exact translational (crystalline or helical) structure. Of course, the size of the original increment is an important decision: too small has insufficient data for a good alignment, too big includes enough curvature to lose accuracy. There is another decision: whether the final corrections will incorporate linear or curvilinear corrections. There has been a progressive tendency to use the simpler linear ones, probably because these suffice for the smallest distortions, which leave the highest-resolution images waiting to be unveiled.

18.2 Crystalline Sheets

18.2.1 Fourier Transforms

Two dimensional crystals are very thin and therefore easily distorted even within the grid plane. A uniform, linear distortion would merely change the crystal's lattice into a new one. This change would be undetectable unless the original lattice had symmetry that was lost by distortion. However, a non-linear distortion no longer gives sharp diffraction spots, but stretched ones that stretch even more at higher resolution. The resolution can be much improved by correcting the distortion, but this involves finding the distortion shifts, i.e. the current (actual) positions of lattice points, minus the ideal lattice positions. We can always estimate the ideal positions from some kind of matching pattern whose position is adjusted to find the exact position of the local lattice. This method was developed in successive studies, and using different kinds of search pattern. At first Crowther & Sleytr (1977) used a pattern that was simply circularly-symmetrical, and the search was done by FT filtering.

18.2.2 Cross-Correlation Functions

When more detailed search patterns were later developed, the search technique changed from reciprocal to real space, and the search patterns were matched with the local lattice by *correlation* methods. If a distorted lattice is to be scanned by a 'reference patch', then that patch must be small to accommodate the distortions. However, it must be large enough to contain enough information to specify the unit cell. Its size (a compromise between these two factors) can vary. Saxton & Baumeister (1982) used five unit cells in a study of negatively-stained bacterial cell wall images. The match positions of the 'reference patch' give the current positions of the 'lattice points' of the distorted lattice; these are the experimental data we need.

We find them by matching small regions of the crystal with the 'correct' structure or, more precisely, with our best estimate for this. We obtain this estimate by computer-filtering the image FT, using reciprocal-space (diffraction-plane) apertures of carefully controlled size, bearing in mind that the purpose of the filtered image is not to represent the overall average, but to find differences due to local distortion. We then take the central part of the filtered image as the comparison structure, and the size of its area must also be chosen carefully. (Henderson et al. (1986) found a suitable reference area contains about 50 unit cells.) We must then find the present 'lattice points' from the CCF of the image with this comparison structure.

As was explained in §3.4.6, the CCF can be calculated more quickly by using the Fast Fourier Transforms (FFT). We here recall this useful trick, supposing that p is the microscope image (picture), c is the comparison structure, and that each has an FT: $c \leftrightarrow C$ and $p \leftrightarrow P$. Then we want $\mathrm{CCF}(c,p) = c \odot p = c_r \star p$, where c_r is the 'reversed function' of c ($= c$ upside-down, in 2D; the same as $C(-)$), so that $c_r \leftrightarrow C^*$. Now $c_r \star p \leftrightarrow C^*P$, so we do the calculation by getting C^*, multiplying it in turn with each picture's FT ($=P$) and then inverting the product. This allows us to speed the calculation of each image's FT, and of the final reverse FT of the product, by using the FFT algorithm (§4.3.2).

The CCF gives us the coordinates of the present 'lattice points'. The average shows us their ideal or correct positions, and the displacement of each ideal position to its actual one is the local distortion vector. Over large distances, where the distortion builds up to more than a unit cell, there is the problem of pairing ideal and actual lattice positions. New peak positions were found from the cross-correlation map with a stepwise algorithm (Crepeau & Fram, 1981) which learns the distortion progressively, starting at the origin peak. (Thus – assuming a rectangular lattice with a sides 'a' along x – to find the 10th peak along x, it starts at $x = 0$, then moves a distance $\Delta x = a$ to look for the next peak, and repeats this until it reaches the 10th peak.)

Each lattice point must have a distortion vector (even if it is in the correct position, there is a null distortion vector). We can plot all these vectors as lines connecting the ideal and actual lattice positions. The resulting collection of hundreds of lines constitutes a 'vector field[1]' (§11.2.1). This vector field is usually relatively simple, allowing us to compress the information into the parameters of an interpolation curve. Henderson et al. (1986) used bicubic splines (see, e.g., section 3.3 of Press et al., 1992). This defines a smoothed vector field which gives us a more accurate picture of the distortion than we get from the original, noisy one. The limited information content of a single unit cell is multiplied by their large number.

Thus the spline formula allows us to calculate the actual position of any desired unit cell. So we run through all the unit cells, as listed by available (h,k) indices. For each cell, we

---

[1] The classic vector field is the pattern of electric or magnetic field lines, which are perpendicular to potential lines, the two sets thus forming coordinate lines that preserve triangle angles ('conformal').

calculate the position of its center and of other standard points where the density is needed. At these points, we interpolate the available densities to give the required densities. So we finally get the corrected unit cell densities of all available (h,k) indices. Then we calculate its FT, which should have sharp spots.

Finally, the variations between different unit cells can be analyzed by multivariate statistical analysis (MSA), throwing light on the source of variation in a cryo-microscope at liquid helium temperatures (Frank et al., 1993).

18.3 Helices

18.3.1 Linear Distortions

Minor distortions are easier to correct, so we start by looking at the very smallest: infinitesimal. These are also the simplest, since infinitesimal changes can be represented as a simple matrix[2]. Such a matrix can be split (rather like 1D functions: §2.2) into 'even' and 'odd' components. The 'odd' component simply describes a small rotation of the distorted part, while the 'even' component is a *symmetric matrix* which describes (§11.2.1) a stretching or compression along two perpendicular directions (the matrix's eigenvectors) (see Fig. 18.2.). When viewed from other directions, however, that distortion would be described as **shear** (Fig. 18.3). So shear is also an infinitesimal distortion, a combination of the types shown in Fig. 18.2 (b) and (c). As already mentioned, the small uniform distortions in Fig. 18.2 have the advantage that we can easily find their FTs. As shown in Chapter 6 (§6.1.2), the FT of rotation (b) is the same rotation, and the FT of uniform distortion (c) is the perpendicular distortion.

18.3.2 Three-Dimensional Linear Distortion

The surface-lattice of a large diameter helix forms a tube. The effects of its distortion are shown in Fig. 18.4. The circumferential vector c cannot change under a continuous distortion, but a basic vector b can change either or both its components, h and Ω. However, this projection uses an angular coordinate ϕ, whereas a real helix has a radius r that *can* change under a continuous distortion. Thus just three parameters can change: r, h and Ω. Changes in r or h stretch or compress the helical lattice along or perpendicular to its axis. But changing Ω is an angular shear that twists the helix, tightening or relaxing its various subsidiary helices, and causing alterations of the helical repeat.

Besides affecting the 2D helical lattice, distortions can affect the entire 3D structure. We have already examined the 2D version of a small first-order distortion (§18.3.1). We found

---

[2] Small distortions can be represented (see Chapter 20) as $(1 + \delta)$, where 1 means the operation that leaves the lattice unaltered (unit matrix) while δ is an infinitesimal matrix with all the distortion parameters (and only zeroes along the leading diagonal). This distortion is linear since two successive distortions have the effect: $(1 + \delta)(1 + \varepsilon) = 1 + (\delta + \varepsilon) + \delta\varepsilon \approx 1 + (\delta + \varepsilon)$ because δ and ε are both small. (See Chapter 20 and Chapter VI of Nye, 1985.)

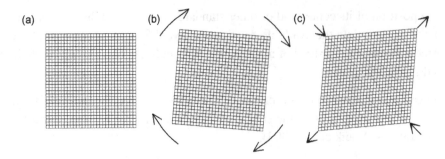

Figure 18.2: Classification of small uniform linear distortions, shown by the distorted lattices. **(a)** Undistorted lattice. **(b)** Rotation. **(c)** Stretch/compression about oblique axes (45°).

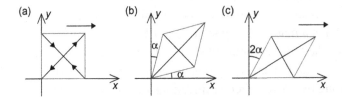

Figure 18.3: Shear is a combination of a diagonal stretching/compression distortion (**(a)**–**(b)**) with rotation (**(b)**–**(c)**).

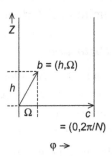

Figure 18.4: Parameters of a helix that can change after minor distortion of the surface-lattice. The radial (φ, z) projection shows two defining vectors: the circumferential vector c and a basic vector b.

that it is a combination of two different processes: a rotation (that leaves the lattice with the same shape and size); and some combination of stretch and compression that distorts it. In 3D, the same division can be accomplished though, of course, both the rotation and the stretch/compression combination occur in 3D. Thus the rotation can have its axis pointing in any direction, as in Fig. 12.11 (but the rotation angle ω must be small). But what is the three-dimensional generalization of stretch/compression?

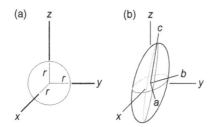

Figure 18.5: General 3D distortion represented by its effect on a sphere. **(a)** Sphere, radius *r*, at origin of (*x*,*y*,*z*). **(b)** Sphere after undergoing a general distortion with three independent stretches or compressions, changing *r* to *a*, *b* or *c* (the lengths of the mutually perpendicular vectors **a**, **b** and **c**). The orientation of **a**, **b** and **c** is arbitrary.

In 2D, a symmetric matrix causes a stretching and compression in perpendicular directions. In 2D, there *are* only two perpendicular directions, but there are three in 3D. So the 3D generalization of stretch/compression can have up to three different stretches or compressions, provided they point in mutually perpendicular directions. We represented the 2D composite distortion by its effect on a circle, as in Fig. 11.9 (a) and (c); it turns the circle into an ellipse, defined by two axes. In 3D, we see the distortion's effect on a *sphere*, which becomes an **ellipsoid** defined by three perpendicular axes **a**, **b** and **c** (Fig. 18.5). This requires six parameters (three to define the orientation of **a**, **b**, **c**, and three to define the axis lengths). As explained in §18.3.1, rotation combines with distortion to yield shear. The most obvious shear is in the (*x*,*y*)-plane.

A tubular helix lying on a flat grid, as in Fig. 15.21, has inherent symmetry that affects the distortion ellipsoid. Its axes will be aligned with the helix axis and the grid plane. Along the helix axis, attachment to the grid will minimize distortion, while the lack of support perpendicular to the grid will maximize it. If the distortion occurs after the helix is embedded in solid stain or ice, this distortion is almost invisible as it only changes the mass distribution down the optic axis. (But, if it occurs before embedding, helix flattening increases its radius without affecting its surface lattice; this *does* affect the FT.) In any case, lateral stretching or compression of the helix is a linear distortion needing correction.

18.3.3 Segmental Correction Using Fourier Transforms

Beroukhim & Unwin (1997) originated a method of correction that has since been widely adopted. They extracted, from the image helix, segments short enough to have the same (linear) distortion; corrected this distortion; and rejoined the segments, not just in their original fashion, but in the arrangement that best enhanced the symmetry. This contrasts with the earlier methods of fitting a bent helix to a curve, and resembles fitting a curve with a series of straight lines.

The segment length is important: too short, and information is lacking; too long, and distortion starts to have an effect. So this choice depends on the specimen. There is some advantage in making segments a repeat (or half-repeat) long, but that is a secondary consideration.

When segments are re-aligned after distortion correction, the new alignment can be used to adjust helical parameters. Thus the axial ('roll') rotation ρ (Fig. 15.21) mimics the twist angle Ω, and the axial shift Δz alters h.

Such simplification is useful, since we must optimize the number of parameters to fit. While too many big distortions imply a specimen not worth image-processing, every image processed for high resolution needs many (small) distortions to be corrected simultaneously, and a high resolution specimen may provide sufficient data for multi-parameter optimization. The chosen parameters should cover the greatest variation independent of each other.

Beroukhim & Unwin chose the following. (i) 2D distortion parameters: r, h and Ω. Of these, Ω was effectively replaced by ρ (see below). (ii) 3D distortion parameters: only frontal shear was considered, but only as a criterion for particle rejection. (iii) Reorientation parameters of the corrected segment: two translation parameters (Fig. 15.21), Δx and Δz (Δy corresponds to defocus) and three rotation angles: ρ, σ, and τ. Thus only seven correction parameters were needed: their r_{scale}, 'repeat', φ, θ, ω, x and z, corresponding to our r, h, ρ, σ, τ, Δx and Δz. They used the 'repeat' (our h) as the major helical parameter because they were concerned to obtain a complete set of orientations contributing to each FT. (The dependence of the helix repeat on Ω was avoided by independent correction for $\Omega \approx \rho = \varphi$, leaving the repeat solely dependent on h.)

Beroukhim & Unwin did their comparisons in reciprocal space, which in any case had to be used for correcting the Contrast Transfer Function (CTF). Moreover, their helix specimens had the advantage of dihedral symmetry which made the FT real in many planes. (As in crystallography – Chapter 13 – this makes phase-determination much more precise.) This distortion-correction was one of the methods used to obtain the structure of the acetylcholine receptor at 4 Å (0.4 nm) resolution (Miyazawa et al., 2003; Unwin, 2005) and of the bacterial flagellum (Yonekura et al., 2003).

18.3.4 Helices: Projection Matching in Real Space

There is a fine choice between the advantages of reciprocal and real space. The former allows the advantages of FTs for structures (including helices) with translational symmetry, and it is natural for CTF correction; but the latter is natural for model-building, and is often preferred for structure averaging (where it was used by Miyazawa et al., 2003). A judicious choice of the space appropriate for each stage of an analysis requires the existence of suitable programs.

For helices, there exist a set of standard reciprocal-space programs (Crowther et al., 1996). Real-space programs have been developed more recently, starting particularly with the successful IHRSR (Iterative Helical Real-Space Reconstruction) program of Egelman (2000), who combined the helical segmental division method of Beroukhim & Unwin (1997) with the technique of real-space projection-mapping developed for single-particle analysis. The beginning and end of the analysis is a helical reference structure, whose symmetry and structure are improved iteratively. From it, many projections (at small φ-increments, perpendicular to z) are calculated, at each cycle iteration, for comparison with the raw image-data. The (2T + R) in-plane orientation parameters of each image are adjusted so that they can generate, by back-projection, the 3D reconstruction of a helical segment. Its helical symmetry parameters (h and Ω) are found as a least-squares fit of the segment's center, and the resulting 3D helical structure is the new reference for another iteration.

IHRSR was developed for helices that are distorted or difficult to index, but (as always) the best applications emerge from experience. We consider two applications to helices with very different characters: TMV (Sachse et al., 2007), one of the best-ordered helices; and HBsAg viral tubes (Short et al., 2009), which are not well ordered.

The essential indexing of TMV was finally fixed by Franklin & Holmes (1958) from the X-ray pattern, so Sachse et al. (2007) only needed to make minor adjustments to h and Ω. The point of their study was to attain a high resolution structural image of a particle which Jeng et al. (1989) had already shown could be reconstructed from electron micrographs by standard Fourier methods to at least 10 Å, showing the four α-helices. From each of 135 particle images, segments were taken with 90% overlap, care being taken to avoid any interpolation (which introduces errors), and the particle symmetry was also exploited to include each segment fifty times. Helix tilt and CTF were corrected by FT methods. Projections were taken at 360 1°-intervals of ρ, and twelve 1°-intervals of τ (Fig. 15.21). After a modified IHRSR projection-mapping gave the alignment parameters, a real-space 3D reconstruction gave a calculated structure that was then symmetrized, by averaging over the known TMV helical symmetry. In these steps, care was again taken to minimize interpolation. The final resolution, measured by FSC (§12.3.2) and the visibility of known features, was better than 5 Å.

The particles studied by Short et al. (2009) were not homogeneous in structure, sometimes even within a single particle, and the resolution of particle FTs was only about 50 Å. Here the real-space IHRSR program was used to obtain indexing as well as structure, to compare with that given by conventional FT methods. It seems, from the resulting agreement, that the real-space method's main value is as a complement to conventional indexing methods, rather than a replacement of them.

18.3.5 Distortion Correction of 'Crystallographic' Structures

It is a paradox that big particles with translational symmetries, which are the best candidates for the 'crystallographic' FT image-analysis methods, are also the most prone to distortion, whose correction requires real-space correlation methods. By contrast, small particles with only rotational symmetry are among those most free of distortion; and yet they are described as 'single particles' and sometimes reconstructed by 'single particle' methods. So there is a tendency for FT-determined structures to need refinement by correlation methods. This raises the question: were those Fourier methods, that needed replacement at the very end of the analysis, really essential at its beginning?

Of course, this issue doesn't arise with those symmetrical aggregates that do give high-resolution diffraction patterns, since it should be possible to extract from them high-resolution images that need no distortion correction. But, when the best diffraction patterns are only mediocre, it might well be wondered if FT methods have any inherent advantages. One possibility is the *reliability* with which the symmetry can be found. This is particularly important with helices, where the correct 'indexing' (symmetry) is necessary to get correct averaging. Here the traditional method, although not the only one, has some advantages in displaying the stages of symmetry-determination visually, providing an additional check on correctness. More generally, FTs are a good way of assessing the accuracy of translational symmetries. So there is an argument for doing parallel studies by FT and real-space methods.

However, there are also some more interesting questions. Is it possible to have real-space methods that are not only adequate, but can convincingly demonstrate their adequacy in the same way as a well-indexed crystalline diffraction pattern? Also, when making distortion corrections by real-space correlation methods, how much does it help to know the approximate subunit positions through some other technique? After all, these 'ideal' or starting positions do supply a considerable fraction of the required information, thus limiting the possibilities of an optimization program falling into a false minimum. Of course, only experience will answer any of these questions.

Note Added in Proof

A recent 6.6Å study of actin fibers (Fuji, T., Iwane, A.H., Yanagida, T. & K. Namba (2010). Direct visualization of secondary structures of F-actin by electron cryomicroscopy. Nature 467, 724–728) provides a detailed molecular model of this thin, flexible helix, in a study combining improved specimen-preparation and microscopy techniques with image analysis (somewhat like sachse et al.) using the IHRSR program and published data.

Mathematical Basis

CHAPTER 19

FT Mathematics

Chapter Outline

19.1 Introduction

This is the first of two chapters that introduce the non-mathematical reader to the essential mathematics of Fourier transforms (FTs) in this chapter, and matrices in the following (Chapter 20). The purpose of these short chapters is not to cover even the most essential properties of FTs (as they were covered in Chapter 3), but to introduce the reader to the subject.

We start this chapter with some revision of elementary algebra, and precede that by an intuitive proof of an important theorem, 'Pythagoras' theorem[1]', in Fig. 19.1. We start (a) with any right-angled triangle (grey), with sides a, b, c (in decreasing order of length). Then we put four identical copies of this triangle in a rigid square frame (black) with sides $(b + c)$. The four identical triangles are free to slide about within this frame but, since the frame and

---

[1] This proof (see R.C. Henry, The Mental Universe, *Nature* **436**, 7 July 2005, p. 29) is apparently based on the Chinese mathematical work 'Chóu-Peï Suan-king', perhaps 1105 BC (see Smith, 1923, p. 30). This would have been roughly 500 years earlier than the famous Pythagoras of Samos.

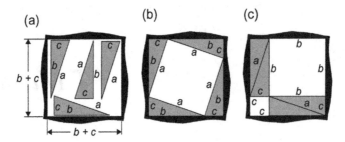

Figure 19.1: Entirely intuitive proof of Pythagoras' theorem. (See text for explanation.)

the triangles have fixed areas, the remaining area (white) is always the same. But what is it? The arrangement (b) turns the entire white area into a square, so we can measure its area: a^2. The second arrangement separates the white area into *two* squares, one of area b^2 and the other of area c^2. So the total white area is now $b^2 + c^2$. As we already know that it is a^2, we conclude that $a^2 = b^2 + c^2$.

19.2 Algebra

19.2.1 Elementary Algebra

We start with ordinary addition and multiplication.

(a) Multiplication Notation
$ab = a \times b$, and $abcd = a \times b \times c \times d = abdc$. The order makes no difference, just as $a + b + c + d = a + b + d + c$. Note that sometimes a simple dot is used for multiplication: $a.b = a \times b$ are common notations.

(b) Multiplication of Brackets
Brackets define the order of operations: one should complete the calculation within a bracket before using the result in any further calculation. We can see the implications with algebra:

$(a + b)(c + d) = ab + bc + ad + bd$. Here the terms in brackets are called binomials; see Fig. 19.2 (a).

If we put $c = a$ and $d = b$, we get:

$(a + b)^2 = (a + b)(a + b) = a^2 + 2ab + b^2$.

These brackets are called 'binomials' as they have just two letters inside. We can introduce one minus sign, getting:

$(a + b)(a - b) = a^2 - b^2$, since the other terms cancel.

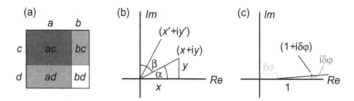

Figure 19.2: (a) Intuitive proof of the product of two binomials. **(b)** Complex numbers in the complex plane, where *Re* and *Im* represent respectively the 'real' and 'imaginary' axes (see section 3.3.2). **(c)** The point representing 1 is rotated by an infinitesimal angle $\delta\varphi$ (see section 19.4.2).

However, when we have two minus signs, we have to decide what happens when they multiply. The result must be consistent, so we try $0 = (1 - 1)^2 = 1^2 + 2(-1) + (-1)^2 = 1 - 2 + (-1)^2$, so $0 = -1 + (-1)^2$, showing that $(-1)^2 = +1$.

Algebra deals only with the *form* of a calculation (i.e., which numbers should be multiplied or added or whatever to which other numbers), so the *letters* have no mathematical significance, and we can re-write $(a + b)(a - b) = a^2 - b^2$ as $(c + d)(c - d) = c^2 - d^2$.

(c) Powers of numbers

$a^2 = a \times a$, $a^3 = a \times a \times a$, so $a^2 \times a^3 = (a \times a) \times (a \times a \times a) = a \times a \times a \times a \times a = a^5$. So we add indices when multiplying $(2 + 3 = 5)$. This rule also applies to negative integers: $a^4 = a \times a \times a \times a = a \times a \times a \times a \times a/a = a^5/a = a^5 \times a^{-1}$, so $1/a = a^{-1}$. Applying the same rule to fractions, $(a^{\frac{1}{2}}) \times (a^{\frac{1}{2}}) = (a^1) = a$. But $(\sqrt{a}) \times (\sqrt{a}) = a$, so $a^{\frac{1}{2}} = \sqrt{a}$. If $b = \sqrt{a}$, then $\sqrt{b} = \sqrt{\sqrt{a}}$, so $(\sqrt{\sqrt{a}})(\sqrt{\sqrt{a}}) = (\sqrt{b})(\sqrt{b}) = b = \sqrt{a}$, and therefore $(\sqrt{\sqrt{a}})(\sqrt{\sqrt{a}})(\sqrt{\sqrt{a}})(\sqrt{\sqrt{a}}) = (\sqrt{a})(\sqrt{a}) = a$. Thus $\sqrt{(\sqrt{a})} = a^{\frac{1}{4}}$.

Thus the rule for powers is $a^m \times a^n = a^{m+n}$. Also, $(a^3)^2 = (a \times a \times a) \times (a \times a \times a) = a \times a \times a \times a \times a \times a = a^6$. So $(a^m)^n = a^{mn}$, and thus:

$$((a^2)^2)^2 = (a^4)^2 = a^8.$$

19.2.2 Exponential Function

'Special functions' need (strictly) infinite numbers of multiplications and additions for perfect accuracy (though of course a quite modest accuracy suffices for practical calculations). A 'scientific' pocket calculator helps us to make the acquaintance of some of these functions. We start with functions that are an extension of squares (raising to the power of 2) or of cubes (raising to the power of 3). However, instead of raising a variable number to a fixed power, we want to raise a *fixed* number (say, 10) to the power of a *variable* number, called x. As a power is also called an 'exponent', this could be called an 'exponential function', n^x, where n is some chosen number (the most convenient being $n = 10$ because of our decimal system). However, nomenclature has settled on another particular value of n for the 'exponential function', $n = e$, where e is a peculiar number, between 2 and 3, that we shall soon investigate. Our study of these 'exponential functions' is helped by the fact that even inexpensive 'scientific' pocket calculators often have two of them: 10^x and e^x. It is hoped that the reader has access to one of these calculators, preferably with the 'trigonometric' functions (sin and cos) as well.

The general exponential function should be familiar from the 'exponential growth' of microbes (or inflation). The defining characteristic of exponential growth is a constant 'doubling time'. If that is 20 minutes, and we started with one microbe, we would have two microbes after 20 minutes, eight after an hour (three 20 minute periods), and a day ($=24 \times 3 = 72$ periods of 20 minutes) would produce 4.7×10^{21} microbes. However, nutrients and oxygen would have exhausted before then, showing that exponential *growth* can only be a transient phase. But the reverse function, which involves successive *halvings*, is far more frequently encountered in practice, e.g. in radioactive decay or cooling (or the value of money).

The number 4.7×10^{21} can be found easily with a pocket calculator. We could double 72 times, but that is tedious and error-prone. Successive squarings are better: $2^8 = ((2^2)^2)^2 = (4^2)^2 = (16)^2 = 256$ so $2^9 = 2 \times 2^8 = 2 \times 256 = 512$. Thus $2^{72} = (2^9)^8 = (512)^8 = ((512^2)^2)^2 = (262144^2)^2 = 4.7 \times 10^{21}$, in two more squarings.

However, it is easier to calculate this with the '10^x' function, which calculates 10 raised to the power of x. Start by finding 10 raised to the power of 0.30103: we[2] get 2. Thus $2^{72} = (10^{0.30103})^{72} = 10^{0.30103 \times 72}$. So we find $72 \times 0.30103 = 21.67416\ldots$ and then immediately press the '10^x' function key, getting 4.7×10^{21}, the answer in two steps. This '10^x' function key gives a kind of 'exponential function' whose properties are interesting. $10^1 = 10$, so entering 1 and pressing '10^x' gives 10. But entering 0 and pressing '10^x' gives 1, which makes sense since we could approach 0 through many successive halvings, i.e. through many successive square-roots, which would bring 10 down to a number little more than 1.

But the calculator also has an 'e^x' function key, which does just the same but with the number e. (This function is called the **exponential function**.) To find what e is, repeat what we did with the '10^x' function key to get 10. So we enter 1 and press 'e^x', getting e. If we try it, we get $2.71828\ldots$, a curious number of no apparent use, so what does it mean? We shall find that in §19.4.1, after looking at some other functions.

19.3 Geometry

19.3.1 Trigonometry

Another set of special functions relates angles to lengths: the trigonometrical functions sine and cosine (plotted in Fig. 3.1). In Fig. 19.2 (b), where the longest side of a right-angled triangle is the hypotenuse, $\sin \beta = y/\text{hypotenuse}$ and $\cos \beta = x/\text{hypotenuse}$. It is useful to remember the definitions of sine, cosine and not to confuse them. The easiest function to

---

[2]   Where did we get 0.30103? Enter 2 and press the 'log' button.

remember would be exactly the same as the angle (not very useful!) but the next easiest is the sine function, which is nearly the same as the angle, when the angle is small. So $\sin(0) = 0$, $\sin(\delta) = \delta$, where δ is very small, and the sine continues increasing up to $\sin(90°) = 1$. (Test this with the keys on a scientific calculator. But note that the calculator expects angles to be in degrees, whereas $\sin(\delta) \approx \delta$ when δ is in radians, where $180°$ (degrees) $= \pi$ radians.)

However, the cosine is the 'complement' ('co-sine') and does the opposite: $\cos(0) = 1$ but $\cos(90°) = 0$. As pointed out in section §3.1, sines and cosines are the same functions, apart from having different starting-points (Fig. 3.10).

Trigonometry is only needed in two or more dimensions, and we now explore some of its applications relevant to FTs.

19.3.2 Complex Numbers (Phasors): Algebra

Recall from Chapter 3 that these use i, the 'square root of -1', which isn't any ordinary ('real') number since $(+1)(+1) = +1$ and also $(-1)(-1) = +1$. Therefore numbers containing i (called 'imaginary') can't be *combined* with ordinary numbers except after squaring or multiplying by i. Thus when we add complex numbers we keep the 'real' and 'imaginary' parts separate: $(a + bi) + (c + di) = (a + c) + (b + d)i$.

Multiplication by a complex number is just like the multiplication of binomials, but with the addition of replacing i^2 by -1:

$$(a + bi)(c + di) = ac + bci + adi + bdi^2 = (ac - bd) + (bc + ad)i. \qquad (19.1)$$

This separation of 'real' and 'imaginary' is central to complex numbers: we must never confuse an imaginary number containing 'i' with a real one that does not. At the end of any calculation involving both, we must segregate the real from the imaginary numbers. Thus we must write complex numbers as separate real and imaginary parts. That is easy enough if we only have addition, subtraction or multiplication. But how do we do it with an expression like $1/(a + ib)$? We use a method involving the 'complex conjugate'.

The complex conjugate of a complex number $(a + bi)$, written $(a + bi)*$, is $(a - bi)$. Therefore:

$(a + bi)(a + bi)* = (a + bi)(a - bi) = a^2 + b^2$, which leaves only the real numbers, which we can use in division. Thus we divide by complex numbers by multiplying top and bottom of a fraction by the complex conjugate of the bottom, $(c - di)$:

$$(a + bi)/(c + di) = ((a + bi)(c - di))/((c + di)(c - di)) = ((ac + bd) + (bc - ad)i)/(c^2 + d^2)$$
$$= ((ac + bd)/(c^2 + d^2)) + ((bc - ad)/(c^2 + d^2))i.$$

(In the first step, we multiplied top and bottom by $(c - d\mathrm{i})$, so we multiplied and divided by the same number. Then we multiplied all brackets; finally, we separated the real and imaginary parts.)

19.3.3 Complex Numbers: Geometry

This separation of 'real' and 'imaginary' resembles the separation of movements in the x-direction from those in the y-direction. Thus complex numbers are a way to do 2D coordinate geometry in what is called the 'complex plane' (see Fig. 19.2 (b)). Imaginary numbers (those containing 'i') are being labeled as 'y-coordinates', the others as 'x-coordinates'.

This geometrical interpretation works well for the following reason. Recall that ordinary ('real') numbers obey the rules $(-1)(+1) = -1$ and $(-1)(-1) = +1$. If we mark all the numbers on the x-axis (Re in Fig. 19.2 (b)) we see that multiplication by -1 involves a rotation of $180°$, from $+1$ to -1 and then back again. Consequently, $\mathrm{i} = \sqrt{-1}$ should be interpreted as a rotation of $90°$ (anti-clockwise is the accepted direction). So all positive real numbers lie on the 'real' $+x$ axis, positive imaginary numbers lie on the 'imaginary' $+y$ axis, and complex numbers have components of both sorts. The 'real' and 'imaginary' axes are conveniently denoted by Re and Im (respectively).

So far, we have considered complex numbers in Cartesian coordinates. We also need to use polar coordinates, and this is where we can introduce trigonometry. The number $(x + \mathrm{i}y)$ in Fig. 19.2 (b) can be expressed as a length r and an angle α. Then $r^2 = x^2 + y^2$, so $r = \sqrt{(x^2 + y^2)}$; and $x = r \cos \alpha$, $y = r \sin \alpha$, so $(x + \mathrm{i}y) = r(\cos \alpha + \mathrm{i} \sin \alpha)$. Suppose there's another number $(u + \mathrm{i}v) = \rho(\cos \beta + \mathrm{i} \sin \beta)$ and we multiply these two, the multiplied polar forms give us $r\rho(\cos \alpha + \mathrm{i} \sin \alpha)(\cos \beta + \mathrm{i} \sin \beta)$. We want to get another polar form, something like $r\rho(\cos \theta + \mathrm{i} \sin \theta)$, so that multiplication by $(\cos \beta + \mathrm{i} \sin \beta)$ would rotate $(x + \mathrm{i}y)$ into $(x' + \mathrm{i}y')$ as in Fig. 19.2 (b). Nevertheless, we would have a lot of trouble getting this with ordinary trigonometrical formulae, and we could easily conclude that the polar form is of little use. However, there is Euler's remarkable formula: $\cos \alpha + \mathrm{i} \sin \alpha = e^{\mathrm{i}\alpha}$ (!) How this could possibly be true only emerges from a much deeper study of the exponential function in the next section.

19.4 Infinitesimals

In section §19.2.2 we had started to become familiar with the 'exponential' functions 10^x and e^x in a pocket calculator, and to have some idea about the 'e^x' key: entering 0 or 1 gives $e^0 = 1$, and $e^1 = 2.71828\ldots$ But what connection does this have with $e^{\mathrm{i}\alpha}$, since one can't enter an imaginary number into a pocket calculator? And if we try entering an angle in radians, like π, we only get the meaningless answer 23.1. However, we shall now find out what it might be good for.

19.4.1 Exponential Function

Infinitesimals are very small numbers, small enough that their squares are negligible in comparison. Thus the square of 0.001 (one thousandth) is 0.000001 (one millionth). By themselves, they're just small but, combined with an ordinary number, the loss of the square (and higher powers) *linearizes* many operations like reciprocals or square roots. We can easily see this effect by doing tests with a pocket calculator. Thus $1/(1 + 0.001) = (1 + 0.001)^{-1}$ $\approx 1 - 0.001 = 0.999...$, and $(1.001)^2 \approx 1 + 2(0.001) = 1.002$. Why not try this 'linearizing' effect with the 'special function' keys, especially the '10^x' and the 'e^x' key?

We should already know what to expect since, if $x > 1$, the effect of raising n to the xth power is something like squaring (n^2 means $x = 2$) and, if we keep squaring a number, it gets bigger. Consequently, if we keep taking its square root ($\sqrt{n} = n^{1/2}$), it must get smaller. But, whereas squaring makes it bigger without limit, 'square-rooting' never makes it smaller than 1, because $\sqrt{1} = 1$ so further square roots make no difference. Thus, if we make x as small as possible, i.e. zero, $n^0 = 1$. Thus $10^0 = 1$ and $e^0 = 1$, so we next explore what happens when we replace zero by a *very small* number, an *infinitesimal*. Starting with a thousandth, 0.001, we find that $10^{0.001} = 1.002305...$, which doesn't have an obvious meaning. But $10^{0.002} = 1.00461...$, where the decimal part .00461 is obviously double .002305... This suggests that $10^{n(0.001)} \approx 1 + n(.002305)$, which turns out to be (approximately) true. Thus $10^{\delta x} \approx 1 + \delta x(2.305...)$, where δx is an infinitesimal. But it would be more elegant if we just got δx without the ugly 2.305... which is always there when we press the '10^x' key. Well, this is quite possible, but only if we press the 'e^x' key instead. (Try it with any *small* number.)

Thus we find by experiment that $e^{\delta x} \approx 1 + \delta x$. This is the special feature of that curious number e that gets it put into a pocket calculator. Indeed, we could take it to be the *definition* of e^x, and use it to find e without depending on the built-in function of the calculator. We do this as follows: $e^{\delta x} \approx 1 + \delta x$ so, raising it to the power m, $(e^{\delta x})^m \approx (1 + \delta x)^m$, or $e^{m\delta x} \approx (1 + \delta x)^m$. We now choose m and δx so that $m\delta x = 1$, so $m = 1/\delta x$ and therefore $e \approx (1 + \delta x)^m$ or $e \approx (1 + 1/m)^m$. We choose $m = 1024$ so that we can raise a number to this power with ten squarings ($2^{10} = 1024$), so $\delta x = 1/1024 = 0.0009765$. Ten squarings of 1.0009765 give 2.7169..., which is quite near the correct value of $e = 2.71828...$

Thus we have found the number e for which $e^{\delta x} \approx 1 + \delta x$, and it corresponds to our pocket calculator key 'e^x'. Notice that $e^{\delta x}e^{\delta y} \approx (1 + \delta x)(1 + \delta y) = 1 + (\delta x + \delta y) + [\delta x \delta y]$, where the quantity between square brackets is quite negligible, so that $e^{\delta x}e^{\delta y} = e^{\delta x + \delta y} \approx 1 + (\delta x + \delta y)$. This confirms that e^x is a function using indices or exponents, so it is called the *exponential function*. This function is much easier to write and read in a different notation: $e^x \equiv \exp(x)$.

19.4.2 Exponential Function in the Complex Plane

A phasor in the complex plane, with real (*Re*) and imaginary (*Im*) axes, is shown in Fig. 3.7, in its Cartesian (c) and polar (b) forms. Consider the polar representation, with magnitude (*M*) constant at 1, so only the angle φ varies and any function of φ (including the exponential function, $\exp(i\varphi)$) lies on a circle of radius 1. We consider the meaning of this function, using its definition in terms of an infinitesimal, the small angle $\delta\varphi$: $\exp(i\delta\varphi) \approx 1 + i\delta\varphi$ (where $i = \sqrt{-1}$). The complex number $(1 + i\delta\varphi)$ is a point close to the real axis (Re) as in Fig. 19.2 (c). Just as we found $e \equiv \exp(1) \approx (1 + \delta x)^m$, where $m\delta x = 1$, so we can now explore the meaning of $\exp(i\varphi)$ for angles like 30°. If, on the circle $\exp(i\varphi)$, we have an intermediate point $x_0 + iy_0$, then a neighboring point at the angle $\delta\varphi$ is $x_1 + iy_1 = (x_0 + iy_0)\exp(i\delta\varphi) = (x_0 + iy_0)(1 + i\delta\varphi) = (x_0 - y_0\delta\varphi) + i(y_0 + x_0\delta\varphi)$. Therefore $x_1 = x_0 - y_0\delta\varphi$ and $y_1 = y_0 + x_0\delta\varphi$, a formula giving successive points round a circle.

But we can also use the quicker method of successive squarings, as in section §19.2.2 above. However, it is more complicated because there are two parameters instead of one, so we shall need a pocket calculator with (ideally) four memory registers. If the current angle φ gives a point with coordinates $(x_0 + iy_0)$, its square has the angle 2φ with coordinates $(x_1 + iy_1)$ where $x_1 = x_0^2 - y_0^2$ and $y_1 = 2x_0y_0$. (This is the 'iteration formula'.) As for the angles, $\delta\varphi$ is measured as the vertical distance (y) on a circle of unit radius; in other words as partial perimeter/radius, the measurement called *radians*, where a complete rotation, with perimeter 2π, has angle 2π radians. So an angle of 30° is $180°/6 = \pi/6$ radians ≈ 0.5236 radians, and a 1/1024th part of that is $0.523599/1024 = 0.0005113...$ radians. So the starting values are $x_0 = 1.0$, $y_0 = 0.00051133... = \varphi$. After ten successive squarings, using the 'iteration formula', we get $\varphi = 0.5236...$, $x_1 = 0.86614...$ and $y_1 = 0.50006...$

These values are very close to $\varphi = \pi/6$ radians, $x = \sqrt{3}/2 = \cos\varphi$, $y = 1/2 = \sin\varphi$ which are the coordinates of the vertex of a triangle whose hypotenuse has the length $\sqrt{[\cos^2\varphi + \sin^2\varphi]} = \sqrt{[(3/4) + (1/4)]} = 1$. The iteration started near the x-axis, and the intermediate parameters followed the anti-clockwise rotation of a point round a circle of unit radius. This is therefore what $\exp(i\varphi)$ implies, so:

$$\exp(i\varphi) = \cos\varphi + i.\sin\varphi \tag{19.2}$$

Also $\exp(-i\varphi) = \cos(-\varphi) + i.\sin(-\varphi) = \cos\varphi - i.\sin\varphi$, showing that $\cos\varphi$ is the even part of $\exp(i\varphi)$, and $\sin\varphi$ is its odd part. Moreover, we can solve the last two equations for $\cos\varphi$ and $\sin\varphi$. By adding the equations, we get:

$$\cos\varphi = \tfrac{1}{2}(\exp(i\varphi) + \exp(-i\varphi)) \tag{19.3}$$

and, by subtracting the equations:

$$\sin\varphi = (\exp(i\varphi) - \exp(-i\varphi))/2i \tag{19.4}$$

After a rotation of $\varphi = \pi$, the point will clearly have reached the x-axis again at $x = -1$. So $\exp(i\pi) = e^{i\pi} = -1$, a remarkable connection between four basic numbers.

19.5 Calculus

To understand FTs we need the calculus, a set of methods for calculating the results of continuously changing processes. (We met a simple, approximate, introductory version of it in Chapter 5, §5.5.) Observing such processes requires film or video, which record very similar images in rapid succession. The sense of motion is conveyed by very slight changes between successive images, so these changes involve very small movements or infinitesimal shifts. We write them as δx, implying that $(\delta x)^2$ is negligible compared with δx. Thus equations in infinitesimals have only the first power, so they are called linear. Linearization is a great simplification, and linear equations are the simplest possible of their type. This simplicity makes the *differential calculus* the study of very simple equations. But this simplification has eventually to be paid for. To make the answers relevant, we must finally add up all the infinitesimals, the part called *integral calculus*. As one might expect, it is hard to turn differentials into integrals.

19.5.1 Areas and Gradients

To handle areas of *curves*, we cut them into vertical infinitesimal strips, of height y (which represents the function) and width δx. We introduce this way of handling curves in Fig. 19.3, where part (a) shows a simple straight-line function $y = kx$. The part of the curve up to x is a 45° triangle which is half of a rectangle with horizontal length x and vertical length $y(x)$.

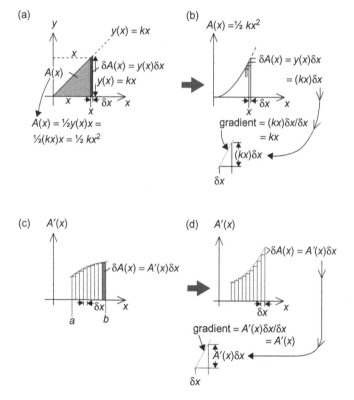

Figure 19.3: (a) Gradient function $y = kx$. **(b)** Its area function $\frac{1}{2}kx^2$. **(c)** General gradient function or derivative. **(d)** Area function of (c).

This is half of the rectangle with sides x and $y(x)$, and which therefore has area $xy(x)$. So the triangle, with half that size, has the area $A(x) = \frac{1}{2}xy(x) = \frac{1}{2}x(kx) = \frac{1}{2}kx^2$. Thus there are two functions implicit in this curve, the curve itself $(y(x))$ and its 'area function' $A(x)$ (§5.5). Part (b) shows a plot of the area function $A(x) = \frac{1}{2}kx^2$.

Note that the area function depends on the starting point chosen for the area (as well as the finishing point). In (a) there was a logical starting point but, if we had chosen the starting point at $x = a$, we would have had $A(x) = \frac{1}{2}kx^2 - \frac{1}{2}ka^2$. Thus, in general, one writes $A(x) = \frac{1}{2}kx^2 + c$ (c is an unknown constant), unless the boundaries have been made explicit. If the area were measured from a to b, it would be $A(x) = \frac{1}{2}kb^2 - \frac{1}{2}ka^2 = \frac{1}{2}k(b^2 - a^2)$.

Thus the simple curve $y = kx$ (left, (a)) has its area function $A(x) = \frac{1}{2}kx^2$ (right, (b)). But this isn't the only connection between these curves. Just as the left curve is divided into infinitesimal strips, so the right curve is built up from corresponding strips, with the same width but with *heights* corresponding to the *areas* of the strips of the left curve. As we see in the little diagram placed just below the right curve (b), its gradient equals kx, the same function as the curve on the left (a).

Summarizing, the right curve (b) is the 'area curve' of the left curve (a); and the left curve is the 'gradient curve' of the right curve. We should fix these important ideas in an appropriate notation. First, the area is a sum of infinitesimal strips, and the mathematical symbol for a sum is taken from its first letter. In the Greek alphabet, this letter (capital or upper-case) is 'sigma': Σ. So we write:

$$A(x) \approx \sum y.\delta x = \sum kx.\delta x \approx \frac{1}{2}kx^2$$

because we know the area of a triangle.

Also the 'gradient function' is:

$$\text{gradient}(A) = A'(x) \approx \delta A(x)/\delta x = \delta(\frac{1}{2}kx^2)/\delta x \approx kx$$

where $A'(x)$ means the gradient of the 'area curve' $A(x)$. So (as in (c), (d)) the two equations can be combined as:

$$A(x) \approx \sum \delta A(x) = \sum(\delta A(x)/\delta x).\delta x \approx \sum A'(x).\delta x \qquad (19.5)$$

Note that we usually start with a function $y(x)$, which we here call $A'(x)$. To use this entire system, we need to know the function $A(x)$ of which $y(x)$ is the gradient function $A'(x)$. If we don't know that, we shall have to find $A(x)$ by using a computer to add all the infinitesimal strips.

19.5.2 Calculus: Essential Details

Constants
The constant k is the same factor for every infinitesimal strip, so we can write $\sum kx.\delta x = k\sum x.\delta x$. Again, because $\frac{1}{2}k$ is a constant factor, we can write $\delta(\frac{1}{2}kx^2)/\delta x \approx \frac{1}{2}k\delta(x^2)/\delta x$.

Finding Gradients

For example, to find $(\delta(x^2)/\delta x)$ algebraically, we first notice that $\delta(x^2)$ means 'the small increase in x^2 when x has a small increase, δx'; so it is $(x + \delta x)^2 - x^2$. Now, from the earlier example $(1.001)^2 \approx 1 + 2(0.001)$ we get the equation $(x + \delta x)^2 \approx x^2 + 2x.\delta x$, from which we subtract x^2 from both sides to get $(x + \delta x)^2 - x^2 \approx 2x.\delta x$, or $\delta(x^2) \approx 2x.\delta x$, or $\delta(x^2)/\delta x \approx 2x$. We shall later give other examples, based on the same idea that $\delta(x^2) = (x + \delta x)^2 - x^2$, i.e. $\delta[f(x)] = f(x + \delta x) - f(x)$.

Area Boundaries (Limits)

Many mathematical curves extend to infinity in both directions, $x = -\infty$ to $x = +\infty$. But we are usually interested in areas of only some *restricted* portion of the curve, say from $x = a$ to $x = b$. Then we write (for x^2) the limits (a to b) as follows:

$$\sum_{a}^{b} \delta(x^2) = \left[x^2\right]_a^b = b^2 - a^2 \dots \tag{19.6}$$

Limits

So far, we have some very useful equations, but they are only *approximate* (hence $\approx$ instead of $=$), like all equations involving infinitesimals. This caused much argument after the invention of calculus, as there seemed to be only two equally bad alternatives. We could continue regarding infinitesimals as *very small* numbers, in which case we only have approximate equations, an ugly blotch on an exact subject. Or, to make the equations more accurate, we could make the infinitesimals *actually* zero; but then we would have absurdities like $\delta(x^2)/\delta x \approx 2x$ becoming $0/0 = 2x$. Eventually, however, mathematicians developed a third, more subtle, alternative. They start with infinitesimals as very small numbers, but then make them smaller and smaller, noticing what the equation would be when the infinitesimals eventually became zero. So we take *limiting cases*, and adopt a different notation for them. Infinitesimals like δx become written as dx, (called a **differential**) and sums like $\Sigma \delta A(x)$ become written as $\int dA(x)$, called an **integral**. (The 'integral sign' $\int$ is also the first letter for 'sum', but written in 17th century handwriting, an idea of Leibniz.)

So the equation (19.5) can be written as:

$$A(x) = \int dA(x) = \int (dA(x)/dx).dx = \int A'(x).dx \dots \tag{19.7}$$

Here the gradient $A'(x) = dA(x)/dx$ is also known as the **differential coefficient** (or **derivative**) of $A(x)$.

19.5.3 Calculus: Important Examples

'Standard Integrals'

Recall that we found $(1.001)^2 \approx 1 + 2(0.001)$. So we might suspect that $(1.001)^n \approx 1 + n(0.001)$, and we would be right, as can be checked for $n = 3, 4, \dots$ by multiplication or, for

$n = \frac{1}{2}$, by taking the square root. These examples suggest that $(1 + \delta x)^n \approx 1 + n\delta x$. We can derive this for the case when $n =$ is a positive integer as follows: $(1 + \delta x)^n = (1 + \delta x)(1 + \delta x)$ ...$(1 + \delta x) \approx 1 + n\delta x$. As each bracket contributes a δx, and there are n brackets. But we can extend this to $(x + \delta x)^n = (x + \delta x)(x + \delta x)...(x + \delta x) \approx x^n + nx^{n-1}\delta x$. Now $\delta(x^n) = (x + \delta x)^n - x^n = nx^{n-1}\delta x$. So we have $\delta(x^n) = nx^{n-1}\delta x$. In the limit, we get the differential:

$$d(x^n) = nx^{n-1}dx, \text{ or } d(x^n)/dx = nx^{n-1} \tag{19.8}$$

This is a 'standard differential' which is useful for differentiating more complicated expressions.

We can also incorporate this into an equation like (19.7). We put $A'(x) = dA(x)/dx = d(x^n)/dx = nx^{n-1}$, so $A(x) = x^n$ and $A'(x) = nx^{n-1}$. So now equation (19.7) becomes $x^n = \int d(x^n) = \int (d(x^n)/dx).dx = \int nx^{n-1}.dx = n\int x^{n-1}.dx$, so $x^n = n\int x^{n-1}.dx$ and we finally we get the integral $\int x^{n-1}.dx = x^n/n$. Replacing $n - 1$ by m, $n = m + 1$ and:

$$\int x^m.dx = x^{m+1}/(m + 1)... \tag{19.9}$$

This formula is probably the most famous example of a 'standard integral', one of the tools for evaluating integrals by purely mathematical means.

Exponential Integral

Recall that $\delta[f(x)] = f(x + \delta x) - f(x)$, so $\delta(\exp(x)) = \exp(x + \delta x) - \exp(x)$. Now $\exp(x + \delta x) = \exp(x).\exp(\delta x) = \exp(x).(1 + \delta x) = \exp(x) + \delta x.\exp(x)$. So $\delta(\exp(x)) = \exp(x + \delta x) - \exp(x) = \delta x.\exp(x)$. That means that $\delta(\exp(x))/\delta x = \exp(x)$. In the limit, this becomes $d(\exp(x)) = \exp(x).dx$, or:

$$d(\exp(x))/dx = \exp(x) \tag{19.10}$$

Thus the gradient of the exponential function is the exponential function itself. But there's something else whose gradient equals the thing itself (here dots imply ordinary multiplication):

$$f(x) = 1 + x/1 + x^2/(1.2) + x^3/(1.2.3) + x^4/(1.2.3.4) + ... \tag{19.11}$$

Now we try differentiating the entire equation. The effect on $f(x)$ is to get $f'(x)$. The number 1 is a constant, so it can't change with x, so its derivative is zero. Using equation 19.8, we get the derivative of $x^2 = 2x$, that of $x^3 = 3x^2$, and that of $x^4 = 4x^3$. So we get:

$$f'(x) = 0 + 1 + 2x/(1.2) + 3x^2/(1.2.3) + 4x^3/(1.2.3.4) + ...$$

When we cancel the numbers 2 (and 3 and ...) that multiply the powers of x, we finally end with exactly the same series that gave $f(x)$! Thus $f'(x) = f(x)$, which is exactly like equation (19.10). So the series in (19.11) is simply e^x, which fits with our earlier finding that $e^{\delta x} \approx 1 + \delta x$.

$$\exp(x) = 1 + x/1 + x^2/(1.2) + x^3/(1.2.3) + x^4/(1.2.3.4) + \ldots \tag{19.12}$$

If the reader tries replacing x in (19.12) by ix, and every i^2 is replaced by -1, the power series will split into real and imaginary terms, with the real terms giving:

$$\cos x = 1 - x^2/(1.2) + x^4/(1.2.3.4) - \ldots \tag{19.13}$$

and the imaginary terms giving:

$$\sin x = x - x^3/(1.2.3) + x^5/(1.2.3.4.5) - \ldots \tag{19.14}$$

(Of course, the angle x must be in radians, where $180° = \pi$ radians. Also, (1.2.3) is usually written 3!, and similarly for 2, 4,…)

However, when we calculate FTs, we shall need $\delta(\exp(ax))$, where a is a constant. To get this, we start with $\delta(\exp(y))/\delta y = \exp(y)$. Then we let $y = ax$, which gives us $\delta(y)/\delta x \approx a$ or $\delta(ax)/\delta x \approx a$. Then we substitute ax for y in the first equation, getting $\delta(\exp(ax))/\delta(ax) = \exp(ax)$. Finally, we multiply by $\delta(ax)/\delta x \approx a$; $(\delta(\exp(ax))/\delta(ax))(\delta(ax)/\delta x) \approx (\exp(ax))(a) = \delta(\exp(ax))/\delta(x)$. Thus:

$$\mathrm{d}(\exp(ax))/\mathrm{d}x = a.\exp(ax) \tag{19.15}$$

We can also incorporate this into equation (19.7). We put $A'(x) = \mathrm{d}A(x)/\mathrm{d}x = \mathrm{d}(\exp(ax))/\mathrm{d}x = a.\exp(ax)$, so $A(x) = \exp(ax)$ and $A'(x) = a.\exp(ax)$. So now equation (19.7) becomes $\exp(ax) = \int \mathrm{d}(\exp(ax)) = \int(\mathrm{d}(\exp(ax))/\mathrm{d}x).\mathrm{d}x = \int a.\exp(ax).\mathrm{d}x = a\int \exp(ax).\mathrm{d}x$, so $\exp(ax) = a\int \exp(ax).\mathrm{d}x$ and we finally get the integral:

$$\int \exp(ax).\mathrm{d}x = \exp(ax)/a + c \tag{19.16}$$

19.5.4 The Fourier Transform

Consider the complex exponential $y(x) = \exp(2\pi \mathrm{i}xX)$. If we substitute $(x + n/X)$ for x, $y(x + n/X) = \exp[2\pi \mathrm{i}X(x + n/X)] = \exp[2\pi \mathrm{i}Xx + 2\pi \mathrm{i}Xn/X] = \exp[2\pi \mathrm{i}Xx + 2\pi \mathrm{i}n] = \exp(2\pi \mathrm{i}n)\exp(2\pi \mathrm{i}xX) = \exp(2\pi \mathrm{i}n).y(x) = 1^n.y(x) = y(x)$. (Recall that $\exp(2\pi \mathrm{i}) = 1$ from Euler's formula.) So $\exp(2\pi \mathrm{i}xX)$ repeats with a period of n/X. If we take equation (19.16) and put $a = 2\pi \mathrm{i}X$, then $\int \exp(ax).\mathrm{d}x = \int \exp(2\pi \mathrm{i}Xx).\mathrm{d}x = \exp(2\pi \mathrm{i}Xx)/(2\pi \mathrm{i}X)$. So:

$$\int_{-b}^{b} \exp(2\pi \mathrm{i}Xx).\mathrm{d}x = [\exp(2\pi \mathrm{i}Xx)/(2\pi \mathrm{i}X)]_{-b}^{b} = (\exp(2\pi \mathrm{i}Xb)/(2\pi \mathrm{i}X)) - (\exp(-2\pi \mathrm{i}Xb)/(2\pi \mathrm{i}X))$$

$$= b(\exp(2\pi \mathrm{i}Xb) - \exp(-2\pi \mathrm{i}Xb))/(2\pi \mathrm{i}Xb) = b(\exp(\mathrm{i}u) - \exp(-\mathrm{i}u))/2\mathrm{i}u, \text{ where } u = (2\pi Xb).$$

Now $(\exp(\mathrm{i}u) - \exp(-\mathrm{i}u))/2\mathrm{i} = \sin u$, (see equation (19.4)) so $(\exp(\mathrm{i}u) - \exp(-\mathrm{i}u))/2\mathrm{i}u = (\sin u)/u = \mathrm{sinc}(u) = \mathrm{sinc}(2\pi Xa)$. Thus:

$$\int_{-b}^{b} \exp(2\pi \mathrm{i}Xx).\mathrm{d}x = b.\mathrm{sinc}(2\pi \mathrm{i}Xb) \tag{19.17}$$

This is the FT of a rectangle from $-b$ to $+b$ (see §5.2.1 and Fig. 5.1). If we define rect(w) as a rectangle function of unit height and width w, centered on the origin, then:

$$\int_{-b}^{b} \exp(2\pi i Xx).\mathrm{d}x = \int_{-\infty}^{\infty} \mathrm{rect}(2b).\exp(2\pi i Xx).\mathrm{d}x$$

so the integral on the right is the FT of rect($2b$). Thus:

$$\int_{-\infty}^{\infty} \mathrm{rect}(2b).\exp(2\pi i Xx).\mathrm{d}x = b.\mathrm{sinc}(2\pi i Xb) \qquad (19.18)$$

is the FT of rect($2b$). In general, the FT of $f(x)$ is:

$$F(x) = \int_{-\infty}^{\infty} f(x).\exp(2\pi i Xx).\mathrm{d}x \qquad (19.19)$$

This is the standard 1D form of the Fourier transform, studied intuitively in Chapter 3. The inversion formula is:

$$f(X) = \int_{-\infty}^{\infty} F(x).\exp(-2\pi i Xx).\mathrm{d}X \qquad (19.20)$$

The changed sign of $2\pi i Xx$ is the basis of the asymmetric arrows in Figs 3.14 and 3.15.

Elementary Matrices

Chapter Outline

The main applications of matrices are in geometry (e.g. adjusting molecular orientations to fit something else) and in statistics (e.g. least-squares data-fits, as in Chapter 11 and in MSA; see Chapter 17).

20.1 Introducing Matrices

20.1.1 From One-Dimension to Two-Dimensions: Vectors

We start by looking at the problem of extending mathematics from 1D to 2D. The obvious change from one direction to two means that, whereas 1D algebra uses single variables (called **scalars**), 2D algebra must unavoidably use number-pairs. But there are alternative ways of forming these pairs. We already saw how multiplication by $i = \sqrt{-1}$ leads to one useful representation of 2D space, with *complex numbers* like $a + ib$, where the real and imaginary parts, a and b, act as the x- and y-coordinates.

Complex numbers form a complete system of algebra in 2D, and we might expect to generalize them for 3D. However, despite the development of 'quaternions' (the 3D equivalent of complex numbers) that proved very difficult, and a different method was used. Obviously, it still needs number-pairs for 2D, but they're called **vectors**, written as (a, b). Although they follow the same rules as complex numbers when they're added or subtracted, e.g. $(a, b) + (x, y) = (a + x, b + y)$, there's a big change with multiplication. The multiplication of two complex numbers gives another complex number, but the multiplication of two vectors to give another vector doesn't turn out to be very useful[1]. Instead, there's a great need for a rule whereby two vectors multiply to give a scalar (so this rule is called the *scalar product*). As always when extending rules into new territories, we have to make sure that the new rule is in harmony with the previous rule in the old territory. Now, if 'a' is a scalar (like 2 or 3), $a \times a = a^2$, so we need to ensure that the same applies to $(a, 0)$, which is entirely on the x-axis. Thus the scalar product is defined so that $(a, 0) \cdot (a, 0) = a^2$, and also $(0, b) \cdot (0, b) = b^2$, so $(a, b) \cdot (a, b) = a^2 + b^2$, and also $(a, b) \cdot (x, y) = ax + by$. It can be shown that the scalar product of two vectors equals the product of their lengths, times the cosine of the angle between them. This works when they're parallel, as the angle is zero so its cosine is one; when they're perpendicular, like (a, b) and $(b, -a)$; and for other simple situations which can be tested by trying different values of x, y in $(a, b) \cdot (x, y)$.

20.1.2 Equation-Pairs and Matrices

We shall therefore hope to do the same sort of 1D to 2D extension with simple equations like $ax = b$, where the solution is $x = b/a$ or ba^{-1} or $a^{-1}b$. (A quicker way to get this is to multiply both sides of $ax = b$ by a^{-1}.) How do we need to extend this when b is not a scalar but a vector, a number-pair? A number-pair gives us an equation-pair (plotted in frame (a) of Fig. 20.1 where line aa is equation (a) below, and line bb is equation (b). So we examine these to see how a better notation develops from left to right, *via* the scalar product version in the middle.

$$\begin{array}{lll} \text{(a)} & 7x + 4y = 15; & (7, 4) \cdot (x, y) = 15; \\ \text{(b)} & 4x + y = 6; & (4, 1) \cdot (x, y) = 6; \end{array} \quad \begin{bmatrix} 7 & 4 \\ 4 & 1 \end{bmatrix}\begin{bmatrix} x \\ y \end{bmatrix} = \begin{bmatrix} 15 \\ 6 \end{bmatrix}$$

The last, best, version $\begin{bmatrix} 7 & 4 \\ 4 & 1 \end{bmatrix}\begin{bmatrix} x \\ y \end{bmatrix} = \begin{bmatrix} 15 \\ 6 \end{bmatrix}$

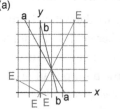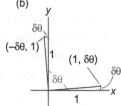

Figure 20.1: (a) Plot of the equations (a, b) as lines aa, bb. **(b)** Rotation of unit vectors by an infinitesimal matrix.

---

[1] Although a *vector product* does exist, it is far less useful than the *scalar product*.

has a matrix: $A = \begin{bmatrix} 7 & 4 \\ 4 & 1 \end{bmatrix}$

multiplying the vector $x = \begin{bmatrix} x \\ y \end{bmatrix}$

to give another vector: $b = \begin{bmatrix} 15 \\ 6 \end{bmatrix}$.

20.1.3 Matrix Multiplication

Before we solve these equations, we should say something about matrix multiplication. The pair of equations at the beginning of the line could be written in matrix form as:

$$\begin{bmatrix} 7x + 4y \\ 4x + y \end{bmatrix}.$$

In the equation $\begin{bmatrix} 7 & 4 \\ 4 & 1 \end{bmatrix}\begin{bmatrix} x \\ y \end{bmatrix} = \begin{bmatrix} 7x + 4y \\ 4x + y \end{bmatrix} = \begin{bmatrix} 15 \\ 6 \end{bmatrix}$,

we see that we form scalar products of the rows of the matrix with the column-vector

$$\begin{bmatrix} x \\ y \end{bmatrix},$$

since $(7, 4) \cdot (x, y) = 7x + 4y$ and $(4, 1) \cdot (x, y) = 4x + y$.

We can multiply matrices using the same rule, so

$$\begin{bmatrix} 7 & 4 \\ 4 & 1 \end{bmatrix}\begin{bmatrix} 2 & -1 \\ 1 & 2 \end{bmatrix} = \begin{bmatrix} 18 & 1 \\ 9 & -2 \end{bmatrix}.$$

Here the 18 comes from the scalar product of the upper row [7, 4] from the left matrix, and the left column [2, 1] from the right matrix; so: $(7, 4) \cdot (2, 1) = (7 \times 2) + (4 \times 1) = 14 + 4 = 18$.

Thus, if the multiply two (2×2) matrices AB, we consider A to consist of two row-vectors a_1, a_2; and B to consist of two column-vectors b_1, b_2; and we form the scalar products as follows:

$$AB = \begin{bmatrix} a_1 \\ a_2 \end{bmatrix}[b_1, b_2] = \begin{bmatrix} a_1 \cdot b_1, a_1 \cdot b_2 \\ a_2 \cdot b_1, a_2 \cdot b_2 \end{bmatrix} \tag{20.1}$$

It is easily shown that this rule has the inconvenience that AB differs from BA, i.e. the order of multiplication affects the result (a feature known as **non-commutativity**). Therefore, when we multiply any equation by a matrix, we must specify the multiplication sequence by saying 'left-multiplication' or 'right-multiplication'.

20.2 Matrix Inversion

20.2.1 Solving a Simple Matrix Equation

Thus we now have the 2D equivalent, $Ax = b$, to the simple 1D equation $ax = b$, whose solution is $x = a^{-1}b$. We would be able to solve $Ax = b$ if we could calculate the inverse matrix A^{-1}. There are of course rules for doing this, but it is an operation almost always left to a computer, for which many 'matrix inversion' subroutines exist. (See e.g. Press et al., 2007, chapter 2.) To solve the original equation-pair, we need a suitable matrix B, and post-multiply by A as shown below.

$$BA = \begin{bmatrix} -1/9 & 4/9 \\ 4/9 & -7/9 \end{bmatrix} \begin{bmatrix} 7 & 4 \\ 4 & 1 \end{bmatrix} = \begin{bmatrix} 1 & 0 \\ 0 & 1 \end{bmatrix}$$

So: $\begin{bmatrix} -1/9 & 4/9 \\ 4/9 & -7/9 \end{bmatrix} = \begin{bmatrix} 7 & 4 \\ 4 & 1 \end{bmatrix}^{-1} = A^{-1}$

Thus we can multiply equation $Ax = b$ by A^{-1}, getting:

$$\begin{bmatrix} -1/9 & 4/9 \\ 4/9 & -7/9 \end{bmatrix} \begin{bmatrix} 7 & 4 \\ 4 & 1 \end{bmatrix} \begin{bmatrix} x \\ y \end{bmatrix} = \begin{bmatrix} -1/9 & 4/9 \\ 4/9 & -7/9 \end{bmatrix} \begin{bmatrix} 15 \\ 6 \end{bmatrix} = \begin{bmatrix} 9/9 \\ 18/9 \end{bmatrix} = \begin{bmatrix} 1 \\ 2 \end{bmatrix}$$

The left side is: $\begin{bmatrix} 1 & 0 \\ 0 & 1 \end{bmatrix} \begin{bmatrix} x \\ y \end{bmatrix} = \begin{bmatrix} 1 \\ 2 \end{bmatrix}$, so $x = 1$ and $y = 2$.

The matrix: $\begin{bmatrix} 1 & 0 \\ 0 & 1 \end{bmatrix}$ is called the **unit matrix**, the matrix equivalent of 1 (in 2D), since multiplication by it leaves any other 2D matrix unchanged. It is usually given the symbol I.

This is how pairs of equations are solved in matrix notation, given a means of calculating the inverse matrix.

20.2.2 *Error-Sensitivity*

In the 1D equation $ax = b$, with the solution $x = a^{-1}b$, we suppose both a and b are experimental measurements and subject to error. How badly will these errors affect the solution? An error in b causes a proportionate error in x, but the effect of an error in a depends on the actual size of a. There can be a serious problem if a is very small for, if an error makes it zero, x could become infinite. Thus the reliability of the solution depends on a being much bigger than any likely error.

We would expect a similar situation to apply in the 2D case, so A should be bigger than any likely error. But A is a matrix with four numbers (elements), so what is meant by its 'size'? The four numbers form two vectors, and the matrix's 'size' should measure the area they enclose. For the matrix

$$A = \begin{bmatrix} a & b \\ c & d \end{bmatrix},$$

that area is $(ad - bc)$ which is called the **matrix's determinant**, written

$$\begin{vmatrix} a & b \\ c & d \end{vmatrix}.$$

The determinant of a matrix appears in the denominator in its inverse. Thus, in the previous section, the matrix

$$A = \begin{bmatrix} 7 & 4 \\ 4 & 1 \end{bmatrix}$$

has the determinant

$$|A| = \begin{vmatrix} 7 & 4 \\ 4 & 1 \end{vmatrix} = 7 - 16 = -9,$$

and the inverse matrix has elements that are written as ninths.

After that numerical example, we consider the general case of a (2D) matrix inverse. If

$$A = \begin{bmatrix} a & b \\ c & d \end{bmatrix}, \ |A| = ad - bc,$$

what is A^{-1}?

Note that: $\begin{bmatrix} a & b \\ c & d \end{bmatrix}\begin{bmatrix} d & -b \\ -c & a \end{bmatrix} = \begin{bmatrix} ad - bc & 0 \\ 0 & ad - bc \end{bmatrix} = (ad - bc)\begin{bmatrix} 1 & 0 \\ 0 & 1 \end{bmatrix} = |A|I.$

So:

$$A\begin{bmatrix} d & -b \\ -c & a \end{bmatrix} = |A|I, \text{ so } A^{-1} = \frac{\begin{bmatrix} d & -b \\ -c & a \end{bmatrix}}{|A|},$$

(20.2)

where

$$|A| = ad - bc$$

(20.3)

So we can test if a matrix is 'effectively zero' (when it is said to be **singular**) by seeing if its determinant is zero. In that case, and if it has eigenvectors (§11.2.1, §20.3.1) at all, at least one of them will be zero. A simple example of a singular matrix is:

$$\begin{bmatrix} 1 & 2 \\ 3 & 6 \end{bmatrix},$$

whose determinant is

$$\begin{vmatrix} 1 & 2 \\ 3 & 6 \end{vmatrix} = (1 \times 6) - (3 \times 2) = 0.$$

It also has a zero eigenvalue, since

$$\begin{bmatrix} 1 & 2 \\ 3 & 6 \end{bmatrix}\begin{bmatrix} x \\ y \end{bmatrix} = 0$$

implies both $x + 2y = 0$ and $3x + 6y = 0$, which are the same equation (the second simply being the first multiplied throughout by 3), so the vector

$$\begin{bmatrix} -2y \\ y \end{bmatrix} \text{ or } \begin{bmatrix} -2x \\ x \end{bmatrix}$$

is its eigenvector (i.e. the straight line $y = -x/2$).

20.2.3 *Infinitesimal Matrices*

Symmetric matrices, although important, are only one type of matrix. Returning for a moment to an earlier topic, we consider the question of infinitesimal matrices. Recall that the interest of infinitesimal objects is not that they scarcely exist, but that they represent only very small perturbations of the *status quo*. So they're very small in isolation, whether added or (even smaller after) multiplication. But they become significant when added to a large quantity like 1, so we consider quantities like $1 + \delta x$. This is also the type of infinitesimal matrix we want to consider: a matrix very close to the *identity matrix* §20.2.1 that leaves everything the same (the matrix equivalent to 1). In 2D, that matrix (also called the unit matrix) is

$$\begin{bmatrix} 1 & 0 \\ 0 & 1 \end{bmatrix}.$$

If its perturbation is going to be *nearly* the same, we need only change the zeroes to infinitesimals. But there are two zeroes and, in general, they won't become the same infinitesimal. So we set them to ε_1 and ε_2, and the 'infinitesimal matrix' is

$$\begin{bmatrix} 1 & \varepsilon_1 \\ \varepsilon_2 & 1 \end{bmatrix}.$$

It represents the effect of a *very small* distortion. Now, we can split this matrix into its even and odd components, as we did with functions in §2.2:

$$\begin{bmatrix} 1 & \varepsilon_1 \\ \varepsilon_2 & 1 \end{bmatrix} = \frac{1}{2} \begin{bmatrix} 1 & \varepsilon_1 + \varepsilon_2 \\ \varepsilon_2 + \varepsilon_2 & 1 \end{bmatrix} + \frac{1}{2} \begin{bmatrix} 1 & \varepsilon_1 - \varepsilon_2 \\ \varepsilon_2 - \varepsilon_1 & 1 \end{bmatrix} \tag{20.4}$$

where the first component is even and the second one is odd. (This equation can be checked by adding the even and odd components, which will give the matrix on the left.) But in what respect are these components 'even' and 'odd'? The first matrix on the right is symmetrical, like all the matrices we have been considering so far in this chapter. But the second matrix is anti-symmetrical, as $\varepsilon_1 - \varepsilon_2 = -(\varepsilon_2 - \varepsilon_1)$; so we can write it as

$$\begin{bmatrix} 1 & -\delta\theta \\ \delta\theta & 1 \end{bmatrix}.$$

We see in the next section what this anti-symmetrical matrix does.

20.2.4 *Rotation Matrices*

We test the unsymmetrical matrix by using it to multiply unit vectors on the *x*- and *y*-axes.

$$\begin{bmatrix} 1 & -\delta\theta \\ \delta\theta & 1 \end{bmatrix} \begin{bmatrix} 1 \\ 0 \end{bmatrix} = \begin{bmatrix} 1 \\ \delta\theta \end{bmatrix} \tag{20.5}$$

and:

$$\begin{bmatrix} 1 & -\delta\theta \\ \delta\theta & 1 \end{bmatrix}\begin{bmatrix} 0 \\ 1 \end{bmatrix} = \begin{bmatrix} -\delta\theta \\ 1 \end{bmatrix}$$

These points are shown in Fig. 20.1 (b) p. 402, and it's clear that the matrix is rotating the points anti-clockwise by an angle $\delta\theta$ radians.

We note that its effect on the x-axis vector $(1, 0)$ is to produce the vector $(1, \delta\theta)$, which is exactly like multiplication by $(1 + i\delta\theta)$. And indeed the effect of multiplying this infinitesimal matrix by itself many times to get a substantial rotation, is the same as multiplying the complex exponentials $\exp(i\delta\theta)$ by itself. So we can rotate the vector (x, y) by an angle θ radians, by multiplying the complex number $(x + iy)$ by $\exp(i\theta)$, as follows: $\exp(i\theta)(x + iy) = (\cos\theta + i \sin\theta)(x + iy) = (x \cos\theta - y \sin\theta) + i(x \sin\theta + y \cos\theta)$. So the new x-coordinate of the rotated point is $x' = x \cos\theta - y \sin\theta$, and the new y-coordinate of the rotated point is $y' = x \sin\theta + y \cos\theta$. These equations can be written in matrix form as:

$$\begin{bmatrix} x' \\ y' \end{bmatrix} = \begin{bmatrix} \cos\theta & -\sin\theta \\ \sin\theta & \cos\theta \end{bmatrix}\begin{bmatrix} x \\ y \end{bmatrix} \tag{20.6}$$

or $x' = Rx$. The matrix R is the simplest form of a **rotation matrix**, a special type of matrix, at least as important as the symmetric matrices we have been considering. Note that it is anti-symmetric, like the infinitesimal version we had in equation (20.5). (However, that feature does not necessarily apply to 3D rotation matrices.)

There's an interesting way to view these equations for x' and y'. We can regard them as the projections of the (x, y)-vector onto two perpendicular unit vectors, one being along the x' axis and with coordinates $(\cos\theta, -\sin\theta)$, the other being along the y' axis and with coordinates $(\sin\theta, \cos\theta)$. The projection of a vector onto a line, is the same as its scalar product with a unit vector along that line, as both are the vector-length times the cosine of the angle between them. This viewpoint leads us to expect that the scalar product of the rotation matrix's row-vectors is zero; and $(\cos\theta, -\sin\theta) \cdot (\sin\theta, \cos\theta) = 0$, and the same applies to its column-vectors $(\cos\theta, \sin\theta) \cdot (-\sin\theta, \cos\theta) = 0$. (Such vectors are described as **orthogonal**.) It also requires that both the row- and column-vectors have unit length $\cos^2\theta + \sin^2\theta = 1$. (Such vectors are described as **normalized**; and a matrix with such row/column vectors is itself *orthogonal* and *normalized*, or **orthonormal**.)

Finally, the **transpose** of the rotation matrix (reflecting it in its leading diagonal, i.e. turning rows into columns and vice-versa) yields the inverse matrix:

$$\begin{bmatrix} \cos\theta & -\sin\theta \\ \sin\theta & \cos\theta \end{bmatrix}^T = \begin{bmatrix} \cos\theta & \sin\theta \\ -\sin\theta & \cos\theta \end{bmatrix} = \begin{bmatrix} \cos(-\theta) & -\sin(-\theta) \\ \sin(-\theta) & \cos(-\theta) \end{bmatrix}$$

so the rotation angle is $-\theta$, the inverse operation. (It is easily verified that a matrix product $C = AB$ gives $C^T = B^T A^T$.) Moreover,

$$\begin{bmatrix} \cos\theta & -\sin\theta \\ \sin\theta & \cos\theta \end{bmatrix}^T \begin{bmatrix} \cos\theta & -\sin\theta \\ \sin\theta & \cos\theta \end{bmatrix} = \begin{bmatrix} 1 & 0 \\ 0 & 1 \end{bmatrix},$$

the unit matrix, usually written I. This feature follows from a requirement of a rotation matrix, that it should leave unchanged the length of any vector, e.g. $(x^2 + y^2)$ for the vector $x = (x, y)$. Now[2]

$$x^2 + y^2 = (x, y) \cdot (x, y) = [x, y] \cdot \begin{bmatrix} x \\ y \end{bmatrix} = x^T x,$$

$x' = Rx$, and so $(x')^T = x^T R^T$, and $x'^T x' = x^T R^T R x = x^T x$ if $R^T R = I$, the unit matrix. But, when $R^T R = I$, we can multiply both sides on the right by R^{-1}. The left side of the equation becomes $R^T R\, R^{-1} = R^T$, and the right side of the equation becomes $I R^{-1} = R^{-1}$. So we have $R^T = R^{-1}$; the transpose is the inverse.

20.3 Eigenvectors

20.3.1 Eigenvalues and -vectors

We have seen (in §11.2.1) that, for symmetric matrices, there exist special vectors called eigenvectors. If v_1 is an eigenvector of A, then $Av_1 = \lambda_1 v_1$ which is a vector parallel to v_1 but λ_1 times as long. The scale factor λ_1 is the corresponding eigenvalue. The symmetric matrix A above, being 2D (two rows and two columns) has two eigenvector/-value pairs. One is $(2y, y)$:

$$\begin{bmatrix} 7 & 4 \\ 4 & 1 \end{bmatrix} \begin{bmatrix} 2y \\ y \end{bmatrix} = \begin{bmatrix} 18y \\ 9y \end{bmatrix} = 9 \begin{bmatrix} 2y \\ y \end{bmatrix}$$

So $\begin{bmatrix} 2y \\ y \end{bmatrix}$ is the eigenvector and 9 is its corresponding eigenvalue. The other is:

$$\begin{bmatrix} 7 & 4 \\ 4 & 1 \end{bmatrix} \begin{bmatrix} x \\ -2x \end{bmatrix} = \begin{bmatrix} -x \\ 2x \end{bmatrix} = - \begin{bmatrix} x \\ -2x \end{bmatrix}$$

so the eigenvalue is -1.

---

[2] This is the simplest example of what is called a **quadratic form**, i.e. a sum of squares like x^2, often including similar terms like xy, where each term is often multiplied by a different constant. Quadratic forms are very important, and symmetric matrices play a big role in understanding them.

The eigenvector $\begin{bmatrix} 2y \\ y \end{bmatrix}$ refers to a series of points such as (2, 1) (if $y = 1$), or (4, 2) (if $y = 2$),

etc. These fit the straight line $y = x/2$. Similarly,

$$\begin{bmatrix} -x \\ 2x \end{bmatrix}$$

refers to points such as $(-1, 2), (-2, 4), \ldots$ that fit the straight line $y = -2x$. These lines are plotted as EE in Fig. 20.1 (a) (p. 402); they are perpendicular. Indeed, it can be shown that the eigenvectors of a symmetric matrix are always perpendicular (provided the eigenvalues differ).

20.3.2 Matrix Diagonalization

It is perhaps a surprising situation to encounter a rotation matrix when finding the eigenvectors of eigenvalues. The previous example showed that

$$\begin{bmatrix} 7 & 4 \\ 4 & 1 \end{bmatrix}\begin{bmatrix} 2y \\ y \end{bmatrix}$$

has two eigenvalues, 9 and -1; and the corresponding eigenvectors are

$$\begin{bmatrix} 2y \\ y \end{bmatrix} \text{ and } \begin{bmatrix} x \\ -2x \end{bmatrix}.$$

Here x and y are adjustable and only affect the eigenvectors' lengths; so we adjust them to give them both unit lengths (i.e. we normalize them). Thus, for the first eigenvector, $4y^2 + y^2 = 5y^2 = 1$, so $y = 1/\sqrt{5}$ and the first eigenvector is $(2/\sqrt{5}, 1/\sqrt{5})$. Similarly, we find that the second eigenvector is $(1/\sqrt{5}, -2/\sqrt{5})$. A check will show that they're also orthogonal, so the matrix

$$U = \begin{bmatrix} 2/\sqrt{5} & -1/\sqrt{5} \\ 1/\sqrt{5} & 2/\sqrt{5} \end{bmatrix}$$

is orthogonal and normalized (i.e., orthonormal). Pre-multiplying by A gives:

$$AU = \begin{bmatrix} 7 & 4 \\ 4 & 1 \end{bmatrix}\begin{bmatrix} 2/\sqrt{5} & -1/\sqrt{5} \\ 1/\sqrt{5} & 2/\sqrt{5} \end{bmatrix} = \begin{bmatrix} 18/\sqrt{5} & 1/\sqrt{5} \\ 9/\sqrt{5} & -2/\sqrt{5} \end{bmatrix}$$

$$= \begin{bmatrix} 2/\sqrt{5} & -1/\sqrt{5} \\ 1/\sqrt{5} & 2/\sqrt{5} \end{bmatrix}\begin{bmatrix} 9 & 0 \\ 0 & -1 \end{bmatrix} = U \cdot \text{diag}(9, -1)$$

(The last notation, applying only to diagonal matrices where all other elements are zero, just lists the diagonal elements.) Multiplying both sides on the left by the transpose of $U = U^T$, we get:

$$\begin{bmatrix} 2/\sqrt{5} & 1/\sqrt{5} \\ -1/\sqrt{5} & 2/\sqrt{5} \end{bmatrix}\begin{bmatrix} 7 & 4 \\ 4 & 1 \end{bmatrix}\begin{bmatrix} 2/\sqrt{5} & -1/\sqrt{5} \\ 1/\sqrt{5} & 2/\sqrt{5} \end{bmatrix} = \begin{bmatrix} 9 & 0 \\ 0 & -1 \end{bmatrix}$$

Thus:

$$U^T A U = U^{-1} A U = \text{diag}(9, -1) \tag{20.7}$$

The entire process is called the **diagonalization** of A. The left side has a special form, found in many applications of matrices. For example, suppose we want to make five 5-fold copies of a molecular group about a 5-fold axis in an arbitrary direction, but we only have a program (A) to do this about the y-axis. So we rotate everything to bring that direction along the y-axis (U), then apply our program (A), and finally back-rotate everything to the original position (U^{-1}).

20.4 Least-Squares Fits

20.4.1 Rectangular Matrices

Matrix multiplication is built from the scalar products of vectors, in which the vectors pair like chromosomes and corresponding elements multiply together. In order to pair, the vectors must have the same number of elements (the same length), which can be as small as one in the product

$$\begin{bmatrix} x \\ y \end{bmatrix}[a, b] = \begin{bmatrix} xa & xb \\ ya & yb \end{bmatrix}.$$

To match, the number of columns in the first must match the number of rows in the second; here the (rows $\times$ columns) are $(2 \times 1)(1 \times 2) = (2 \times 2)$; when two 2D matrices multiply, (rows $\times$ columns) are $(2 \times 2)(2 \times 2) = (2 \times 2)$. And, when we form the scalar product of two vectors, $(1 \times 2)(2 \times 1) = (1 \times 1) = $ scalar. In each case the rule is $(r_1 \times c_1)(c_1 \times c_2) = (r_1 \times c_2)$ and, provided this applies, the other numbers, r_1 and c_2, can be adapted to convenience. This means that they should fit the data-structure of the numbers we need to process.

We have also developed a notation for writing out these vectors in a compact form. A 2D vector can be written as a row $[x, y]$, or as a column

$$\begin{bmatrix} x \\ y \end{bmatrix}.$$

In section §20.1.2 we wrote

$$\begin{bmatrix} 7 & 4 \\ 4 & 1 \end{bmatrix} \begin{bmatrix} x \\ y \end{bmatrix}$$

in the compact form as $\mathbf{Ax}$, implying that

$$\mathbf{x} = \begin{bmatrix} x \\ y \end{bmatrix},$$

the column form. Consequently the row form $[x, y]$, the transpose of the column, is written as $\mathbf{x}^\mathrm{T}$. So, with

$$\mathbf{x} = \begin{bmatrix} x \\ y \end{bmatrix}, \ \mathbf{x}^\mathrm{T} = [x, \ y],$$

we write

$$x^2 + y^2 = (x, \ y) \cdot (x, \ y) = [x, \ y]\begin{bmatrix} x \\ y \end{bmatrix} = \mathbf{x}^\mathrm{T}\mathbf{x},$$

as in section §20.2.4.

This algebraic machinery is used for statistical calculations, especially those related to principal component analysis (section §11.1.6). But, as a simple introduction to these methods, we start with a simple straight-line fit.

20.4.2 Straight-Line Fits

Suppose we have pairs of experimental data values, (x, y), which show a scattering of points around a straight line whose parameters we want to calculate. So we want to fit all data pairs (x_j, y_j) to the same equation $y_j = ax_j + b$, $(j = 1, 2, ...)$ and thus need the constants a and b. There are just two unknowns (a and b), but there could be hundreds of data pairs (x_j, y_j), so an exact fit is impossible. Thus there are residuals ε_j to make each equation balance: $y_j = ax_j + b + \varepsilon_j$. So (for three measurements) the equations are:

$$y_1 = ax_1 + b + \varepsilon_1$$

$$y_2 = ax_2 + b + \varepsilon_2$$

$$y_3 = ax_3 + b + \varepsilon_3$$

In matrix form this is:

$$\begin{bmatrix} y_1 \\ y_2 \\ y_3 \end{bmatrix} = \begin{bmatrix} x_1 & 1 \\ x_2 & 1 \\ x_3 & 1 \end{bmatrix} \begin{bmatrix} a \\ b \end{bmatrix} + \begin{bmatrix} \varepsilon_1 \\ \varepsilon_2 \\ \varepsilon_3 \end{bmatrix}, \text{ or } y = Xa + \varepsilon \qquad (20.8)$$

The adjustable parameters are in a, and they determine the size of ε, which we want to minimize; or rather to minimize the sum of squares $\varepsilon^T\varepsilon$. If the situation were that $a = b = 0$, $\varepsilon = y$, so $\varepsilon^T\varepsilon = y^Ty$, which is large. In this case, ε would contain a large contribution from the signal in y; so we would want to minimize this contribution by adjusting a, b to make ε uncorrelated with either y or X. If our adjustment should make ε uncorrelated with X, then (on the average) $X^T\varepsilon = 0$, so $X^Ty = X^TXa$. (Check the rows/columns: y is (3×1), X is (3×2) and a is (2×1); so $y = Xa$ gives (3×1) = (3×2)(2×1), $X^Ty =$ gives (2×3)(3×1) = (2×1) and X^TXa gives (2×3)(3×2)(2×1) = (2×1).) Now X^TX is a square (2×2) matrix, which normally has an inverse $(X^TX)^{-1}$ which is used to multiply the entire equation, giving:

$$a = (X^TX)^{-1}(X^Ty) \qquad (20.9)$$

What are these terms? First,

$$(X^TX) = \begin{bmatrix} x_1 & 1 \\ x_2 & 1 \\ x_3 & 1 \end{bmatrix}^T \begin{bmatrix} x_1 & 1 \\ x_2 & 1 \\ x_3 & 1 \end{bmatrix} = \begin{bmatrix} x_1 & x_2 & x_3 \\ 1 & 1 & 1 \end{bmatrix} \begin{bmatrix} x_1 & 1 \\ x_2 & 1 \\ x_3 & 1 \end{bmatrix} = \begin{bmatrix} \Sigma x_j^2 & \Sigma x_j \\ \Sigma x_j & \Sigma 1 \end{bmatrix},$$

where the sum includes the total number of measurements (here 3). So $\Sigma 1 = 1 + 1 + 1 = 3$; in general, it is just the number of measurements. In the case of a 2D problem (just two unknowns a and b), we can find the inverse $(X^TX)^{-1}$ by using equations (20.2) and (20.3).

Also,

$$X^Ty = \begin{bmatrix} x_1 & x_2 & x_3 \\ 1 & 1 & 1 \end{bmatrix} \begin{bmatrix} y_1 \\ y_2 \\ y_3 \end{bmatrix} = \begin{bmatrix} \Sigma x_j y_j \\ \Sigma y_j \end{bmatrix}$$

So:

$$a = \begin{bmatrix} \Sigma x_j^2 & \Sigma x_j \\ \Sigma x_j & \Sigma 1 \end{bmatrix}^{-1} \begin{bmatrix} \Sigma x_j y_j \\ \Sigma y_j \end{bmatrix} \qquad (20.10)$$

20.4.3 Principal Component Analysis

We need to use some of these rectangular matrices in order to calculate the variance of the projected points in Fig. 11.5 in section §11.1.5. Each point, defined by its (x, y) coordinates, is to be projected onto a line passing through the origin and with an orientation defined by the unit vector $\mathbf{u}^{\text{T}} = [u, v]$. Now one of the points, with coordinates $\mathbf{x_1}^{\text{T}} = [x_1, y_1]$, gives a projection

$$x_1 u + y_1 v = [x_1, \ y_1]\begin{bmatrix} u \\ v \end{bmatrix} = \mathbf{x_1}^{\text{T}}\mathbf{u}.$$

We need to minimize the sum of all the squares of projections, and the square of this projection is

$$(\mathbf{x_1}^{\text{T}}\mathbf{u})^2 = (\mathbf{x_1}^{\text{T}}\mathbf{u})^{\text{T}}(\mathbf{x_1}^{\text{T}}\mathbf{u}) = (\mathbf{u}^{\text{T}}\mathbf{x_1})(\mathbf{x_1}^{\text{T}}\mathbf{u}) = [u, \ v]\left(\begin{bmatrix} x_1 \\ y_1 \end{bmatrix}[x_1, \ y_1]\right)\begin{bmatrix} u \\ v \end{bmatrix} = \mathbf{u}^{\text{T}}(\mathbf{x_1}\mathbf{x_1}^{\text{T}})\mathbf{u}.$$

(We have used the rule that a matrix product $C = AB$ gives $C^{\text{T}} = B^{\text{T}}A^{\text{T}}$, and also $(B^{\text{T}})^{\text{T}} = B$.) So we have the result that one of the squared projections,

$$(x_1 u + y_1 v)^2 = [u, \ v]\left(\begin{bmatrix} x_1 \\ y_1 \end{bmatrix}[x_1, \ y_1]\right)\begin{bmatrix} u \\ v \end{bmatrix}$$

which is

$$[u, \ v]\begin{bmatrix} x_1^2 & x_1 y_1 \\ x_1 y_1 & y_1^2 \end{bmatrix}\begin{bmatrix} u \\ v \end{bmatrix} = [u, \ v]\begin{bmatrix} x_1^2 u + x_1 y_1 v \\ x_1 y_1 u + y_1^2 v \end{bmatrix} = x_1^2 u^2 + 2x_1 y_1 uv + y_1^2 v^2.$$

It is easily checked that the last expression is the same as $(x_1 u + y_1 v)^2$.

All this algebra would be unnecessary just for one squared projection, but it is easily generalized to three of them (from three points), and from there to any number. Using the symbol 'Σ' (meaning 'sum') to abbreviate the algebra, the sum of three squared projections is
$\Sigma(\mathbf{x_j}^{\text{T}}\mathbf{u})^2 = (\mathbf{x_1}^{\text{T}}\mathbf{u})^2 + (\mathbf{x_2}^{\text{T}}\mathbf{u})^2 + (\mathbf{x_3}^{\text{T}}\mathbf{u})^2 = (x_1 u + y_1 v)^2 + (x_2 u + y_2 v)^2 + (x_3 u + y_3 v)^2 =$

$$\begin{bmatrix} x_1 u + y_1 v \\ x_2 u + y_2 v \\ x_3 u + y_3 v \end{bmatrix}^{\text{T}}\begin{bmatrix} x_1 u + y_1 v \\ x_2 u + y_2 v \\ x_3 u + y_3 v \end{bmatrix} = \left(\begin{bmatrix} x_1 & y_1 \\ x_2 & y_2 \\ x_3 & y_3 \end{bmatrix}\begin{bmatrix} u \\ v \end{bmatrix}\right)^{\text{T}}\left(\begin{bmatrix} x_1 & y_1 \\ x_2 & y_2 \\ x_3 & y_3 \end{bmatrix}\begin{bmatrix} u \\ v \end{bmatrix}\right) = [u, \ v]\begin{bmatrix} x_1 & x_2 & x_3 \\ y_1 & y_2 & y_3 \end{bmatrix}\left(\begin{bmatrix} x_1 & y_1 \\ x_2 & y_2 \\ x_3 & y_3 \end{bmatrix}\begin{bmatrix} u \\ v \end{bmatrix}\right)$$

$$= [u, \ v]\begin{bmatrix} x_1^2 + x_2^2 + x_3^2 & x_1 y_1 + x_2 y_2 + x_3 y_3 \\ x_1 y_1 + x_2 y_2 + x_3 y_3 & y_1^2 + y_2^2 + y_3^2 \end{bmatrix}\begin{bmatrix} u \\ v \end{bmatrix} = [u, \ v]\begin{bmatrix} \Sigma x_j^2 & \Sigma x_j y_j \\ \Sigma x_j y_j & \Sigma y_j^2 \end{bmatrix}\begin{bmatrix} u \\ v \end{bmatrix}.$$

In summary: $\Sigma\left(x_j^T u\right)^2 = u^T C u.$ (20.11)

Here $C = \begin{bmatrix} \Sigma x_j^2 & \Sigma x_j y_j \\ \Sigma x_j y_j & \Sigma y_j^2 \end{bmatrix}$

is a symmetric matrix that has two eigenvalue–vector pairs. The equation describes an ellipse (Fig. 11.5) and the eigenvectors are the directions of its axes, the eigenvalues being related to the axis lengths.

References

Abbe, E., 1873. Beiträge zur Theorie des Mikroskops und der mikroskopischen Wahrnehmung. Arch. Mikr. Anat. 9, 413–468.

Abramowitz, M., Stegun, I.A. (Eds.), 1964. Handbook of Mathematical Functions with Formulas, Graphs, and Mathematical Tables, Washington, D.C., National Bureau of Standards, Applied Mathematics Series, 55. (Also reprinted by Dover Publications, New York.) [The National Institute of Standards and Technology, the Digital Library of Mathematical Functions; see dlmf.nits.gov.]

Acton, F.S., 1970. Numerical Methods that Work. Harper & Row, N.Y.

Alber, F., Dokudovskaya, S., Veenhoff, L.M., Zhang, W., Kipper, J., Devos, D., et al., 2007. Determining the architectures of macromolecular assemblies. Nature 450, 683–694.

Allen, M.P., Tildesley, D.J., 1987. Computer Simulation of Liquids. Clarendon Press, Oxford.

Amos, L.A., Henderson, R., Unwin, P.N.T., 1982. Three-dimensional structure determination by electron microscopy of two-dimensional crystals. Progr. Biophys. Mol. Biol. 39, 183–231.

Arnold, J.T., Dharmatti, S.S., Packard, M.E., 1951. Chemical effects on nuclear induction signals from organic compounds. J. Chem. Phys. 19, 507.

Aue, W.P., Bartholdi, E., Ernst, R.R., 1976. Two-dimensional spectroscopy. Application to nuclear magnetic resonance. J. Chem. Phys. 64 (5), 2229–2246.

Baker, T.S., Caspar, D.L.D., Murakami, W.T., 1983. Polyoma virus 'hexamer' tubes consist of paired pentamers. Nature 303, 446–448.

Baker, T.S., Cheng, R.H., 1996. A model-based approach for determining orientations of biological macromolecule images by cryoelectron microscopy. J. Struct. Biol. 116, 120–130.

Baker, T.S., Olson, N.H., Fuller, S.D., 1999. Adding the third dimension to viral life cycles: three-dimensional reconstruction of icosahedral viruses from cryo-electron micrographs. Microbiol. Molec. Biol. Rev. 63, 862–922.

Baker, T.S., Henderson, R., 2001. Electron cryomicroscopy (Chapter 19.6). In: Rossmann, M.G. (Ed.), International Tables for Crystallography, Crystallography of Biological Macromolecules, vol. F. Kluwer Academic Publishers, Dordrecht, The Netherlands, pp. 451–463.

Baldwin, J.M., Henderson, R., 1984. Measurement and evaluation of electron diffraction patterns from two-dimensional crystals. Ultramicroscopy 14, 319–336.

Baldwin, J.M., Henderson, R., Beckman, E., Zemlin, F., 1988. Images of purple membrane at 2.8 Å resolution obtained by cryo-electron microscopy. J. Mol. Biol. 202, 585–591.

Barlow, W., 1901. I. Crystal symmetry. The actual basis of the thirty-two classes. Philosophical Magazine 1 (6), 1–36.

Barrett, A.N., Leigh, J.B., Holmes, K.C., Leberman, R., Mandelkow, E., Sengbusch, P.V., 1972. An electron-density map of tobacco mosaic virus at 10Å resolution. Cold Spring Harbor Symp. Quant. Biol. 36, 433–448.

Belnap, D.M., Olson, N.H., Cladel, N.M., Newcomb, W.W., Brown, J.C., Kreider, J.W., et al., 1996. Conserved features in papillomavirus and polyomavirus capsids. J. Mol. Biol. 259, 249–263.

Bernal, J.D., Crowfoot, D., 1934. X-ray photographs of crystalline pepsin. Nature 134, 794–795.

Bernal, R.A., Hafenstein, S., Olson, N.H., Bowman, V.D., Chipman, P.R., Baker, T.S., et al., 2003. Structural studies of bacteriophage α3 assembly. J. Mol. Biol. 325, 11–24.

Beroukhim, R., Unwin, N., 1997. Distortion correction of tubular crystals: improvements in the acetylcholine receptor structure. Ultramicroscopy 70, 57–81.

Blow, D., Crick, F.H.C., 1959. The treatment of errors in the isomorphous replacement method. Acta Crystallogr. 12, 794–802.

Blow, D., 2002. Outline of Crystallography for Biologists. University Press, Oxford.

Bluemke, D.A., Carragher, B., Josephs, R., 1988. The reconstruction of helical particles with variable pitch. Ultramicroscopy 26, 255–270.

Bokhoven, C., Schoone, J.C., Bijvoet, J.M., 1951. Fourier synthesis of the crystal structure of strychnine sulphate pentahydrate. Acta Crystallogr. 5, 275–280.

Born, M., Wolf, E., 2002. Principles of Optics: Electromagnetic Theory of Propagation, Interference and Diffraction of Light, first–seventh ed. University Press, Cambridge.

Böttcher, B., Wynne, S.A., Crowther, R.A., 1997. Determination of the fold of the core protein of hepatitis B virus by electron cryomicroscopy. Nature 386, 88–91.

Bracewell, R.N., 1989. The Fourier transform. Sci. Am. 260 (6), 62–69.

Bracewell, R.N., 1965–2000. The Fourier Transform and its Applications. McGraw-Hill Book Co., Singapore.

Bragg, W.L., 1914. The diffraction of short electromagnetic waves by a crystal. Proc. Cambridge Phil. Soc. 17, 43–57.

Bragg, W.L., Stokes, A.R., 1945. X-ray analysis with the aid of the 'fly's eye. Nature 156, 332–333.

Bragg, W.L., Perutz, M.F., 1952. The structure of haemoglobin. Proc. Roy. Soc. A 213, 425–435.

Brenner, S., Horne, R.W., 1959. A negative staining method for high resolution electron microscopy of viruses. Biochim. Biophys. Acta 34, 103–110.

Bricogne, G., 2001. Fourier transforms in crystallography: theory, algorithms and applications (Chapter 1.3), pp. 25–98. In: Shmueli, V. (Ed.), International Tables for Crystallography, Reciprocal Space, vol. B. Kluwer Academic Publishers, Dordrecht, The Netherlands for the International Union of Crystallography.

Brigham, E.O., 1974. The Fast Fourier Transform and its Applications. McGraw-Hill, New York.

Brisson, A., Unwin, P.N.T., 1984. Tubular crystals of acetylcholine receptor. J. Cell Biol. 99, 1202–1211.

Buerger, M.J., 1959. Vector Space and its Application in Crystal Structure Investigation. J. Wiley, New York.

Bullough, P., Henderson, R., 1987. Use of spot-scan procedure for recording low-dose images of beam-sensitive specimens. Ultramicroscopy 21 (3), 223–230.

Calladine., C.R., 1978. Change of waveform in bacterial flagella: the role of mechanics at the molecular level. J. Mol. Biol. 118, 457–479.

Cantele, F., Lanzavecchia, S., Bellon, P.L., 2003. The variance of icosahedral virus models is a key indicator in the structure determination: a model-free reconstruction of viruses, suitable for refractory particles. J. Struct. Biol. 141, 84–92.

Caspar, D.L.D., 1956. Structure of bushy stunt virus. Nature 177, 4756.

Caspar, D.L.D., Klug, A., 1962. Physical principles in the construction of regular viruses. Cold Spring Harbor Symp. Quant. Biol. 27, 1–24.

Ceska, T.A., Henderson, R., 1990. Analysis of high-resolution electron diffraction patterns from purple membrane labelled with heavy-atoms. J. Mol. Biol. 213, 539–560.

Chapman, S.K., 1986. Maintaining and Monitoring the Transmission Electron Microscope. University Press, Oxford.

Cheng, L., Fang, Q., Shah, S., Atanasov, I.C., Zhou, Z.H., 2008. Subnanometer-resolution structures of the grass carp reovirus core and virion. J. Mol. Biol. 382, 213–222.

Chiu, W., 2007. Icosahedral particles. Chapter 13 of Glaeser et al. (2007) *Electron Crystallography of Biological Macromolecules*. New York: Oxford University Press.

Clemons, W.M., Broderson, D.E., McCutcheon, J.P., May, J.L., Carter, A.P., Morgan-Warren, R.J., et al., 2001. Crystal structure of the 30S ribosomal subunit from *Thermus thermophilus*: purification, crystallization and structure determination. J. Mol. Biol. 310, 827–843.

Cochran, W., Crick, F.H.C., Vand, V., 1952. The structure of synthetic polypeptides. I. The transform of atoms on a helix. Acta Crystallogr. 5, 581–586.

Cohan, N.V., 1958. The spherical harmonics with the symmetry of the icosahedral group. Proc. Cambridge Phil. Soc. 54, 28–38.

Conway, J., Cheng, N., Wingfield, P.T., Stahl, S.J., Steven, A.C., 1997. Visualization of a 4-helix bundle in the hepatitis B virus capsid by cryo-electron microscopy. Nature 386, 91–94.

Cooley, J.W., Tukey, J.W., 1965. An algorithm for the machine calculation of complex Fourier series. Math. Comput. 19, 297–301.

Cowley, J.M., 1975. Diffraction Physics. North-Holland, Amsterdam.

Cowley, J.M., 2001. Electron diffraction and electron microscopy (Chapter 2.5.2). In: Shmueli, V. (Ed.), International Tables for Crystallography, Reciprocal Space, vol. B. Kluwer Academic Publishers, Dordrecht, The Netherlands, for the International Union of Crystallography, pp. 277–285.

Coxeter, H.S.M., 1989. Introduction to Geometry, second ed. J. Wiley and Sons, New York and London.

Crepeau, R.H., Fram, E.K., 1981. Reconstruction of imperfectly ordered zinc-induced tubulin sheets using cross-correlation and real space averaging. Ultramicroscopy 6, 7–17.

Crick, F.H.C, 1953. Ph.D. Thesis, Cambridge University.

Crick, F.H.C., Watson, J.D., 1956. The structure of small viruses. Nature 177, 473–475.

Crowther, R.A., Amos, L.A., Finch, J.T., DeRosier, D.J., 1970a. Three-dimensional reconstructions of spherical viruses by Fourier synthesis from electron micrographs. Nature 226, 421–425.

Crowther, R.A., DeRosier, D.J., Klug, A., 1970b. The reconstruction of three-dimensional structure from projections and its application to electron microscopy. Proc. Roy. Soc. Lond. A 317, 319–340.

Crowther, R.A., 1971. Procedures for three-dimensional reconstruction of spherical viruses by Fourier synthesis from electron micrographs. Phil. Trans. Roy. Soc. Lond., B 261, 221–230.

Crowther, R.A., Amos, L.A., 1971. Harmonic analysis of electron microscope images with rotational symmetry. J. Mol. Biol. 60, 123–130.

Crowther, R.A, 1972. In: Rossmann, M.G. (Ed.), The Molecular Replacement Method. Gordon & Breach, New York, pp. 173–178.

Crowther, R.A., Sleytr, U.B., 1977. An analysis of the fine structure of surface layers from two strains of Clostridia, including correction for distorted images. J. Ultrastr. Res. 58, 41–49.

Crowther, R.A., Padron, R., Craig, R., 1985. Arrangement of the heads of myosin in relaxed thick filaments from tarantula muscle. J. Mol. Biol. 184, 429–439.

Crowther, R.A., Kiselev, N.A., Böttcher, B., Berriman, J.A., Borisova, G.P., Ose, V., et al., 1994. Three-dimensional structure of hepatitis B virus core particles determined by electron cryomicroscopy. Cell 77, 943–950.

Crowther, R.A., Henderson, R., Smith, J.M., 1996. MRC image processing programs. J. Struct. Biol. 116, 9–16.

Crowther, R.A., 2008. Microscopy goes cold: frozen viruses reveal their structural secrets. (The Leeuwenhoek lecture 2006.). Phil. Trans. R. Soc. B 363, 2441–2451.

DeRosier, D.J., Klug, A., 1968. Reconstruction of three dimensional structures from electron micrographs. Nature 217, 130–134.

DeRosier, D.J., Moore, P.B., 1970. Reconstruction of three-dimensional Images from electron micrographs of structures with helical symmetry. J. Mol. Biol. 52, 355–369.

DeRosier, D.J., 2007. Electron crystallography of helical structures. Chapter 12 of Glaeser et al. (2007) *Electron Crystallography of Biological Macromolecules*. New York: Oxford University Press.

Dietrich, I., Fox, F., Knapek, E., Lefranc, G., Nachtrieb, K., Weyl, R., et al., 1977. Improvements in electron microscopy by application of superconductivity. Ultramicroscopy 2, 241–249.

Downing, K.H., Glaeser, R.M., 1986. Improvement in high-resolution image quality of radiation sensitive specimens achieved with reduced spot size of the electron beam. Ultramicroscopy 20, 269–278.

Dube, P., Tavares, P., Lurz, R., Heel, M. van, 1993. The portal protein of bacteriophage SPP1: a DNA pump with 13-fold symmetry. EMBO J. 12, 1303–1309.

Dubochet, J., Chang, J.-J., Freeman, R., Lepault, J., McDowall, A.W., 1982. Frozen aqueous suspensions. Ultramicroscopy 10, 55–62.

Dubochet, J., Adrian, M., Chang, J.-J., Homo, J.-C., Lepault, J., McDowall, A.W., et al., 1988. Cryo-electron microscopy of vitrified specimens. Quart. Rev. Biophys. 21, 129–228.

Dykstra, M.J., Renss, L.E., 2003. Biological Electron Microscopy: Theory, Technologies and Troubleshooting, second ed. Kluwer Academic/Plenum Publications, New York.

Egelman, E.H., 1986. An algorithm for straightening images of curved filamentous structures. Ultramicroscopy 19, 367–373.

Egelman, E.H., 2000. A robust algorithm for the reconstruction of helical filaments using single particle methods. Ultramicroscopy 85, 225–234.

Ernst, R.R., Anderson, W.A., 1966. Application of Fourier Transform spectroscopy to magnetic resonance. Rev. Sci. Instrum. 37, 93–102.

Feynman, R.P., Leighton, R.B., Sands, M., 1963. The Feynman Lectures in Physics: Mainly Mechanics, Radiation, and Heat. Vol. 1, Addison-Wesley Publishing Company, Inc., Reading, Massachusetts.

Feynman, R.P., Leighton, R.B., Sands, M., 1965. The Feynman Lectures in Physics: Quantum Mechanics. Vol. 3, Addison-Wesley Publishing Company, Inc., Reading, Massachusetts.

Finch, J.T., Holmes, K.C., 1967. Structural studies of viruses. In: Maramorosch, K., Koprowski, H. (Eds.), Methods in Virology, vol. III. Academic Press, New York and London, pp. 351–474.

Frank, J., 1990. Classification of macromolecular assemblies studied as 'single particles'. Quart. Rev. Biophysics 23, 281–329.

Frank, J., Chiu, W., Henderson, R., 1993. Flopping polypeptide chains and Suleika's subtle imperfections: analysis of variations in the electron micrograph of a purple membrane crystal. Ultramicroscopy 49, 387–396.

Frank, J., 2006. Three-dimensional Electron Microscopy of Macromolecular Assemblies: Visualization of Biological Molecules in their Native State. Oxford University Press, New York.

Frank, J., 2007. Single particles. Chapter 14 of Glaeser et al. (2007) *Electron Crystallography of Biological Macromolecules*. New York: Oxford University Press.

Franklin, R.E., Klug, A., 1955. The splitting of layer lines in X-ray fibre diagrams of helical structures: application to tobacco mosaic virus. Acta Crystallogr. 8, 777–781.

Franklin, R.E., Holmes, K.C., 1958. Tobacco mosaic virus: application of the method of isomorphous replacement to the determination of the helical parameters and radial density distribution. Acta. Crystallogr. 11, 213–220.

Freitag, B., Kujawa, S., Mul, U., Ringuelda, J., Tiemeijer, P.C., 2005. Breaking the spherical and chromatic aberration barrier in transmission electron microscopy. Ultramicroscopy 102, 209–214.

Fuller, S.D., 1987. The T = 4 envelope of Sindbis virus is organized by interactions with a complementary T = 3 capsid. Cell 48, 923–934.

Fuller, S.D., Butcher, S.J., Cheng, R.H., Baker, T.S., 1996. Three-dimensional reconstruction of icosahedral particles – the uncommon line. J. Struct. Biol. 116, 48–55.

Giacovazzo, C., Monaco, H.L., Viterbo, D., Scordari, F., Gilli, G., Zanotti, G., et al., 1992. Fundamentals of Crystallography. International Union of Crystallography. University Press, Oxford.

Gilbert, P.F.C., 1972. The reconstruction of a three-dimensional structure from projections and its application to electron microscopy. II. Direct methods. Proc. Roy. Soc. Lond. B 182, 89–102.

Glaeser, R.M., Taylor, K.A., 1978. Radiation damage related to transmission electron microscopy of biological specimens at low temperatures: a review. J. Microsc. 112, 127–138.

Glaeser, R.M., Downing, K., DeRosier, D.J., Chiu, Wah, Frank, J., 2007. Electron Crystallography of Biological Macromolecules. Oxford University Press, New York.

Goldstein, H., 1980. Classical Mechanics, second ed. Addison-Wesley, Reading, MA.

Gonen, T., Sliz, P., Kistler, J., Cheng, Y., Walz, T., 2004. Aquaporin-0 membrane junctions reveal the structure of a closed water pore. Nature 429, 193–197.

Goodman, J.W., 1968–2004. Introduction to Fourier Optics, third ed. Roberts & Company, Englewood, Colorado (Earlier editions: San Francisco: McGraw-Hill Book Company.).

Gordon, R., Bender., R., Herman, G.T., 1970. Algebraic reconstruction techniques (ART) for three-dimensional electron microscopy and X-ray photography. J. Theor. Biol. 29, 471–481.

Green, D.W., Ingram, V.M., Perutz, M.F., 1954. Structure of haemoglobin IV: sign determination by the isomorphous replacement method. Proc. Roy. Soc. Lond. A 225, 287–307.

Grigorieff, N., Ceska, T.A., Downing, K.H., Baldwin, J.M., Henderson, R., 1996. Electron-crystallographic refinement of the structure of bacteriorhodopsin. J. Mol. Biol. 259, 393–421.

Hahn, T., 2002. International Tables for Crystallography: Space-group symmetry, fifth ed., vol. A. Kluwer Academic Publishers, Dordrecht, The Netherlands, for the International Union of Crystallography.

Hall, C.E., 1955. Electron densitometry of stained virus particles. J. Biophys. Biochem. Cytol. 1, 1–12.

Harburn, G., Taylor, C.A., Welberry, T.R., 1975. An Atlas of Optical Transforms. Bell & Hyman, London.

Harauz, G., van Heel, M., 1986. Exact filters for general geometry three dimensional reconstruction. Optik 73, 146–156.

Harris, H., 1999. The Birth of the Cell. Yale University Press, New Haven & London.

Hartline, H.K., 1959. Receptor mechanisms and the integration of sensory information in the eye. In: Biophysical Sciences – A Study Program. J. Wiley, New York, pp. 515–523.

Heide, H.G., 1982. Design and operation of cold stages. Ultramicroscopy 10, 125–154.

Heidenreich, R.D., 1964. Fundamentals of Transmission Electron Microscopy. John Wiley & Sons, New York.

Henderson, R., Moffat, J.K., 1971. The difference Fourier technique in protein crystallography: errors and their treatment. Acta Cryst. B 27, 1414–1420.

Henderson, R., Unwin, P.N.T., 1975. Three-dimensional model of purple membrane obtained by electron microscopy. Nature 257, 28–32.

Henderson, R., Glaeser, R.M., 1985. Quantitative analysis of image contrast in electron micrographs of beam-sensitive crystals. Ultramicroscopy 16, 139–150.

Henderson, R., Baldwin, J.M., Downing, K.H., Lepault, J., Zemlin, F., 1986. Structure of purple membrane from *Halobacterium halobium*: recording, measurement and evaluation of electron micrographs at 3.5 Å resolution. Ultramicroscopy 19, 147–178.

Henderson, R., Baldwin, J.M., Ceska, T.A., Zemlin, F., Beckmann, E., Downing, K.H., 1990. Model for the structure of bacteriorhodopsin based on high-resolution electron cryo-microscopy. J. Mol. Biol. 213, 899–929.

Henderson, R., 1992. Image contrast in high-resolution electron microscopy of biological macromolecules: TMV in ice. Ultramicroscopy 46, 1–18.

Henderson, R., 1995. The potential and limitations of neutrons, electrons and X-rays for atomic-resolution microscopy of unstained biological molecules. Q. Rev. Biophys. 28, 171–193.

Henderson, R., 2004. Realizing the potential of electron cryo-microscopy. Q. Rev. Biophys. 37, 3–13.

Herman, G.T., 1980. Image Reconstruction from Projections. Academic Press, New York.

Herrmann, T., Güntert, P., Wüthrich, K., 2002. Protein NMR structure determination with automated NOE assignment using the new software CANDID and the torsion angle dynamics algorithm DYANA. J. Mol. Biol. 319, 209–227.

Hilbert, D., Cohn-Vossen, S., 1952. Geometry and the Imagination. Chelsea Publishing Company, New York (Chapter II).

Hite, R.K., Li, Z., Walz, T., 2010. Principles of membrane protein interactions with annular lipids deduced from aquaporin-O 2D crystals. EMBO J., 1–7 (13 April 2010).

Holmes, K.C., Mandelkow, E., Leigh, J.B., 1972. Determination of heavy atom positions in tobacco mosaic virus from double heavy atom derivatives. Naturwiss 59, 247–254.

Holser, W.T., 1958. Point groups and plane groups in a two-sided plane and their sub-groups. Z. Kristallogr. 110, 268–281.

Hosemann, R., Bagchi, S.N., 1962. Direct Analysis of Diffraction by Matter. Amsterdam, North Holland

Hotelling, H., 1933. Analysis of a complex of statistical variables into principal components. J. Educ. Psychol. 24, 417–441 & 498–520.

Hunter, G.K., 2004. Light is a Messenger: The life and science of William Lawrence Bragg. University Press, Oxford.

Huxley, H.E., 1956. Some observations on the structure of tobacco mosaic virus. Proceedings of the Stockholm Conference on Electron Microscopy. Almquist & Wiksell, Stockholm, pp. 260–261.

Huxley, H.E., 1968. Structural difference between resting and rigor muscle; evidence from intensity changes in the low-angle equatorial X-ray diagram. J. Mol. Biol. 37, 507–520.

Jeffreys, H., Jeffreys, B.S., 1962. Methods of Mathematical Physics, third ed. University Press, Cambridge.

Jelsch, C., Teeter, M.M., Lamzin, V., Pichon-Pesma, V., Blessing, R.H., Lecomte, C., 2000. Accurate protein crystallography at ultra-high resolution: valence electron distribution in crambin. Proc. Natl. Acad. Sci., USA 97, 3171–3176.

Jeng, T.-W., Crowther, R.A., Stubbs, G., Chiu, W., 1989. Visualization of alpha-helices in tobacco mosaic virus by cryo-electron microscopy. J. Mol. Biol. 205, 251–257.

Jiang, W., Chiu, W., 2006. Cryo-electron microscopy of isosahedral virus particles. In: Kui, J. (Ed.), Methods in Molecular Biology. Humana Press, N.J.

Kabius, B., Hartel, P., Haider, M., Müller, H., Uhlemann, S., Loebau, U., et al., 2009. First application of C_c-corrected imaging for high-resolution and energy-filtered TEM. J. Electron Microsc. 58, 147–155.

Keller, A., O'Connor, A., 1958. Study of single crystals and their associations in polymers. Disc. Farad. Soc. 25, 114.

Kendrew, J.C., Dickerson, R.E., Strandberg, B.E., Hart, R.G., Davies, D.R., Phillips, D.C., et al., 1960. Structure of myoglobin: a three-dimensional Fourier synthesis at 2 Å resolution. Nature 185, 422–427.

Keynes, J.M., 1921. Treatise on Probability. Macmillan and Company, Limited, London.

Kiselev, N.A., Klug, A., 1969. The structure of viruses of the papilloma-polyoma type. V. Tubular variants built of pentamers. J. Mol. Biol. 40, 155–171.

Klug, A., Crick, F.H.C., Wyckoff, H.W., 1958. Diffraction by helical structures. Acta. Crystallogr. 11, 199–212.

Klug, A., Berger, J.E., 1964. An optical method for the analysis of periodicities in electron micrographs, and some observations on the mechanism of negative straining. J. Mol. Biol. 10, 505–569.

Kühlbrandt, W., Wang, D.N., Fujiyoshi, Y., 1994. Atomic model of plant light-harvesting complex by electron crystallography. Nature 367, 614–621.

Lake, J.A., 1976. Ribosome structure determined by electron microscopy of *Escherichia coli* small subunits, large subunits and monomeric subunits. J. Mol. Biol. 105, 131–159.

Langer, R., Frank, J., Feltynowski, A., Hopoe, W., 1970. Anwendung des Bilddifferenzverfahrens auf die Untersuchung von Strukturänderungen dünner Kohlefolien bei Elektronenbestrahlung. Ber. Bunsenges. Phys. Chem. 74, 1120–1126.

Leisegang, S., 1954. Versuch mit einer Kühlbaten Objectpatrone. In: Proceedings of the 3rd International Conference on Electron Microscopy, London. Royal Microscopy Society, pp. 176–184.

Liljas, A., Liljas, L., Piskur, J., Lindblom, G., Nissen, P., Kjeldgaard, M., 2009. Textbook of Structural Biology. World Scientific Publishing Co., Singapore.

Lipson, H., Taylor, C.A., 1958. Fourier Transforms and X-ray Diffraction. G. Bell & Sons, London.

Lipson, S.G., Lipson, H., Tannhauser, D.S., 1995. Optical Physics. Cambridge University Press.

Liu, H., Cheng, L., Zeng, X., Cai, C., Zhou, Z.H., Yang, Q., 2008. Symmetry-adapted spherical harmonics method for high-resolution 3D single-particle reconstructions. J. Struct. Biol. 161, 64–73.

Luzzati, V., 1953. Resolution d'un structure crystalline lorsque les positions d'une partie des atoms sont connues: traitement statistique. Acta Crystallogr. 6, 142–152.

Markham, R., Frey, S., Hills, G.J., 1963. Methods for the enhancement of image detail and accentuation of structure in electron microscopy. Virology 20, 88–102.

Mathews, J., Walker, R.L., 1964. Mathematical Methods of Physics. W.A. Benjamin, Inc., New York.

McCammon, J.A., Harvey, S.C., 1987. Dynamics of Proteins and Nucleic Acids. University Press, Cambridge.

Maxwell, J.C., 1854. On the transformation of surfaces by bending. Cambridge Phil. Soc. Trans. 9, 455–469 (Reprinted, 1890, in The Scientific Papers of J.C. Maxell, vol. 1, ed. W.D. Niven, pp. 81–114. University Press, Cambridge.).

Mertz, L., 1965. Transforms in Optics. J. Wiley, New York.

Mitsuoka, K., Hirai, T., Murata, K., Miyazawa, A., Kidera, A., Kimura, Y., et al., 1999. The structure of bacteriorhodopsin at 3.0 Å resolution based on electron crystallography: implication of the charge distribution. J. Mol. Biol. 286, 861–882.

Miyazawa, A., Fuyiyoshi, Y., Stowell, M., Unwin, N., 1999. Nicotinic acetylcholine receptor at 4.6 Å resolution: transverse channels in the channel wall. J. Mol. Biol. 288, 765–786.

Miyazawa, A., Fuyiyoshi, Y., Unwin, N., 2003. Structure and gating mechanism of the acetylcholine receptor pore. Nature 423, 949–955.

Moody, M.F., 1965. The shape of the T-even bacteriophage head. Virology 26, 567–576.

Moody, M.F., 1967. Structure of the sheath of bacteriophage T4. I. Structure of the contracted sheath and polysheath. J. Mol. Biol. 24, 167–200.

Moody, M.F., 1973. Structure of the sheath of bacteriophage T4. III. Contraction mechanism deduced from partially contracted sheaths. J. Mol. Biol. 80, 613–635.

Moody, M.F., 1990. Image analysis of electron micrographs. In: Hawkes, P.W., Valdrè, U. (Eds.) Biophysical Electron Microscopy. Academic Press, London, pp. 145–287.

Moody, M.F., 1999. Geometry of phage head construction. J. Mol. Biol. 293, 401–429.

Namba, K., Stubbs, G., 1986. Structure of tobacco mosaic virus at 3.6 Å resolution: implications for assembly. Science 231, 1401–1406.

Natterer, F., 1986. The Mathematics of Computerized Tomography. John Wiley & Sons, Ltd., Chichester.

Navaza, J., 2003. On the three-dimensional reconstruction of icosahedral particles. J. Struct. Biol. 144, 13–23.

Nogales, E., Wolf, S., Downing, K.H., 1998. Structure of the $\alpha\beta$ tubulin dimer by electron crystallography. Nature 391, 199–203.

Nye, J.F., 1985. Physical Properties of Crystals: Their Representation by Tensors and Matrices, second ed. Clarendon Press, Oxford.

Oesterhelt, D., Stoeckenius, W., 1971. Rhodopsin-like protein from the purple membrane of *Halobacterium halobium*. Nature New Biol. 233, 149–152.

Patterson, A.L., 1935. A direct method for the determination of the components of interatomic distances in crystals. Z. Kristallogr. 90, 517–542.

Pauling, L., Corey, R.B., Branson, H.R., 1951. The structure of proteins: two hydrogen-bonded helical configurations of the polypeptide chain. Proc. Nat. Acad. Sci. USA 37, 205–211.

Penczek, P.A., Radermacher, M., Frank, J., 1992. Three-dimensional reconstruction of single particles embedded in ice. Ultramicroscopy 40, 33–53.

Press, W.H., Teukolsky, S.A., Vetterling, W.T., Flannery, B.P., 2007. Numerical Recipes in Fortran: The Art of Scientific Computing, third ed. Cambridge University Press, Cambridge.

Provencher, S.W., Vogel, R.H., 1988a. Three-dimensional reconstruction from electron micrographs of disordered specimens. I. Method. Ultramicroscopy 25, 209–222.

Provencher, S.W., Vogel, R.H., 1988b. Three-dimensional reconstruction from electron micrographs of disordered specimens. II. Implementation and results. Ultramicroscopy 25, 223–240.

Radermacher, M., Wagenknecht, T., Verschoor, A., Frank, J., 1987. Three-dimensional reconstruction from a single-exposure, random conical tilt series applied to the 50S ribosomal subunit of *Escherichia coli*. J. Microsc. 146, 113–136.

Radermacher, M., 1988. Three-dimensional reconstruction of single particles from random and non-random tilt series. J. Elect. Microsc. Tech. 9, 359–394.

Radon, J., 1917. Über die Bestimmung von Funktionen durch ihre Integralwerte längs gewisser Mannigfaltigkeiten. Ber. Sachs. Akad. Wiss. Leipzig. Kl. 69, 262–277.

Ray, S., Meyhofer, E., Milligan, R.A., Howard., J., 1993. Kinesin follows the microtubule's protofilament axis. J. Cell Biol. 121, 1083–1093.

Rayleigh, Lord, 1896. On the theory of optical images, with special reference to the microscope. Phil. Mag. 42, 167–195.

Reimer, L., 1989, 1997. Transmission Electron Microscopy, Physics of Image Formation and Micro-analysis. Springer-Verlag, Berlin Springer Series in Optical Sciences, vol. 36. Springer-verlag, Berlin.

Rhodes, G., 2006. Crystallography Made Crystal Clear: A Guide for Users of Macromolecular Models, third ed. Academic Press, Burlington, Mass.

Robertson, J.M., 1943. Interpretation of Patterson diagrams. Nature 152, 411.

Rosenthal, P.B., Henderson, R., 2003. Optimal determination of particle orientation, absolute hand, and contrast loss in single-particle electron cryomicroscopy. J. Mol. Biol. 333, 721–745.

Rossmann, M.G., Blow, D.M., 1962. The detection of sub-units within the crystallographic asymmetric unit. Acta Crystallogr. 15, 24–31.

Rossmann, M.G., Blow, D.M., 1963. Determination of phases by the conditions of non-crystallographic symmetry. Acta Crystallogr. 16, 39–45.

Rupp, B., 2010. Biomolecular Crystallography: Principles, Practice and Application to Structural Biology. Garland Science, Taylor & Francis Group.

Sachse, C., Chen, J.Z., Coureux, P.-D., Stroupe, M.E., Fändrich, M., Grigorieff, N., 2007. High-resolution electron microscopy of helical specimens: a fresh look at tobacco mosaic virus. J. Mol. Biol. 371, 812–835.

Saxton, W.O., Frank, J., 1977. Motif detection in quantum noise-limited electron micrographs by cross-correlation. Ultramicroscopy 2, 219–227.

Saxton, W.O., Baumeister, W., 1982. The correlation averaging of regularly arranged bacterial cell envelope protein. J. Microsc. 127, 127–138.

Schatz, M., van Heel, M., 1990. Invariant classification of molecular views in electron micrographs. Ultramicroscopy 32, 255–264.

Scherzer, O., 1936. Über einige Fehler von Elektronenlinsen. Z. Physik. 101, 593–603.

Scherzer, O., 1949. The theoretical resolution limit of the electron microscope. J. Appl. Phys. 20, 20–29.

Schmeing, T.M., Ramakrishnan, V., 2009. What recent ribosome structures have revealed about the mechanism of translation. Nature 461, 1234–1242.

Shannon, C.E., 1949. Communication in the presence of noise. Proc. Inst. Radio Eng. New York 37, 10–21.

Short, J.M., Chen, S., Roseman, A.M., Butler, P.J.G., Crowther, R.A., 2009. Structure of hepatitis B surface antigen from subviral tubes determined by electron cryomicroscopy. J. Mol. Biol. 390, 135–141.

Schröder, G.F., Levitt, M., Brunger, A.T., 2010. Super-resolution biomolecular crystallography with low-resolution data. Nature 464, 1218–1222.

Smith, D.E., 1923. History of Mathematics. Dover Publications, New York.

Smith, M.F., Langmore, J.P., 1992. Quantitation of molecular densities by cryo-electron microscopy. Determination of the radial density distribution of tobacco mosaic virus. J. Mol. Biol. 226, 763–774.

Sommerville, J., Scheer, U., 1987. Electron Microscopy in Molecular Biology. IRL Press, Oxford.

Spence, J.C.H., 1981–2003. Experimental High-resolution Electron Microscopy, first–third ed. Clarendon Press, Oxford.

Stewart, I., Golubitsky, M., 1992. Fearful symmetry: Is God a geometer? Penguin Books, Harmondsworth (England).

Stewart, M., 1988. Computer image processing of electron micrographs of biological structures with helical symmetry. J. Electron Microsc. Tech. 9, 325–358.

Stoeckenius, W., Kunau, W.H., 1968. Further characterization of particulate fractions from lysed cell envelopes of *Halobacter halobium* and isolation of gas vacuole membranes. J. Cell Biol. 38, 337–357.

Taylor, C.A., Hinde, R.M., Lipson, H., 1951. Optical methods in X-ray analysis. I. The study of imperfect structures. Acta Crystallogr. 4, 261.

Taylor, C.A., Lipson, H., 1964. Optical Transforms. G. Bell & Sons, London.

Taylor, K.A., Glaeser, R.M., 1974. Electron diffraction of frozen, hydrated protein crystals. Science 186, 1036–1037.

Taylor, K.A., Glaeser, R.M., 1976. Electron microscopy of frozen hydrated biological specimens. J. Ultrastruct. Res. 55, 448–456.

Thon, F., 1966. Zur Defokussierungsabhängigkeit des Phasenkontrastes bei der elektronenmikroskopischen Abbildung. Z. Naturforschg 21a, 476–478.

Thon, F., 1968. Zur Deutung der Bildstrukturen in hochaufgellösten elektronenmikroskopischen Aufnahmessn dünner amorpher Objekte. Ph.D. Dissertation, University of Tübingen.

Thouvenin, E., Laurent, S., Madelaine, M.-F., Rasschaert, D., Vautherot, J.-F., Hewat, E.A., 1997. Bivalent binding of a neutralizing antibody to a calcivirus involves the torsional flexibility of the antibody hinge. J. Mol. Biol. 270, 238–246.

Thuman-Commike, P.A., Chiu, W., 2000. Reconstruction principles of icosahedral virus structure determination using electron cryomicroscopy. Micron. 31 (6), 687–711.

Toyoshima, C., Unwin, N., 1988. Contrast transfer for frozen-hydrated specimens: determination from pairs of defocused images. Ultramicroscopy 25, 279–292.

Toyoshima, C., 2000. Structure determination of tubular crystals of membrane proteins. 1. Indexing of diffraction patterns. Ultramicroscopy 84, 1–14.

Trachtenberg, S., DeRosier, D.J., Zemlin, F., Beckmann., 1998. Non-helical perturbations of the flagellar filament: *Salmonella typhimurium* SJW117 at 9.6 Å resolution. J. Mol. Biol. 276, 759–773.

Unwin, P.N.T., Henderson, R., 1975. Molecular structure determination by electron microscopy of unstained crystalline specimens. J. Mol. Biol. 94, 425–440.

Unwin, N., 2005. Refined structure of the nicotinic acetylcholine receptor at 4 Å resolution. J. Mol. Biol. 346, 967–989.

Valpuesta, J.-M., Carrascosa, J.L., Henderson, R., 1994. Analysis of electron microscope images and electron diffraction patterns of thin crystals of Φ29 connectors in ice. J. Mol. Biol. 240, 281–287.

van Heel, M., Keegstra, W., Schutter, W., van Bruggen, E.F.J., 1982. Arthropod hemocyanin structures studied by image analysis. In: Wood, E.J. (Ed.), Life Chemistry Reports, Supplement 1, The Structure and Function of Invertebrate Respiratory Proteins. EMBO Workshop, Leeds, pp. 69–73.

van Heel, M., 1987. Similarity measures between images. Ultramicroscopy 21, 95–100.

van Heel, M., Gowen, B., Matadeen, R., Orlova, E.V., Finn, R., Pape, T., et al., 2000. Single-particle electron cryomicroscopy: towards atomic resolution. Quart. Rev. Biophysics 33, 307–369.

van Heel, M., Portugal, R., Schatz, M., 2009. Multivariate statistical analysis in single particle (cryo) electron microscopy. 3D-EM in Life Sciences (Sixth Framework Paper). (DVD). <www.single_particles.org>

van Vlijmen, H.W.T., Karplus, M., 2005. Normal mode calculations of icosahedral viruses with full dihedral flexibility by use of molecular symmetry. J. Mol. Biol. 350, 528–542.

Vinothkumar, K.R., Henderson, R., 2010. Structures of membrane proteins. Quart. Rev. Biophys. 42, 1–93.

Walker, J.S., 1996. Fast Fourier Transforms. CRC Press, Boca Raton.

Ward, A., Moody, M.F., Sheehan, B., Milligan, R.A., Carragher, B., 2003. Windex: a toolset for indexing helices. J. Struct. Biol. 144, 172–183.

Ward, J.H., 1963. Hierarchical grouping to optimize an objective function. Am. Statist. Assoc. J. 58, 236–244.

Waser, J., 1955. Fourier transforms and scattering of tubular objects. Acta. Crystallogr. 8, 142–150.

Watson, G.N., 1958. A Treatise on the Theory of Bessel Functions, second ed. University Press, Cambridge (reprinted 2006).

Watson, J.D., Crick, F.H.C., 1953. Molecular structure of nucleic acids. A structure for deoxyribose nucleic acid. Nature 171, 737–738.

Weyl, H., 1952. Symmetry. University Press, Princeton.

White, H.E., Saibil, H.R., Ignatiou, A., Orlova, E.V., 2004. Recognition and separation of single particles with size variation by statistical analysis of their images. J. Mol. Biol. 336, 453–480.

Whittaker, E.T., 1915. On the functions which are represented by the expansions of the interpolatory theory. Proc. Roy. Soc. Edinburgh 35, 181–194.

Wiley, D.C., Lipscomb, W.N., 1968. Crystallographic determination of symmetry of aspartate transcarbamylase: studies of trigonal and tetragonal crystalline forms of aspartate transcarbamylase show that the molecule has a three-fold and a two-fold symmetry axis. Nature 218, 1119–1121.

Williams, R.C., Fisher, H.W., 1970. Electron microscopy of tobacco mosaic virus under conditions of minimal beam exposure. J. Mol. Biol. 52, 121–123.

Wilson, A.J.C., 1942. Determination of absolute from relative X-ray intensity data. Nature 150, 152.

Woolfson, M.M., 1997. An Introduction to X-Ray Crystallography, second ed. University Press, Cambridge.

Wüthrich, K., 1986. NMR of Proteins and Nucleic Acids. Wiley, New York.

Wynne, S.A., Crowther, R.A., Leslie, A.G., 1999. The crystal structure of the human hepatitis B virus capsid. Mol. Cell. 3, 771–780.

Yonekura, K., Maki-Yonekura, S., Namba, K., 2003. Complete atomic model of the bacterial flagellar filament by electron cryomicroscopy. Nature 424, 643–650.

Zemlin, J., Zemlin, F., 2002. Diffractogram tableaux by mouse click. Ultramicroscopy 93, 77–82.

Zhang, X., Fang, Q., Hui, W.H., Zhou, Z.H., 2010. 3.3 Å cryo-EM structure of a nonenveloped virus reveals a priming mechanism for cell entry. Cell 141, 1–11.

Zhou, Z.H., Chiu, W., 2002. Determination of icosahedral virus structures by electron cryomicroscopy at subnanometer resolution. Adv. Protein Chem. 64, 93–124.

Index

Printed in the United States
By Bookmasters